Manufacturing Handbook of Best Practices

An Innovation, Productivity, and Quality Focus

The St. Lucie Press/APICS Series on Resource Management

Titles in the Series

Applying Manufacturing Execution Systems
by Michael McClellan

Back to Basics: Your Guide to Manufacturing Excellence
by Steven A. Melnyk and R.T. "Chris" Christensen

Enterprise Resources Planning and Beyond: Integrating Your Entire Organization
by Gary A. Langenwalter

ERP: Tools, Techniques, and Applications for Integrating the Supply Chain
by Carol A. Ptak with Eli Schragenheim

Integrated Learning for ERP Success: A Learning Requirements Planning Approach
by Karl M. Kapp, with William F. Latham and Hester N. Ford-Latham

Integral Logistics Management: Planning and Control of Comprehensive Business Processes
by Paul Schönsleben

Inventory Classification Innovation: Paving the Way for Electronic Commerce and Vendor Managed Inventory
by Russell G. Broeckelmann

Lean Manufacturing: Tools, Techniques, and How To Use Them
by William M. Feld

Macrologistics Management: A Catalyst for Organizational Change
by Martin Stein and Frank Voehl

Restructuring the Manufacturing Process: Applying the Matrix Method
by Gideon Halevi

Basics of Supply Chain Management
by Lawrence D. Fredendall and Ed Hill

Supply Chain Management: The Basics and Beyond
by William C. Copacino

Handbook of Supply Chain Management
by Jim Ayers

Manufacturing Handbook of Best Practices

An Innovation, Productivity, and Quality Focus

Edited by

Jack B. ReVelle, Ph.D.

ST. LUCIE PRESS

A CRC Press Company

Boca Raton London New York Washington, D.C.

Library of Congress Cataloging-in-Publication Data

Manufacturing handbook of best practices : an innovation, productivity, and quality focus / edited by Jack B. ReVelle
p. cm. -- (St. Lucie Press/APICS series on resource management)
Includes bibliographical references and index.
ISBN 1-57444-300-3
1. Technological innovations--Management. 2. Product management. 3. Quality control. I. ReVelle, Jack B. II. Series.

HD45 .M3295 2001
658.5--dc21
2001048504

Direct all inquiries to CRC Press LLC, 2000 N.W. Corporate Blvd., Boca Raton, Florida 33431.

St. Lucie Press is an imprint of CRC Press LLC

No claim to original U.S. Government works
International Standard Book Number 1-57444-300-3
Library of Congress Card Number 2001048504
Printed in the United States of America 2 3 4 5 6 7 8 9 0
Printed on acid-free paper

Table of Contents

Preface

By Jack B. ReVelle

Sometimes it seems as though there is no end to the number of new or nearly new manufacturing methods that are now available. The primary objective for bringing together this book is for it to become your single-source reference to what's currently happening in modern manufacturing.

Whether your goal is to improve organizational responsiveness, product quality, production scheduling, or sensitivity to customer expectations, or to reduce process cycle time, cost of quality, or variation in products or processes, there is a methodology waiting to be discovered and introduced to enhance your operations.

In an effort to facilitate your use of this book, it has been organized in two ways: alphabetically, to ease the location of a specific topic; and by application, to indicate primary usage. No matter how the topics are enumerated or organized, there is seemingly no end to the scope of tools and techniques available to the well-informed manufacturing manager. The topics addressed in this book have been classified and then subclassified according to their major applications in Table 1.

The next few pages are dedicated to briefly describing each of these topics.

- An *agile enterprise* is adept at rapidly reorganizing its people, management, physical facilities, and operating philosophy to be able to produce highly customized products and services that satisfy a new customer or a new market.
- *Design for manufacture and assembly (DFMA) and design for six sigma (DFSS)* are complementary approaches to achieve a superior product line that maximizes quality while minimizing cost and cycle time in a manufacturing environment. DFMA stresses the achievement of the simplest design configuration. DFSS applies statistical analysis to achieve nearly defect-free products.
- *Design of experiments (DOE)* is the statistical superstructure upon which DFMA and DFSS are based. By analyzing the results of a predetermined series of trial runs, the optimal levels or settings for each critical parameter or factor are established.
- *Integrated product and process development (IPPD)* is a cross-functional, team-oriented approach to maximize concurrent development of both a product design and the means to produce the design.
- *ISO 9000:2000* is the most recent version of the international standard for quality management systems (QMS). Originally approved in 1987 and revised in 1994, this is the most recent version of ISO 9000. Because of substantial changes, even persons familiar with earlier versions of this standard need additional training.
- *ISO 14001* is the international standard for environmental management systems (EMS) and their integration into overall management structures.

- *Lean manufacturing* is an integrated collection of tools and techniques, traceable back to the Toyota production system, that focuses on the elimination of waste from the production process.
- *Manufacturing controls integration* brings together a collection of related systems such as enterprise resource planning (ERP) and manufacturing resource planning (MRP) to manage their internal operations and establish the demands of their supply chains.
- *Measurement systems analysis (MSA)* is the examination and understanding of the entire measurement process as well as its impact on the data it generates. The process includes procedures, gauges, software, personnel, and documentation.
- *Process analysis* is the mapping, input–output analysis, and detailed examination of a process including each of its sequential steps.
- *Quality function deployment (QFD)* is a matrix-based approach to acquisition and deployment of the "voice of the customer" throughout an organization to ensure that customer expectations, demands, and desires are thoroughly integrated into products and services. The initial QFD matrix is widely known as the House of Quality (HOQ).
- *Robust design* of a product or a process is the logical search for its optimal design (the levels or settings for each controllable parameter or factor) when considering the negative effect of the most critical uncontrollable/noise factors.
- *Six sigma* is a financially focused, highly structured approach to advancing the objectives of continuous improvement. The first of two chapters addresses the benefits resulting from the application of Six Sigma quality, while the second chapter focuses on the Six Sigma problem-solving methodology.
- *Statistical quality/process control (SQC/SPC)* was initially developed in the 1920s, but was substantially enhanced in the 1970s and 1980s by W. Edwards Deming and Joseph Juran and in the 1990s through the use of personal computers. This chapter emphasizes when and how to use SQC/SPC to improve products and processes as well as how this collection of tools differs from other statistical techniques.
- *Supply chain management (SCM)* is the control of the network used to deliver products and services from raw materials to end consumers through an engineered flow of information, physical distribution, and cash. The first of two chapters addresses the basics of SCM, while the second chapter focuses on SCM applications.
- The concepts known as the *theory of constraints (TOC)* and the *critical chain* were developed by Eli Goldratt. They represent a major expansion of the existing methodology known as critical path planning or the activity network diagram.
- *TRIZ* (a Russian acronym also known as *the theory of innovative problem solving [TIPS])* is a highly integrated collection of facts regarding physical, chemical, electrical, and biological principles that are used to predict where future breakthroughs are likely to occur and what they are likely to be.

Our contributing authors are all seasoned manufacturing veterans who have a particular interest in and extensive understanding of the topics about which they have written. In many cases the editor has worked directly with these authors at one point or another in their careers, so he can attest to their knowledge and willingness to share this knowledge with those who want to learn more about their profession. However, the idea to create this book, the choice of topics, and the selection of contributing authors are all mine and so, as editor, I accept full responsibility for any shortcomings you may find.

At this point it should be evident that this book is intended to provide information for both novice and experienced manufacturing managers. If a particular topic is of special interest to you for purposes of review or to initiate your understanding of its "fit" within the broad spectrum of tools and techniques that are a regular part of today's manufacturing venue, you will have immediate access to the basics as well as a bridge to more advanced information regarding that topic.

Remember, this is a handbook, not a textbook. Although you may wish to read the entire book from front to back, it is not necessary to do so. Simply search out the topic(s) of interest to you and begin your journey into the future of manufacturing.

TABLE 1
Topical Classification by Major Usage

	Design		Operations	
Topic	**Product**	**Process**	**Produce**	**Support**
Agile Enterprises		x		
Design for Manufacture & Assembly/Design for Six Sigma (DFMA/DFSS)	x			
Design of Experiments (DOE)				x
Integrated Product and Process Development (IPPD)	x	x		
ISO 9000:2000		x		
ISO 14000		x		
Lean Manufacturing		x		
Manufacturing Controls Integration				x
Measurement Systems Analysis (MSA)				x
Process Analysis		x		
Quality Function Deployment (QFD)				x
Robust Design	x	x		
Six Sigma Benefits Resulting from Six Sigma Quality				x
Six Sigma Problem Solving	x	x		
Statistical Quality/Process Control (SQC/SPC)			x	
Supply Chain Management Basics				x
Supply Chain Management Applications				x
Theory of Constraints/Critical Chain			x	
TRIZ/Theory of Innovative Problem	x			
Solving (TIPS)	x			

Acknowledgments

The team of authors, editor, and publisher that helped us to convert the original concept for a highly focused manufacturing handbook into this final product deserves public recognition. My thanks are extended to all the contributing authors who produced their respective chapters. Special thanks and appreciation are due to Drew Gierman, our publisher at St. Lucie Press, who pushed and pulled us to ensure that this handbook would eventually become a reality. Maria Muto of Muto Management Associates, our Phoenix-based editor, deserves more than thanks and appreciation: she has earned our enduring respect for her tenacity and professionalism. Without her intervention and involvement, we would still be running the race trying to bring everything together for you, our readers. And of course, her check is in the mail.

Editor

Dr. Jack B. ReVelle, The Wizard of Odds, provides his advice and assistance to his clients located throughout North America. In this capacity, he helps his clients to better understand and continuously improve their processes through the use of a broad range of Six Sigma, Total Quality Management, and continuous improvement (Kaizen) tools and techniques. These include process mapping, cycle time management, quality function deployment, statistical quality control, the seven management and planning tools, design of experiments, strategic planning (policy deployment), and integrated product and process development. In May 2001, Dr. ReVelle completed instructing "An Introduction to Six Sigma," a Web-based graduate course on behalf of California State University, Dominguez Hills.

Previously, he was Director of the Center for Process Improvement for GenCorp Aerojet in Azusa and Sacramento, CA, where he provided technical leadership for the Operational Excellence program. This included support for all the Six Sigma, Lean/Agile Enterprise, Supply Chain Management, and High Performance Workplace activities. Prior to this, Dr. ReVelle was the leader of Continuous Improvement for Raytheon (formerly Hughes) Missile Systems Company in Tucson, AZ. During this period, he led the Hughes teams that won the 1994 Arizona Pioneer Award for Quality and the 1997 Arizona Governor's Award for Quality. He also established the Hughes team responsible for obtaining ISO 9001 registration in 1996. On behalf of Hughes, Dr. ReVelle worked with the Joint Arizona Consortium-Manufacturing and Engineering Education for Tomorrow (JACME^{2}T) as the leader of the Quality Curriculum Development Group and as the lead TQM trainer.

Dr. ReVelle's previous assignments with Hughes Electronics were at the corporate offices as Manager, Statistical and Process Improvement Methods, and as Manager, Employee Opinion Research and Training Program Development. Prior to joining Hughes, he was the Founding Dean of the School of Business and Management at Chapman University in Orange, CA.

Currently, Dr. ReVelle is a member of the Board of Directors, Arizona Governor's Award for Quality (1999–2000). Previously, he was a member of the Board of Examiners for the Malcolm Baldrige National Quality Award (1990 and 1993), a judge for the Arizona Governor's Award for Quality (1994–1996), a member of the Awards Council for the California Governor's Award for Quality (1998–1999), and a judge for the RIT — *USA Today* Quality Cup (1994–2001).

Following publication of his books, *Quantitative Methods for Managerial Decisions* (1978) and *Safety Training Methods* (1980, revised 1995), Dr. ReVelle authored chapters for *Handbook of Mechanical Engineering* (1986, revised 1998), *Production Handbook* (1987), *Handbook of Occupational Safety and Health* (1987), and *Quality Engineering Handbook* (1991). His most recent texts are *From Concept to Customer: The Practical Guide to Integrated Product and Process Development and Business*

Process Reengineering (1995) and *The QFD Handbook* (1998). Dr. ReVelle led the development of two innovative, expert-system software packages, **TQM ToolSchool™** (1995) and **QFD/Pathway™** (1998). His latest text is *What Your Quality Guru Never Told You* (2000).

Dr. ReVelle is a fellow of the American Society for Quality, the Institute of Industrial Engineers, and the Institute for the Advancement of Engineering. He is listed in *Who's Who in Science and Engineering, Who's Who in America, Who's Who in the World,* and as an outstanding educator in *The International Who's Who in Quality.*

Dr. ReVelle is a recipient of the Distinguished Economics Development Programs Award from the Society of Manufacturing Engineers 1990, the Taguchi Recognition Award from the American Supplier Institute 1991, the Akao Prize from the QFD Institute 1999, and the Lifetime Achievement Award from The National Graduate School of Quality Management 1999. He is one of only two persons ever to receive both the Taguchi Recognition Award (for his successful application of Robust Design) and the Akao Prize (for his outstanding contribution to the advancement of quality function deployment).

Dr. ReVelle's award-winning articles have been published in *QUALITY PROGRESS, INDUSTRIAL ENGINEERING, INDUSTRIAL MANAGEMENT,* and *PROFESSIONAL SAFETY* magazines. During 1994 and 1995, Dr. ReVelle created and hosted a series of monthly satellite telecasts, "*Continuous Improvement Television" (CITV)*, for the National Technological University.

Dr. ReVelle received his B.S. in chemical engineering from Purdue University and both his M.S. and Ph.D. in industrial engineering and management from Oklahoma State University. Prior to receiving his Ph.D., he served 12 years in the U.S. Air Force. During that time, he was promoted to the rank of major and was awarded the Bronze Star Medal while stationed in the Republic of Vietnam as well as the Joint Services Commendation Medal for his work in quality assurance with the Nuclear Defense Agency.

Dr. ReVelle was a Senior Vice President and Treasurer of the Institute of Industrial Engineers (IIE), Director of the Aerospace and Defense Division of the IIE, a Co-Chair of the Total Quality Management (TQM) Committee of the American Society for Quality (ASQ), and a member of the Board of Directors of the Association for Quality and Participation (AQP).

Other professional memberships include the American Statistical Association (ASA) and the American Society of Safety Engineers (ASSE). Dr. ReVelle's national honor society memberships include Sigma Tau (all engineering), Alpha Pi Mu (industrial engineering), Alpha Iota Delta (decision sciences), and Beta Gamma Sigma (business administration).

Contributors

Jonathon L. Andell
Andell Associates
Phoenix, AZ

Douglas Burke
General Electric
Gilbert, AZ

Adi Choudri
GenCorp Aerojet
Folsom, CA

R.T. "Chris" Christensen
University of Wisconsin
Madison, WI

Charles A. Cox
Compass Organization, Inc.
Gilbert, AZ

Syed Imtiaz Haider
Gulf Pharmaceutical Industries
United Arab Emirates

John W. Hidahl
GenCorp Aerojet
Rancho Cordova, CA

Robert Hughes
Ethicon
Cincinnati, OH

Paul A. Keller
Quality America/Quality Publishing
Tucson, AZ

Edward A. Peterson
GenCorp Aerojet
Auburn, CA

Jack B. ReVelle
ReVelle Solutions, LLC
Tustin, CA

Lisa J. Scheinkopf
Chesapeake Consulting, Inc.
Tempe, AZ

Steven F. Ungvari
Consultant
Brighton, MI

Dedication

This handbook is dedicated to

- Bren, my wife of 33 years and the love of my life. No significant decision can or should be made without her counsel.
- Karen, our daughter who has become a lovely young lady and an exceptional commercial artist.
- Manufacturing vice presidents, directors, managers, engineers, specialists, and technicians around the world. This is your book; let it help you focus on innovation, productivity, and quality in manufacturing.

About APICS

APICS, The Educational Society for Resource Management, is an international, not-for-profit organization offering a full range of programs and materials focusing on individual and organizational education, standards of excellence, and integrated resource management topics. These resources, developed under the direction of integrated resource management experts, are available at local, regional, and national levels. Since 1957, hundreds of thousands of professionals have relied on APICS as a source for educational products and services.

- **APICS Certification Programs**—APICS offers two internationally recognized certification programs, Certified in Production and Inventory Management (CPIM) and Certified in Integrated Resource Management (CIRM), known around the world as standards of professional competence in business and manufacturing.
- *APICS Educational Materials Catalog*—This catalog contains books, courseware, proceedings, reprints, training materials, and videos developed by industry experts and available to members at a discount.
- *APICS—The Performance Advantage*—This monthly, four-color magazine addresses the educational and resource management needs of manufacturing professionals.
- *APICS Business Outlook Index*—Designed to take economic analysis a step beyond current surveys, the index is a monthly manufacturing-based survey report based on confidential production, sales, and inventory data from APICS-related companies.
- **Chapters**—APICS' more than 270 chapters provide leadership, learning, and networking opportunities at the local level.
- **Educational Opportunities**—Held around the country, APICS' International Conference and Exhibition, workshops, and symposia offer you numerous opportunities to learn from your peers and management experts.
- **Employment Referral Program**—A cost-effective way to reach a targeted network of resource management professionals, this program pairs qualified job candidates with interested companies.
- **SIGs**—These member groups develop specialized educational programs and resources for seven specific industry and interest areas.

- **Web Site**—The APICS Web site at http://www.apics.org enables you to explore the wide range of information available on APICS' membership, certification, and educational offerings.
- **Member Services**—Members enjoy a dedicated inquiry service, insurance, a retirement plan, and more.

For more information on APICS programs, services, or membership, call APICS Customer Service at (800) 444-2742 or (703) 354-8851 or visit http://www.apics.org on the World Wide Web.

1 The Agile Enterprise

Adi Choudri

1.1 INTRODUCTION

An agile enterprise is adept at reorganizing its people, management, physical facilities, and operating philosophy very quickly to produce highly customized products and services to satisfy a new customer or a new market. Agility is the deliberate, strategic response for survival in today's market conditions.

A company that knows how to be agile:

- **Strategizes** to fragment mass markets into niche markets
- **Competes** on the basis of customer-perceived value
- **Produces** multiple products and services in market-determined quantities
- **Designs** solutions interactively with customers
- **Organizes** for proficiency at change and rapid response
- **Manages** through leadership, motivation, support, and trust
- **Exploits** information and communication technologies to the fullest
- **Leverages** all its capabilities, resources, and assets regardless of location
- **Works** through entrepreneurial and empowered teams
- **Partners** with other companies as a strategy of choice, not of last resort
- **Thrives** and is widely **imitated**

As we transition into the 21st century, radical changes are taking place that are reshaping every aspect of a business, including the way we produce goods and services. With the advent of Internet and high-speed communication, the marketplace has truly become global and fragmented. Customers are requiring smaller quantities and more customized products quickly. Traditional manufacturing, with its large batch approach, extensive inventories, and static organizational style, simply cannot compete in this marketplace. The notion of "economies of scale" becomes almost obsolete in such a changing and fragmented market. In the 1980s and '90s we learned lean manufacturing techniques, reduced cycle time and cost, and strived to become world-class. We introduced just-in-time (JIT) techniques such as one-piece part flow and quick changeover, and practiced team-based continuous improvement. Yet our customers pressed for even more flexibility, shorter lead times, and more varied products and services. Lean manufacturing is about being very good at doing things we can control. Agility of an enterprise gives the ability to deal with things it cannot control. Agility means not only accommodating change but also relishing the opportunities inherent within a turbulent environment.

Here are some of the axioms of agile manufacturing: Mass production is moribund. Mass customization requires that each customer be treated as an individual.

1-57444-300-3/02/$0.00+$1.50

This leads to a people-intensive, relationship-driven operation. Increasingly, a company ceases to sell products but rather sells its ability to fulfill customers' needs, utilizing its information and people skills. New information technology such as the ability to leverage the Internet and a highly educated, skilled workforce becomes the real asset base for the corporation. This allows local decision-making by people who understand the company's vision, principles, customer requirements, and products and services. They must know how to create cooperative alliances across the supply chain, how to reconfigure products and production facilities, and how to combine expertise to satisfy the changing marketplace. Agile companies put enormous emphasis on training and developing their people. For example, Saturn Corporation requires their employees to take no less than 96 hours of training every year. The latest information technology such as Internet and object-oriented programming can provide a tremendous amount of information and computer system flexibility in the hands of a highly trained workforce. Forming virtual teams within the supply chain (sometimes even with a competitor) to satisfy a customer need becomes commonplace with agile enterprises. Internet and information technology become key enablers.

Many industries and markets are increasingly requiring much greater flexibility and timeliness from their manufacturers and service providers. These changes are taking place very fast in some industries and more slowly in others. But the companies that will meet the challenges of the ever-changing global marketplace of the 21st century must go beyond lean and become agile in every aspect of their business. Agility is not a magic wand to solve all ills. But without agility, survivability in the 21st century will be questionable for many corporations. However, agility must be built on the firm foundation of world-class or lean manufacturing methods and high-quality Six Sigma processes, coupled with an organization that is physically, technologically, and managerially and culturally flexible enough to capitalize on rapid and unpredictable change.

1.2 TRADITIONAL MANUFACTURING

Why does traditional batch-and-queue manufacturing seem right intuitively, yet carry so much waste? We human beings are into a mental world of "functions" and "departments" and have a commonsense conviction that activities ought to be grouped by type so they can be performed more efficiently and managed more easily. Intuitively, this makes sense if the activity contains some form of "set-up" activity. For example, making numerous trips to the supermarket to get groceries one item at a time would be tremendously wasteful, and our intuition would be right in this case. So it is natural for us to take this intuitive sense of efficiency and extend it to an enterprise where processes are not independent, and we start thinking that to get tasks done more efficiently within departments we must perform like activities in batches. In the paint department we tend to paint all the cars green and then shift to red, then to white, in between creating as large a batch size as possible regardless of the need. Batches, it turns out, always mean long delays as the product sits patiently awaiting the department's changeover to the type of activity the product needs next. But this approach keeps the department and its people and equipment busy and gives a sense of "efficiency" because everyone and everything is working hard. This comes from our lack of "systems" thinking that is actually counter-intuitive. We must see this from the perspective of the part flowing through the system rather than the viewpoint of the individual process. Taiichi Ohno, the father of the Toyota production system, blamed this batch-and-queue mode of thinking on civilization's first farmers, who he claimed lost the one-thing-at-a-time wisdom of the hunter as they became obsessed with batches (once-a-year harvest) and inventory (the grain depository). Or perhaps we are simply born with batch thinking in our heads, along with many other commonsense illusions. For example, time seems constant rather than relative or the sun seems to revolve around Earth and not the other way around. But we all need to fight departmentalized batch thinking because tasks can almost always be accomplished more efficiently and accurately when the product is worked continuously from raw materials to finished goods. In short, things work better when you focus on the product and its needs rather than the organization, the equipment, or the people, so that all the activities needed to design, manufacture, and ship a product occur in a continuous flow.

Henry Ford and his associates were the first people to fully realize the benefit of flow thinking. Ford reduced the amount of effort required to assemble a Model T Ford by 90% during the fall of 1913 by switching to continuous flow in final assembly. Subsequently, he lined up all the machines needed to produce the parts for the Model T in the correct sequence and tried to achieve flow all the way from raw materials to shipment of the finished car, achieving a similar productivity leap. But he discovered only the special case. His method worked only when production volumes were high enough to justify high-speed assembly lines, when every product used exactly the same parts and when the same model was produced for many years.

After World War II, Taiichi Ohno and his technical collaborators, including Shigeo Shingo, concluded that the real challenge was to create continuous flow in small-lot production, when dozens or hundreds of copies of a product were needed — not millions. They achieved continuous flow by learning to quickly change over tools from one product to the next and by rightsizing the machines so that processing steps of different types could be conducted immediately adjacent to each other with the product being kept in continuous flow. These concepts led to what is now known as lean manufacturing.

1.3 EVOLUTION FROM LEAN TO AGILE ENTERPRISE

> When change is discontinuous, the success stories of yesterday have little relevance to the problems of tomorrow; they might even be damaging. The world at every level has to be reinvented to some extent.
>
> Charles Handy, *Beyond Certainty*, Arrow Business Books, 1996

As we approached the new millennium, companies started to build upon those improvements gained through application of lean manufacturing principles.

Most of the things presented as agile practices are in fact lean production practices. The agile enterprise is concerned with a post-lean production paradigm. Lean production is one of yesterday's success stories, although because ideas diffuse very slowly, many companies are still in the process of implementing it. And because lean is so popular and easy to understand, it's a common mistake to assume that lean and agile are the same. They are not.

With the emerging collapse of mass/lean production-oriented competitive conditions, a need has arisen to develop new types of enterprises capable of dealing with and thriving in a complex and ever-changing business environment — enterprises that can continually reinvent themselves. The strategic vision is therefore the development of enterprises totally committed to embracing the emerging business environment. This involves creating a strategy that moves enterprises forward in three interrelated areas:

The niche enterprise — develops and exploits capabilities to thrive and prosper in the face of increasing diversity (arising from individual customers as well as different markets) and to deal with the wider issues of a fragmenting and diverse world.

The knowledge-based enterprise — develops and exploits capabilities to use knowledge and information for sustainable competitive advantage (in effect acknowledging information and knowledge as a source of wealth).

The agile (or adaptive) enterprise — develops and exploits capabilities to thrive and prosper in a changing, nonlinear, uncertain, and unpredictable business environment.

Agile manufacturing takes its name from the last of these three interrelated areas. However, agility is just one component of a 21st century manufacturing enterprise strategy. The issues of knowledge-based and niche enterprises need to be considered and, most importantly, the interrelationships among the three elements should be addressed.

Many companies have moved forward in the area of niche enterprise, using concepts and strategies linked to what is called mass customization (individually customized products at mass production prices). However, many have not actively explored the issue of knowledge enterprising, although more and more companies are starting to explore this area and to better define and further develop the concepts. Few companies have fully understood, let alone implemented, agile attributes (meaning that capability to deal with change, uncertainty, and unpredictability). None has linked the three elements together.

Therefore, although much is now known about how to mass customize, very little is known about what creates agile attributes. When companies involved in mass customization are analyzed, the lack of agility is often very apparent, since most of the mass customization techniques assume only limited uncertainty and unpredictability in the business environment. Agility is therefore truly a frontier activity, challenging many of today's "best practices."

The key points to understand are as follows:

Agile manufacturing is a strategy aimed at developing capabilities (the enterprise platform) to prosper in the next century. In this respect it is similar to a manufacturing strategy in that it should support business and marketing strategies. However, these strategies also need to be modified to take advantage of agile manufacturing capabilities.

As a strategy, agile manufacturing is concerned with objectives, structures, processes, and resources and not with individual point solutions, particular technologies, methods, etc. considered in isolation.

The emphasis is on designing the enterprise as a whole so that certain characteristics are achieved and not on the piecemeal adoption of quick fixes, prescriptions, and panaceas.

Agile manufacturing may require some current best practices, lean production concepts, technologies, and taken-for-granted assumptions to be re-evaluated, modified, or even abandoned.

In the same way that mass production resulted in the demise of many craft-based firms, agile manufacturing is likely to lead to the elimination of many mass production firms, even those with lean production enhancements.

One of the biggest problems to overcome is the misunderstanding that lean and agile are synonymous. They are not, although most (as much as 99%) of what is portrayed as agile is in fact lean.

Agile and *lean* are not synonymous terms. One of the biggest differences between the two can be seen in supplier relationships. Lean manufacturers, particularly Japanese automakers, believe that successful relationships must be cultivated over a long (20-year) period. Agile manufacturers believe they can find the best suppliers by searching the market of open competition whenever they need a service.

1.4 AGILE ENTERPRISE FOUNDATION

As organizations become leaner through the relentless pursuit of internal waste reduction, many are beginning to turn their focus to the outside. In the spirit of lean, this is forcing a new way of looking at how we satisfy the value expectations of the individual customer. This requires the organization to adapt to yet new ways of working. There are three key themes that form the foundation of an agile enterprise: customer focus, strategy deployment, and focus on work.

1.4.1 Customer Focus

A company exists to turn what it does for customers into profits for its shareholders. Along the way, it offers a societal context of providing gainful activity for its employees and support for the respective community. However, the fundamental basis for an enterprise is to supply customers with valuable products and services for which they pay. Without this customer focus, benefits and support to other stakeholders cannot occur. For a company to successfully make the agile transition, there cannot be any doubt as to who the customers are and what their expectations are. Quality in this environment can be defined as meeting or exceeding customer expectations. Achieving such quality objectives will bring extraordinary customer loyalty which, in turn, will drive success in terms of market share and the margins a company can draw from sales. However, achieving such objectives goes beyond just the quality of the physical product. The total customer "value" has to be central to every aspect of the company's operation.

The concept of "internal customer" is based on the idea of the employees being made up of interdependent links in a chain known as the service chain. However, every one of those links must know and focus on the impact he or she has on fulfilling the expectations of the external customer. Therefore, it is imperative to align the entire organization around

- Meeting or exceeding customer expectations in everything they do
- Everyone seeking out what the customer's explicit and tacit (unspoken) expectations are
- The organization as an unbreakable "service chain" focused on the external customer
- Measures of success based on how the customer values one's service in terms of loyalty, market share, and margins

1.4.2 Strategy Deployment

Every company usually has some sort of overall strategy for how it intends to turn market opportunities into shareholder value. Sometimes these strategies are fairly complete; at other times the environment demands that they remain flexible, even to the point of being vague. However, these strategies rarely have real meaning at organizational levels where the actual work is done. The top-level strategy must be clearly understood and translated into tasks and actions to which people can relate. If there is lack of alignment, it manifests itself in many ways, including

- Lack of a common vocabulary
- Apparent conflict between development programs and improvement programs
- Difficulty setting priorities among improvement opportunities
- Confusion between tools and results
- "Program-of-the-month" syndrome

Ensuring operational alignment is essential for a solid foundation for an agile enterprise over the long term. Without alignment, the effort to be agile will be reduced to a tools-based exercise with more hype than results. The end product is wasted resources, increased frustration, and a loss of credibility in company leadership. Operational alignment ensures that every individual in the organization knows how he or she can personally have an impact on the strategic objectives of the company. A process must be developed to drive the leadership-derived master strategy down to the base of the organization. This process should force all organization levels to

- Reformulate their own *master* strategy which has to be directly relevant to what they have a direct impact on
- Develop their own *deployment* strategy as to how they are going to make the strategy happen within the area they are responsible for through specific actions and metrics
- Agree on *execution* strategies with each part of their area to act as the input for the next level (including criteria of success)

The process drives down and then it drives back up. This forces real dialogue around business issues and ensures that everyone starts to manage with "strategic intent." Any transition to an agile enterprise requires enormous cultural change for everyone, and a clearly defined and well-communicated strategic intent goes a long way to help that process.

1.4.3 Focus on Work

The two key themes described above are vital for a program to be effective in implementing a new culture within the organization. However, on their own, employees will not change. For the customer to feel a beneficial effect in the value of what he or she receives, real change has to occur where actual work is done, and that change has to be implemented by those who do the work. Value-added work needs to be defined for each job. Usually, it means that agility must be addressed within the following work factors:

- Equipment, tools, and software work as planned and are available at the work location.
- Material and information required as input to the process are available in the quantities, format, and quality needed.
- Workers have the skills required to complete the task to the quality expected by the customer and at the productivity level required by the shareholders.
- Standard work instructions and processes are qualified to provide consistent, acceptable quality.
- Priorities are provided to satisfy customers' differing delivery expectations.

This phenomenon is not restricted to the factory, but is present in every part of the organization. That is why it is recommended that an agile transformation process start by focusing on specific work processes. This focus will immediately demonstrate respect for the customer. It brings immediate improvements in the way work is done and generates visible and measurable improvements in the work lives of everyone, which will fuel a successful agile enterprise.

1.5 AGILE MANUFACTURING

1.5.1 Definition

Agile manufacturing (or *agile competition*) is an umbrella term that embraces a wealth of ideas. It is not some vague concept of how a company should be run 5 to 10 years from now; it is how many businesses are being run today, not only to survive but to excel.

These ideas include

- Innovative alliances among suppliers, customers, and manufacturers in the pursuit of value
- Powerful concepts of technology-enabled agility

The alliances and concepts are integrated with and characterized by flat organizations, team production, empowerment, customization, and concern for social issues.

There are no well-established road maps to achieve agility; however, there are four overarching guidelines to help organizations start the agility journey.

1. Enrich the customer
   - Sell solutions — provide an unlimited variety of products, information, and services
2. Cooperate to enhance competition
   - Internal — cross-functional teams, empowerment
   - External — managing the supply chain
3. Organize to manage change and uncertainty
   - Rapid reconfiguration of plant and facilities
   - Rapid decision making — empowered at all organizational levels
4. Leverage people and information
   - Distribution of authority, resources, and rewards

1.5.2 Agile Manufacturing Challenges in the Automotive Industry

Agile manufacturing is a recent movement. It is viewed by the auto industry, which shares with the consumer electronics industry the distinction of being the pacesetter in manufacturing process innovation, as the next step in its development. It represents the demise of the century-long tradition of manufacturing driven by scale. It aspires to total flexibility without sacrificing quality or adding costs.

Agile manufacturing is contrasted with lean production, Toyota's composite of tools, culture, and organizational philosophy that ensures high quality, low cost, and continuous and sustained improvement. The Japanese Manufacturing 21 (21st century) consortium defines it in terms of nine major challenges to carmakers, one being the 3-day car — 3 days from a customer order for a customized car to dealer delivery. The goal is practical; leading Japanese automakers can deliver the 10-day car now.

U.S. firms moving in the same direction have a strong advantage over Japanese companies in some areas relevant to the nine challenges of Manufacturing 21. Those challenges are listed below.

- Break dependency on scale and economies of scale (reducing setup costs is key).
- Produce vehicles in low volumes at a reasonable cost (Nissan's intelligent body system, a Lego-block approach that favors existing over newly designed body components, leaves tooling as the only major expense for a new model).
- Guarantee the 3-day car.
- Replace the large centralized approach with distributed clusters of mini-assembly plants located near customers (as much as 5 days' time is required to ship cars to dealers; Japan's horrendous traffic congestion has become the weak link in just-in-time inventory management, with suppliers unable to deliver on time).
- Be able to reconfigure components in many different ways.
- Make work stimulating (those who carry out Lego-block production should not be treated as Lego blocks).
- Turn the customer into a "prosumer," an ugly neologism that means proactive something; the idea is that the customer will take an active role in the product design by, for example, configuring options on a computer in a dealer showroom.
- Streamline ordering systems and establish close relationships with suppliers.
- Manage the massive volumes of data generated by the production system to analyze that data quickly and agilely.

Agile production would appear to be the blueprint for future manufacturing and a key strategy for agile enterprises. Managers in every industry would do well to incorporate the essence of the Manufacturing 21 challenges into their agendas. Although these challenges were presented in the automotive context, similar agile challenges exist in almost every industry. Publishing, retailing, and banking are but a few of the industries likely to rally around agility.

1.6 AGILE ENTERPRISE GUIDING PRINCIPLES

1.6.1 Benefits of Being Agile

If successful, the key characteristics of agile manufacturing in your company will be

- Customer-integrated process for designing, manufacturing, marketing, and supporting all products and services
- Decision-making at functional knowledge points, not in centralized management "silos"
- Stable unit costs (low variability) no matter what the volume

- Flexible manufacturing — ability to increase or decrease production volumes at will
- Easy access to integrated data whether it is customer driven, supplier driven, or product and process driven
- Modular production facilities that can be organized into ever-changing manufacturing nodes
- Data that is rapidly changed into information that is used to expand knowledge
- Mass-customized product vs. mass-produced product

1.6.2 What's New or Different?

Agile manufacturers must recognize the volatility of change, and put mechanisms in place to deal with it. They must move from being manufacturing driven to customer driven, and they must also realize that customers won't pay a premium for quality — it's assumed.

Agile manufacturers must partner with customers, suppliers, and competitors (cooperate and compete) and understand that the soft side of business (trust, empowered teams, risk taking, reward, and recognition) must drive the entire process.

In an agile environment, information is the primary enabling resource. Firms must know their customers, products, and competitors.

What's new or different about this list? Not much! Some new terminology perhaps.

What is new, however, is the packaging and intensiveness by which a company tries to reinvent itself. This is *not* 5% a year continuous improvement. Agile manufacturing concepts are the key to future competitiveness, but many of these concepts are still in the development state. No one book or seminar will bring you to the "Fountain of Agility."

We do know, however, that world class manufacturing is the culmination of all these processes. The ultimate compliment is to be named by your sister companies, customers, and competitors as the leader in customer responsiveness, brought about by high quality, low cost, and innovative products and services. How well this is done becomes the measure of profitability.

1.7 AGILE ENTERPRISE TOOLS AND METRICS

1.7.1 Transaction Analyses

Transaction analyses are interview-based studies of how organizations operate. Performing transaction analyses helps recognize the inherent complexities of engineering partnerships and shows the need to develop tools to make the complexities visible and deal with them. Transaction analyses reveal where intensive transaction activity occurs and also permit one to see how activities at one point in the process are linked to activities elsewhere. Actual transactions do not necessarily correspond to official organization charts or approved information transfers, and the degree to which they differ is a good indication of how the participants must skew the official process in order to make progress.

1.7.2 Activity/Cost Chains

Activity/cost chains are an extension of activity-based costing. They are the result of using direct cost measurement techniques during the transaction analyses. In many cases, transactions can be associated with costs, so those cascades of transactions can be linked in order to sum up their component costs. Activity/cost analyses show how much it costs to do some basic activity such as make a design change, adjust a fixture, or tighten a tolerance. Knowing costs can help justify improvements in design and business processes. However, most companies do not know their actual costs to the required accuracy and usually compile costs in functionally defined cost centers rather than associating them with processes, especially when those processes cross alliance or functional boundaries.

1.7.3 Organization Maps

Organization maps show explicitly who does what in the web of suppliers. These maps turn out to be quite complicated, since assemblies and related tooling seem to be divided into very small elements, and each element is contracted out to a different supplier (at least in the car industry). If companies were to make these maps during early product design, they would be able to plan who should be in the partnerships and begin thinking about who should do what. Supplier selection criteria could be formulated based on where suppliers lie on the map and what their parts are in delivering the final customer requirement.

1.7.4 Key Characteristics (KCs)

Key characteristics (KCs) are aspects of the product that require close attention. They are intended to capture customer requirements and express them systematically as design and production metrics. Hundreds of specifications, dimensions, and tolerances typically appear on drawings. The assignment of a KC to a dimension or surface finish, for example, indicates that this particular aspect is the important one to deliver. Different companies have used this idea in different ways. GM distinguishes key product characteristics (KPCs) that the customer is aware of and key control characteristics (KCCs) that the manufacturer must control in order to deliver the KPCs.

1.7.5 Contact Chains

Contact chains link the key characteristics of assemblies of parts and fixtures to each other to describe how fitup is supposed to be achieved. KCs, for example, highlight visible fits such as those around car doors, since fitup dimensions and tolerances are documented by the chains and fitup is a KC for customer satisfaction. A metric that has been proposed is to count how many company or organizational boundaries are crossed by a single contact chain. The assumption is that smaller is better. If companies define these contact chains early in the design, they can assign responsibility explicitly to the different suppliers for their roles in supporting the chains. However, it appears that, although individual engineers commonly calculate these chains for local assembly fitup analyses, the contact chain concept has not been utilized as a way of unifying the work of several cooperating companies. No current computer-aided design (CAD) tools include contact chain representation capability, although the potential to add this capability exists. CAD is commonly used to define parts, less often for assemblies, and hardly at all for assembly fixtures.

1.8 CUSTOMER ORIENTATION

The key to this is to look at the products and services that a company provides in terms of how much value they add to the customers. World-class manufacturers have placed great emphasis on being close to the customers; customer prosperity goes much further and examines how much value is added to the customers by the use of a company's products and services. This requires an intimate understanding of the customers' needs. It requires a short-term, medium-term, and long-term view. It requires a company to understand the customers' use of its products more thoroughly than the customers know themselves. To address the customers' real needs, a company must sell solutions — not products. Selling solutions requires a detailed and thorough understanding of customer needs, and requires the bringing together of a package of products and services to fulfill those needs. A company's product alone may not be enough. It may need to add extra services or technical support or special terms. It may need to add complementary products supplied by other companies — perhaps even by its competitors — to truly satisfy customers' needs.

To be agile, a company will almost certainly need to design or develop products that are focused specifically on an individual customer's requirements. Product design, in most cases, will need to be closely integrated with the production process. The need for fast and effective design means that the traditional approach of having all new products routed through a design area must be eliminated. It always causes delay, misunderstanding, and a lack of cooperation between design and production. The design process must be integrated with the manufacturing process. Often, the manufacturing people in the production cell can be trained to do the majority of the design functions. Often, the products can be modularized to allow configuration rather than the separate design of each product, thus simplifying the design process.

Sometimes automated design systems need to be introduced so that the CAD systems can remove much of the detailed skills from the design process. Sometimes these CAD systems are integrated with CAM (computer-aided manufacturing) systems so that the designs can be automatically fed into the computer-controlled production machines. The design process can be significantly enhanced by having customers fully participate in the effort. With the two companies working together cooperatively, the customer bringing its design skills to bear on the project and your company adding its production skills into the equation, everyone benefits. In some cases the suppliers and outside process vendors can also be integrated into the design process so that the product is designed to meet the customer's needs very effectively. This close cooperation allows for the development of service-rich products that can evolve over time, as the customer and the company work together. This leads to the development of long-term relationships. The products may be designed to not only meet current needs but to be reconfigurable to meet the future needs of the customer.

Attention is paid to configurability, modularity, and design for the longer-term satisfaction of customer requirements. If the product contains software, it can be built to accept software updates over time. If the product is mechanical, it can be designed for easy reconfiguration and upgrades as technologies change, as new features are added, and as the customer's needs change over time.

Honda Motorcycle in Japan has developed a range of machines that have a credit-card sized electronic key. This key serves not only as a security device to unlock the steering mechanism, the electronic fuel pump, and other major components; it also contains information that changes the performance of the machine by adjusting the fuel injection, the timing, the ignition settings, and other controls. The rider can choose between fast, high-performance, economy, town, or mountainous driving, and so forth. The addition of electronic configurability allows the rider to easily reconfigure the machine to meet his or her needs. This flexibility and customer responsiveness were created because Honda has an understanding of the customer's varying needs and saw an information-based method of providing a wide-ranging solution. Increasingly, the company's information and the skill of its people become the premium product. The company ceases to sell products as such; instead, it is selling its ability to fulfill the customer's needs. This knowledge and skill need to be valued, protected, and shared. New information systems technology has made it possible for the company's personnel to be in direct contact with each other wherever they are in the world. This makes information, skills, and knowledge accessible to the people who are the primary providers of customer service. This can be a powerful tool linking people, customers, and other third parties closely together.

1.9 INFORMATION SYSTEM DESIGN

The skills and knowledge of the people within the company become a paramount consideration as a company develops solutions-based selling. This includes product knowledge and experience, but it also includes a rich depth of knowledge of customers' needs, anxieties, and service requirements. The relationships that develop between the company's people and their customers when the company sells solutions instead of products become very much a part of the product itself. Customers need to be treated as individuals, having individual needs and a history of experience with your company. This is very much a part of the agile approach. In some ways it is good, old-fashioned service, but in other ways it is very modern. This level of customer enrichment can only be achieved through the use of knowledge-based systems. Increasingly, the best way to create close customer awareness is to provide the people within the company, and the customers themselves, with a great deal of information. This may be product information, company information, education and training in the use of the company's products, analysis and data, product upgrades, manuals, drawings, instructions, or specifications. These days all this information can reside within the computer systems and be readily available to all authorized users including customers, suppliers, and other third-party partners. In this way, the sales representatives can be highly knowledgeable about the customers, their requirements, their ordering patterns, their payment histories, their use of the technical support or customer service facilities, and so forth. Available, complete, pertinent, and easy-to-access information is fast becoming a key competitive weapon that enables all customer contacts to be thorough and satisfactory.

Leading from this, of course, is the ability to closely link customers' information systems into your company's systems. Orders can be placed automatically from the customer and scheduled within the plant, yielding the customer accurate delivery promises. Design requirements can be automatically picked up in the customers' information systems without drawings or specifications being printed and passed. This enables the company to address customer needs with great agility. Design, delivery information, history, accounts receivable, and customer service contacts can all be integrated and made accessible. Some of the technologies required to achieve this level of information sharing have only recently become available. The wide access to the Internet and the World Wide Web opens up a standard and direct method of accessing information and providing the customers with a standard link into a company's system. For customers to be linked into a company's information systems in the past required a direct link (usually through dialing into the company's computer center). The Internet, as well as other networks, allow the customer to have a simple and standard link to place orders, make inquiries, send messages, and specify its needs. IBM has recently established a worldwide information system for its 350 partner companies. The system, using Lotus Notes communication methods, provides the partners with a window into IBM for technical information, product availability, personal communication, help-desk facilities, trouble-shooting data, and the ability to enter orders and check delivery dates and order status. This kind of information was previously either unavailable or it required the partner company to contact a customer service or sales representative. This open sharing of information is a key aspect of creating an agile operation.

1.10 COOPERATION THROUGH VIRTUAL TEAMS AND CORPORATIONS

The rapid change in technology (and other skills), added to customers' requiring highly specific, customized products, has led to the need for far greater cooperation within and among firms. No company can have all the required skills and knowledge. In high-tech areas it is often the small and virile organizations that develop and harness the latest advances. It is just not possible for one firm to have everything it takes to fully meet customer needs. Additional services, information, or logistics may be required. To meet these diverse and ever-changing needs requires great cooperation within the firm. Often, traditional companies have very little flexibility and cooperation from one department to another. This must be solved, and the various departments or areas of the company must work together for the enrichment of the customers, irrespective of any department's short-term benefit. The customers, suppliers, and other third parties can be brought into the cooperative effort to design a product or develop a value-added service. In some cases the company will need to seek out specific partners with special skills or attributes and create a virtual corporation from several parties to focus on meeting the needs of a customer or a market.

These virtual corporations are opportunistic alliances of core competencies across several firms to provide focused services and products to meet the customer's highly focused needs. With the advent of the information revolution, these various companies can readily communicate and cooperate across long distances and provide products and services that are widely scattered geographically and politically. The beginnings of the information age have made it possible to create diverse virtual corporations that can quickly and effectively address the needs of the customers and the marketplace. The agile organization will choose inter-enterprise cooperation as its first choice.

These cooperative partnerships are not the traditional joint ventures or mergers; they are informally created by companies dedicated to cooperation. Usually there is no complex legal structure. The cooperative arrangements are quickly made, written down so everyone understands their roles and expectations, and then put into practice. Virtual corporations require considerable trust, respect, and openness. Information technologies that allow groups of people to work together effectively, even if they are geographically separated, are tools that enable these kinds of informal, cooperative endeavors to flourish. Before the advent of the Internet, video conferencing, and multilingual systems it was not possible to provide the level of personal contact required to work together effectively and in a timely manner. These new technologies have opened up a world of communications that facilitates cooperative and virtual corporations to meet the needs of specific customers and markets. A notable example of this kind of cooperation is the link that has been forged between IBM, Motorola, and Apple Corporation to develop the new PowerPC chip to compete with the Intel Pentium. The companies, in some aspects competitors with each other, have created a team to design, develop, and manufacture the PowerPC chip. None of them could have done this alone.

An Australian company that was experiencing high costs and problems with the replenishment of materials from their principle suppliers entered a cooperative relationship with a transportation company. The truck drivers were given keys to the company's production plants and trained to identify component parts that were in short supply or had kanban requirements. Now the driver simply enters a requirement message in the computer system and drives to the supplier for replenishment of the item. These transactions occur continually throughout a 24-hour period, even when the plants are closed and empty. This has significantly reduced costs, eliminated the purchasing/order entry role between the customer and the supplier, and solved many of the part shortage problems. Cooperation of this kind requires trust, training, and openness to unorthodox approaches. The difficult aspect of this change was not the organization of the new plan, but the acceptance by company managers that this would even work.

1.11 HIGHLY EDUCATED AND TRAINED WORKFORCE

Everybody recognizes that the next few years will be a time of unprecedented change and uncertainty. But how should an organization be structured to take advantage of this turbulence? There is no clear-cut and simple answer to this question, but there are a number of issues that can be addressed to help a company become change-ready. Change and customer focus require the people closest to the customer to have the authority to change the company's methods to better address the customer's needs. The local people need to have considerable authority. The company needs to have a clearly defined vision of where the company is going, what its objectives are, and how those objectives will be met. This vision must be thoroughly disseminated throughout the organization. Principles of conduct and practice must be laid out so the local people making the decisions and the changes have clear policy guidelines to direct them. But the local people must then have complete authority, within the vision and principles of the company, to address the customer's needs.

For local decision-making to be effective, a company must have a highly educated and trained workforce. They must be people who know and understand the company's vision and principles, the customers' requirements, and the company's products and services. They must also know how to create cooperative alliances, how to reconfigure products, when to "go the second mile," and how to combine expertise to reach a common goal. Added to this, an agile company will often have smaller production and service centers geographically spread out, so that customers can be served locally. Sometimes this need for "local-ness" can be met by appropriate use of information systems, but often the need for very short lead times and customer responsiveness requires physical proximity as well as excellent communications. Saturn Corporation, the U.S. car manufacturer, requires its employees to take no less than 96 hours of training every year. Although training is voluntary, the company's bonus system is set up so that there are strong incentives to achieve or exceed the training requirement. In the early days, the company used training achievement as the only performance measure for the plant people because it was clear that training was the key to quality, timeliness, low cost, teamwork, and the company's other strategies. If the working people are to have considerable authority, then they must also have the resources, the knowledge, and the authority to meet customers' needs. Agile companies put enormous emphasis on the training and development of their people. Some of this is through traditional training classes, books, and seminars. Some of it is through team-based, cross-functional improvement initiatives. Some of it is through the intelligent use of information technologies, making the latest information immediately available for education or for analysis.

Some recent advances in information technology are important to changing readiness. The move to object-oriented programming may seem to be a technical nicety, but in reality it makes computer systems highly flexible. Instead of a program performing certain defined functions and those functions alone, object-oriented technology allows the users to string together the objects (or small, modular business transactions) so that processes are created within the system to address the needs of the organization. In fact, more than one set of object-oriented processes can be present within the system. This enables the company to serve different customers in different ways — according to their needs — but using a single, highly flexible system that can be readily adapted as the needs change.

The company must also become adept at changing the organization. It is not only the ability to make changes that is a critical skill; it is also the ability to recover quickly and effectively from the disruption caused by the changes. Like a lightweight boxer or a graceful gymnast, an agile organization can elegantly recover from any blow or disturbance. Practice at change is essential. Reorganization must become routine. An agile company will often need more than one organizational structure at the same time. Different customers will need to be served differently. These differences will often require different internal structures. These are the challenges of agility. Agility requires significant management skills, wide distribution of expertise and authority, local decision-making to address local customers, and highly skilled and trained people. Leadership, motivation, and trust must replace the traditional management style of command and control.

1.11.1 The Rise of the Knowledge Worker

A major trend underpinning the ability of an agile enterprise is its ability to coordinate through its knowledge workers. This increased need for coordination is necessitated by shorter product life cycles and is reflected in the changing makeup of the workforce. In fact, it seems that the need to coordinate has gone from a pernicious task to be gotten out of the way as quickly as possible, to becoming a central competency.

A study by the Educational Testing Service found that since the 1960s the number of office workers has risen from 30 to 40% of all workers. Greater proportions of these workers are professionals, and a lesser proportion is support staff, e.g., secretaries.

1.12 AGILE ENTERPRISE AND THE INTERNET

Any discussion about the evolution of the agile enterprise and its ultimate impact on our society will not be complete without a discussion of the Internet and how it is changing the rules of the competitive game. Whether a company is in the business of planning, sourcing, making, or delivering products, the Internet is changing the way work is done today. The traditional model of a vertically oriented enterprise is becoming obsolete. To set the valuation of a company, it is more important to know how flexible a company is in leveraging its core capability to other entities surrounding it. Quickly changing demand or rapid product technology turnover means that the company holding the fewest assets and the best information wins. Systems must be looking outwardly rather than focused inwardly. The name of the game is collaborative business processes that help determine demand, coordinate production, and optimize distribution. The increase in product market competition brought on by the accelerating use of the Internet and the World Wide Web has created a buyers' market, with compressed cycle times and unique product designs becoming the norm. Customers will no longer accept mass-produced products that only partially address their needs. Knowing that alternatives exist, customers expect services and products tailored to their specific requirements. As manufacturers struggle to meet this increasing demand for customer-tailored products, they face exponential increases in product complexity, unprecedented competitive pressures to bring products to market faster, and ever-increasing dependence on their supply chains. To compete, an agile enterprise must create an environment where both customers and partners can participate in the innovation process, and where new products can be delivered dynamically as the customer demand requires, all at competitive prices.

Focused on all phases of a product's life cycle from concept and definition to production, service, and retirement, collaborative product commerce (CPC) allows manufacturers to collaborate over the Internet with customers, suppliers, and partners throughout the development and delivery process. In one sense, the evolution of Internet has been compared to the great Industrial Revolution of the 19th century, where tremendous productivity gains were achieved in a relatively short period. Now the industrial economy of the past is giving way to the creative economy, and corporations are at another crossroad. Attributes that made them ideal for the 20th century could cripple them in the 21st. The Darwinian struggle of daily business will be won by the people — and the organizations — that adapt most successfully to the new world that is unfolding.

Converting from a traditional supply-chain concept to a customer-focused value chain in which all resources and processes are optimized toward serving customers faster and better is a challenging task for a company of any size. However, for growth-oriented, small- and medium-sized manufacturers and distributors, the prospect of implementing enterprise and supply-chain information technology solutions may be even more daunting.

The perception among many of these organizations is that making the leap from manual or nonintegrated automated business processes to totally integrated automated processes enabled by information technology (IT) would be too difficult, disruptive to the business, and expensive. Yet, the fact is, with the technology available today, enterprises failing to improve their business processes to deliver greater value to customers will be left in the dust by those who succeed.

For small- and medium-sized companies, the thing to remember is that implementing IT doesn't have to be an all-or-nothing proposition. Given companies' unique challenges, it is true that technology for the low-to-middle market has to be business focused, low maintenance, easy to implement, and easy to learn and use. In addition, implementation must be fairly rapid — and so must return-on-investment (ROI).

Small and midmarket companies seeking to adopt the value chain paradigm can succeed — if they select appropriate solutions in keeping with the scale and scope of their business and if they apply technology intelligently in those areas where it will add the most value.

1.12.1 Supply Chain Challenges

In many growing companies, the application of technology has been an evolution rather than a revolution. Companies typically start out with an off-the-shelf accounting package. Later, they may add software packages for specific functions — inventory management, for example, or bar coding and identification. For some manufacturers, large retail customers such as Wal-Mart and JCPenney may demand compliance with electronic data interchange (EDI) and advance shipping notice

(ASN) capabilities, so these may be implemented. But, by and large, many business processes within the small or medium-sized organization still are performed manually.

However, more and more companies have recognized the value of integrated information and have adopted it in one form or another in applications specifically designed for manufacturing, such as material requirements planning (MRP), manufacturing resources planning (MRP II), and manufacturing execution systems (MES). The problem is that these solutions, while offering a certain level of integration, focus almost exclusively on manufacturing and plant-floor operations rather than overall business processes. While shop-floor solutions often result in significant bottom-line savings through operational efficiencies and reduced waste, they cannot by themselves make the entire enterprise agile nor they can increase the value of the enterprise or build intrinsic value into customer relationships.

1.12.2 Growth and Value

When it comes to factory floor solutions, many small and medium-sized companies already are on their second or third generation software products, yet they are still searching for ways to differentiate themselves in the market and grow. Forward-thinking companies have begun to move in a more strategic direction — beyond the traditional "command-and-control" mentality, which focuses almost exclusively on applying technology for internal monitoring and control to cut costs. In reality, top-line growth is not the result of cost cutting. To achieve sustainable growth, companies need to take the next logical step and focus their efforts on those who can fuel that growth — their customers. Building a value chain involves integrating every aspect of the business to deliver optimum value to its customers. This is the surest path to building sustainable growth and long-term value.

If customer-focused production employees are to coordinate their efforts to serve customers better, there has to be a timely flow of actionable integration across these functional areas. The lack of seamless integration not only hampers efficiency but — more important — it inhibits the coordination of internal and external business processes and business partners that would otherwise add velocity and responsiveness to customer service.

Advances in technology have enabled the development of a whole range of new solution options for the small-to-medium-sized enterprise — from scalable enterprise resources planning (ERP) systems to shop-floor solutions to stand-alone supply chain applications. Many supply chain solutions tend to focus behind the scenes, with such functional modules as advanced planning, scheduling, and warehouse management, when what small and medium-sized manufacturers really need is to find more and better ways to interact with customers and deliver the value that keeps them coming back.

1.12.3 Impact of the Internet on Various Aspects of Agility

In the next few pages we will focus on how the advent of the Web and its ever-increasing nexus of information flow is changing the way future enterprises will be required to function.

1.12.4 Customer Orientation — The Rise of CRM (Customer Relationship Management)

There are enormous advantages in using the Internet to deepen and secure customers relationships such as being more accessible, providing better service, and locking in key relationships. To accomplish this, an enterprise needs to design an information system that is open and that can be integrated with all the supply chain partner's applications. As a result, those businesses that are most flexible and have the quickest response time will succeed.

With the current trend toward consolidation of markets and companies within industries, achieving differentiation in a specific market space has become more critical than ever. In this competitive climate, small and mid-tier companies may find it difficult to figure out just how to differentiate their company and its products.

In some industries, goods have become commoditized to the point where the product is no longer the chief differentiator. Examples of commoditization range from consumer foods and beverages such as cereal, coffee, and beer, to industrial components such as mechanical and electronic parts. As a result of commoditization, profit margins are being squeezed beyond all reason simply because manufacturers believe the only option they have is to compete on price.

In other industries, both individual and business-to-business (B2B) customers have become more sophisticated and demanding. In these areas, price is no longer the prime factor. Realizing that manufacturers are willing to compete mightily for their business, customers are raising the bar on a number of fronts, including customization of products (e.g., multipacks), delivery time, individualized packaging options, and customized transportation choices.

Given such challenges, how can a small or medium-sized company differentiate itself from its competitors? More and more the answer is value-added customer service. A fusion of products and services is occurring, with service becoming the prime differentiator.

Putting customers first is the driving force behind the growing popularity of customer relationship management (CRM) systems. Sometimes referred to as the next generation of sales force automation (SFA), these systems integrate sales and marketing information with all transactional information related to getting products to customers when, where, and how they want them. Rather than being internally focused, CRM focuses on front office, or customer-facing, processes with an emphasis on delivering a high level of personalized customer contact and care.

According to ISM, the Bethesda, Maryland-based research and consulting firm, CRM is big business. Already a $40 billion industry, CRM is expected to grow more than 40% per year for the next 5 years. This alone is ample proof that more companies are beginning to realize the urgency of building a more customer-focused value chain. However, to create value, CRM needs to be customized to the way a company does business and must also be integrated into the enterprise system — tasks requiring more technology infrastructure than most companies can afford.

Agile enterprises of the future will develop "learning relationships" — remembering what the customer wants and making the product and services better as a result. Amazon.com is an Internet pioneer, studying the books its customers buy and making future recommendations based on what they are reading. Dell Computer, which sells built-to-order PCs, remembers what customers have ordered in the past and uses individualized Web pages to make it simpler with every subsequent order to add new computers. Getting real-time customer feedback and acting on that feedback are going to be the norm of an agile enterprise. Everyone in the company should be listening, not just the sales department. Anyone can visit an online chat room and find out what customers are saying about their products.

1.12.4.1 What Will It Take to Keep the Customer in the Future?

- Customized Product: No more off-the-rack items. Customers need products designed to their specs in everything.
- Personalized Marketing: Customers will want ads about the products they want. Send it through E-mail, airwaves, or magazine pages. If it is news they can use, they will pay attention.
- No-Excuses Service: Sales staff must be trained to respond to customer concerns as if they are the most important things in the world. Ban the phrase, "It's not my department."
- Rapid Change: An agile enterprise will not wait to make these shifts. Customers may already be shopping the competition.

1.12.4.2 A Value Chain Proposition

Hand in hand with growing emphasis on CRM is another major trend — E-commerce — that already is redefining the way companies do business. With transactions over the Internet and World Wide Web gaining greater acceptance globally, E-commerce is growing at an amazing rate. According to Framingham, Massachusetts-based International Data Corporaton (IDC), retail Internet revenues will hit $29 billion by 2002, and Web-enabled B2B revenues are expected to soar as high as $66 billion.

E-commerce has two distinct sides — B2B and business-to-consumer (B2C). For midmarket manufacturers, B2B E-commerce is not a new concept. Many of them have engaged in EDI at one level or another to transact business with their suppliers and customers. However, B2B E-commerce has been expanded to include Web-enabled media such as corporate intranets and extranets.

Whether traditional EDI or Internet-enabled, B2B E-commerce can help companies forge closer links across the entire supply chain — from suppliers at one end, though internal processes, to distributors and retailers at the other. Publishing a product catalog, or allowing dealers to check on inventory or order product online can greatly streamline business processes. Product configuration is also on manufacturers' minds today. The challenge for software developers is to build a rules-based configurator that can be integrated with Web sites, and CRM and ERP systems. It is a tall order, one that even the software giants have not yet fully solved, but it is coming.

The benefits of enhancing supply chain visibility and partner communications in this way often include shorter lead times, lower inventories, reduced work-in-process (WIP), more accurate forecasting, more efficient production scheduling, and a higher level of customer responsiveness.

On the B2C side, the growing number of Internet-literate consumers has led hordes of companies — from the largest retailers to the smallest providers of consumer goods and services — to set up Web sites and sell directly to consumers via these electronic storefronts. This is a far different world from B2B E-commerce.

Through their electronic storefronts, some manufacturers are building a community with customers by adding value to the goods they sell. For example, amazon.com will make recommendations on book and music selections you might enjoy, based on tracking your previous purchases. Dell Computer will help you configure a computer system that fits your needs, then take your order through a secure credit card transaction on the spot. This kind of convenient, personalized service — available 24 hours per day, 7 days per week — builds customer loyalty and retention. Virgin Atlantic Airlines is using the Internet to streamline its far-flung operation. For instance, it is tapping into the Net to improve the efficiency of its supply chain. The airline now buys most of its new and used parts online. Whenever mechanics need a part, they log on and place their order — instead of Virgin having to stock a complete array of plane parts. This just-in-time approach has helped the carrier achieve great savings by reducing the amount of inventory it needs to warehouse. If a plane is stranded on the tarmac or in the hangar because of a faulty part, Virgin Atlantic can check the Web for a local supplier that stocks that part and have it sent to the runway in a matter of hours, something that would have been impossible 3 years ago.

Does this mean that every company should rush out and build a Web site based on the assumption of "If we build it, they will come?" Definitely not. On the contrary, before launching an E-commerce project, whether it is B2B or an electronic storefront, much research should be done. Complex Web initiatives can be an expensive proposition.

Even now, while Dell is raking in sales at its computer site, amazon.com has yet to turn a penny of profit, despite large sales volumes. The reason is that this particular E-commerce strategy has to reach critical mass before it becomes profitable. And, in addition to the initial investment, companies typically have to deploy considerable skill sets — usually from outside the organization — to design and set up the site, as well as maintain and refresh it.

For companies considering a Web presence, the first step is to research successful sites and analyze them for design, ease of navigation, usefulness of content, and interactive functions. The next step is to determine realistically whether such an initiative will provide value and ROI, and will further the strategic objectives of the enterprise. The importance of an Internet strategy based on value to customers and value to the company cannot be overemphasized. A failed Web initiative is far worse than no initiative at all.

1.12.4.2.1 Functional Requirements

In manufacturing industries, the concept of the value chain has evolved over time and tends to be defined by retailers and distributor demands for EDI and ASNs, as well as by the need for regulatory compliance, for example, to OSHA and FDA labeling rules.

Today the value chain is also being driven by end-user preferences and demands, as well as by a company's own strategic objectives. For profit-minded businesses, a customer-oriented value chain can be a viable path to sustainable growth through higher productivity and increased market share. To achieve these goals, companies are seeking to reduce operating and inventory costs, streamline production, shorten order fulfillment and time to delivery, and maximize profit margins and return on assets.

To build an effective value chain, it is essential to put process before technology and examine the issue from purely a business standpoint. Initially, the basic decision steps must:

Identify problems: Is the company's weak point excessive inventory or WIP? Material or machine bottlenecks? Inefficient scheduling or poor resource utilization? Longer than average time-to-delivery? Whatever the problem, it must be clearly identified before it can be solved.

Pinpoint the goals: Determine specifically what operational and business improvements you want to achieve in solving the problems, e.g., lower inventories, reduced operating costs, faster turnover and order fulfillment, higher productivity and capacity, and improved return on assets.

Rethink business processes: It does not pay to automate bad processes. If certain business processes are identified as part of the problem, it will be necessary to rethink and perhaps reengineer these processes to bring them into alignment with the industry's best practices. Target processes to check might include order entry, procurement and inventory management, logistics (i.e., transportation and shipping) management, and data collection.

Determine where technology can help: Based on the results of the first three steps, identify IT components that will automate key processes to help the company serve customers faster and better, make it more competitive, and enable it to achieve the targeted goals.

Evaluate marketplace technology solutions: Customer-focused functional requirements are simply pieces of supply chain execution systems that represent how a manufacturer brings its products to its customers. In a scalable enterprise system, these pieces can be unbundled into discrete, yet integrated components that serve a specific business need.

Once a company has taken these steps, it needs to stabilize and automate the business processes that will add the most value for the customer and the enterprise. Reducing costs by increasing production speed and efficiency and by gaining better control of materials and resources is a worthy goal that can contribute to bottom-line savings. However, the real value of an integrated technology solution is the top-line growth created by the ability to serve new markets, gain new customers, and provide existing customers with innovative new products and a high level of service. That is why customer-centric solutions such as CRM and business intelligence are gaining popularity.

1.12.4.2.2 Reaping Business Benefits from IT

In applying technology to build the value chain, a commonsense approach works best — technology is never a solution unto itself. To add value to the enterprise, IT must be applied intelligently and strategically to where it will do the most good. Companies should always start with the core business drivers behind the technology imperative and determine exactly what it is they want to accomplish. To do this, the myth of all-or-nothing thinking must be shattered.

With today's open architecture and communications standards, application integration and scalability are possible. With the right solutions, a company can implement only those modules it needs now, and add functionality as business needs expand and grow. Integration is the key and this is typically the strength of a single-vendor solution. While best-of-breed solutions may be attractive initially, they can be difficult to implement — and even more difficult to interface with legacy data and other information systems.

This is not to imply that a single vendor can necessarily provide all the pieces required to solve all the problems. But if the vendor selects design software with open standards and open architecture, this will allow other third-party solutions to hook in at critical process points to integrate and automate more of the processes.

1.12.4.2.3 Setting the Stage for Success

Once the IT solution has been selected, time and thought should be given to its implementation and the training of users. Many smaller organizations do not fully realize that enterprise applications are not plug-and-play solutions. It is critical to allocate ample time, budget, and internal resources (or external, if required) to ensure the effectiveness of the IT infrastructure and the success of its implementation.

Since most companies don't have a wealth of in-house expertise to call upon, there can be significant value in working with technology partners, such as value-added resellers, to assess the needs, evaluate the hardware and software requirements, implement the solution for maximum value, and train end-users in its use and maintenance.

Taking a business-focused approach to transforming the traditional supply chain into a customer-focused value chain, with an eye toward ROI, is a sound decision in that technology such as Web becomes the enabler it is meant to be. Intelligently applied, integrated software solutions make good processes better, slow processes faster, and valuable information infinitely more accessible to everyone who needs it, across the enterprise and beyond.

1.12.5 The Future of the Agile Enterprise

1.12.5.1 Idea-Centric Society

As the industrial economy of the 20th century gives way to the creative economy of the 21st century, attributes that made enterprises ideal for the 20th century could cripple them in the 21st century. So they will have to change — dramatically. Industrial economies have gotten so efficient at producing food and physical goods that most of the workforce has been freed to provide services or to produce abstract goods: data, software, news. The Bureau of Labor Statistics projects that by 2005 the percentage of workers employed in manufacturing will fall below 20%, the lowest level since 1850. In an economy based on ideas rather than physical capital, the potential for breakaway successes such as Yahoo! is far greater. Power and money will flow to corporations with indispensable intellectual property. The most important intellectual property is inside every employee's head.

In the old economy, shareholders owned assets that were physical, such as coal mines. But when vital assets are people, there can be no true ownership. The best that a corporation can do is to create an environment that makes the best people want to stay. Enduring relations with employees become an enormous asset, because those employees are what connect the company to its partners. Managers in this new economy must go by new set of rules. The Internet gives everyone in an organization the ability to access a mind-boggling array of information — instantaneously, from anywhere. That means that the 21st century corporation must adapt itself to management by the Web. Leading edge technology will enable workers on the bottom rungs of the organization to seize opportunity as it arises. Employees will increasingly feel the pressure to get breakthrough ideas to market first. Thus the corporation will need to nurture an array of formal and informal networks to ensure that these ideas can speed into development. The rapid flow of information will permeate the organization. Orders will be fulfilled electronically without a single phone call or piece of paper. The "virtual financial close" will put real time sales and profit figures at every manager's fingertips via the wireless phone or a spoken computer command.

1.12.5.2 The Agile Enterprises of the Future Will Have Certain Defining Characteristics

1.12.5.2.1 Management by Web

This means not just Web as Internet but also the web shape of successful organizations in the future. Agile enterprises of the 21st century will look like a web, a flat, intricately woven form that links partners, employees, external contractors, suppliers, and customers in various collaborations. Managing this intricate network of partners, spin-off enterprises, contractors, and freelancers will be as important as managing internal operations.

1.12.5.2.2 Information Management

The most profitable enterprises will manage information instead of solely focusing on physical assets. Sheer size will no longer be the hallmark of success; instead, the market will prize the ability to efficiently deploy assets and leverage information. Good information management can enable an upstart to beat an established player. By using information to manage themselves and better serve their customers, companies will be able to do things cheaper, faster, and with far less waste.

1.12.5.2.3 Mass Customization

The past 100 years have been marked by mass production and mass consumption. The company of the future will tailor its products to each individual by turning customers into partners and giving them the technology to design and demand exactly what they want. Mass customization will result in waves of individualized products and services, as well as huge savings for companies, which will no longer have to guess what and how much customers want. The Procter & Gamble spin-off, Reflect.com LLC, an online cosmetics merchant, is a harbinger of things to come. By answering a series of queries ranging from color preferences to skin type, consumers can custom design up to 50,000 different formulations of cosmetics and perfumes. When they are done, they can even design the packaging for the products. Customers given the option of mixing their own shades are not as likely to try comparison shopping.

1.12.5.3 Dependence on Intellectual Capital

The advantage of bringing breakthrough products to market first will be shorter lived than ever, because technology such as the Internet will let competitors match or exceed them almost instantly. To keep ahead of the new steep product curve, it will be crucial for enterprises to attract and retain the best thinkers. Companies will need to build a deep reservoir of talent, drawing on both employees and free agents. They will need to create the kind of cultures and reward systems that keep the best minds engaged. The old command-and-control hierarchies must give way to organizations that empower vast numbers of people and reward them as if they were owners of the enterprise.

1.12.5.4 Global

The agile enterprise of the future will call on talent and resources, especially intellectual capital, wherever they can be found around the globe, just as it will sell its products and services around the globe. The new global corporation might be based in the United States but do its software programming in Sri Lanka, its engineering in Germany, its manufacturing in China. The Net will seamlessly connect every outpost so that far-flung employees and freelancers can work together in real time.

1.12.5.5 Speed

The Internet is a tool, and the biggest impact of that tool is speed. The speed of action and speed of deliberations and speed of information have increased. Speed in every aspect of the product life cycle will be critical.

1.12.5.6 Flexible Facilities and Virtual Organizations

The 21st century corporation will not have one ideal form. Some will be completely virtual, wholly dependent on a network of suppliers, manufacturers, and distributors for their survival. Some of the most successful companies will be very small and very specialized. Some enterprises will last no longer than the time it takes to reach the market. Once it does, these temporary organizations will pass their innovations to host companies that can leverage them more quickly and at less expense. The reason is that every enterprise has capabilities as well as disabilities such as deeply held beliefs, rituals, practices, and traditions that often smother radical thinking.

2 Benefiting from Six Sigma Quality

Jonathon L. Andell

To benefit from *Six Sigma* first requires knowing what it is. There are various definitions of Six Sigma. Table 2.1 presents some of the confusing array of descriptions.

Each of these definitions contains an element of truth. Six Sigma includes quantitative and problem-solving aspects, along with underlying management issues. What makes Six Sigma successful is less about doing anything new than it is of finally following what has been advocated for decades. The alleged failures ascribed to TQM and a variety of other "initiatives" are usually the result of a departure from well-founded counsel.

This chapter starts with a discussion of Six Sigma's historical context, including factors that distinguish the success stories from lesser outcomes. Following this are some thoughts on how Six Sigma benefits the bottom line of an organization when implemented effectively. Finally, the chapter takes a look at what characterizes the so-called Six Sigma organization.

Many references address the need for problem-solving experts, champions, and other specific individuals. Before we discuss this, we compare departmental duties between traditional and Six Sigma organizations, and finally provide some project management guidelines on how to implement a successful Six Sigma effort.

Throughout the discussion are contrasting examples of what happens in an "ideally Six Sigma" vs. an extremely traditional organization. Although no organization personifies every characteristic of either extreme, every example is based on an actual experience or observation.

Discussion of how the problem-solving methodology actually works appears in Chapter 14.

2.1 A BRIEF HISTORY OF QUALITY AND SIX SIGMA

Certain approaches to quality have been around for ages, such as standards for performing work and auditing to evaluate compliance to those standards. However, compliance to standards does not guarantee satisfactory outcomes. For instance, records show that HMS *Titanic* conformed to many rigorous standards.

Most modern quality concepts have originated since the onset of the Industrial Revolution. Prior to that, an effective and dependable product could only be made slowly and painstakingly by hand; quality and economy could not coexist. Though mass production enhanced access to products, their quality was often poor by today's standards.

1-57444-300-3/02/$0.00+$1.50

TABLE 2.1
"Six Sigma Is..."

A management system	No, it's a statistical methodology.
A quality philosophy based on sound . . . . fundamental principles	No, it's an arbitrary defect rate. (3.4 parts per million [ppm]).
A vast improvement over the flawed total quality. . . . management (TQM) system	No, it's new feathers on an old hat: quality tools that have been around for decades.
A comprehensive approach to improving all . . . . aspects of running an organization	No, it's a person with a hammer, trying to treat the entire world like a nail.
A stunning success story	No, it's a stupendous waste of resources.

However, two major contributions early in the 20th century made it not only feasible, but downright indispensable, to merge quality with economy. Sadly, a lingering misconception is this so-called *tyranny of the or*,* the notion that one must choose between quality and cost. We will return to this topic from time to time during this chapter.

One contribution, attributed to Sir Ronald Fisher, is an efficient way of gathering and analyzing data from a process called statistical design of experiments, or DOE. The other is Walter Shewhart's recognition that variation in a process can be attributed primarily to what many modern practitioners call "common" vs. "special" causes. Shewhart developed a data-driven methodology to recognize and respond to such causes, a methodology currently referred to as *statistical process control* (SPC). Both topics are covered as individual chapters of this handbook.

Although DOE was used widely in agriculture, neither technique saw extensive industrial application until the United States entered World War II. To meet armaments manufacturers' urgent requirements for maximum output, dependable performance, and minimal waste, Shewhart and many of his distinguished colleagues brought SPC to shop floors. It would be arrogant to presume that this was the sole reason for America's wartime success, but these methodologies contributed substantially to the unprecedented productivity levels that ensued.

However, after the war ended, the use of these quality management tools diverged widely throughout the world. This divergence had profound implications in subsequent decades.

One extreme took place in the Western world, particularly the United States. During the war, many workers had been part of the armed forces. Many returned to their old jobs, but lacked the SPC skills instilled in the temporary workforce. Simultaneously, the nation's sense of urgency diminished. In fact, buoyed by pride in what had been achieved, manufacturing management became downright complacent. The result was that relatively few managers appreciated the benefits of statistical methods or quality management, and few postwar workers received the training to implement the tools.

* Collins and Porras, *Built to Last,* NY: Harper Business, 1994, 44.

The other extreme took place in those nations defeated in the same war, notably Japan. Determined not to repeat the Versailles blunders following World War I, the Allies strove to secure lasting peace by giving the vanquished nations a fighting chance at prosperity. Among the many decisions to ensue from that policy was a request that Shewhart provide guidance to Japanese manufacturers. Due to advancing age, he recommended instead a "youthful" associate, Dr. W. Edwards Deming.

Deming, Dr. Joseph Juran, and numerous others gave the Japanese some tools to accelerate their economic recovery. Those included SPC and DOE, along with how to use quality as a strategic management tool. As the Japanese grew comfortable applying the methodologies, their own pioneers began to emerge: Taguchi, Shingo, Ishikawa, Imai, and others.

By the late 1970s and early 1980s, Japan's reputation for quality had undergone a remarkable transformation. Their success has been discussed at great length, but a few anecdotal examples warrant mention:

- One Japanese company could build and ship a copy machine to the United States at a lower cost than the inventors of photocopying could deliver a comparable unit to their own shipping dock.
- A typical design cycle for a Japanese automobile was 50 to 60% of the equivalent U.S. cycle, and the resulting vehicles contained discernibly fewer design defects.
- Technical developments patented in the United States frequently were brought to market solely by Japanese firms.

There may have been merit to some claims of dumping — exporting goods with government-subsidized, artificially low prices — but the above facts show that there was vastly more to Japan's success than price cuts alone could accomplish.

Thus, two postwar developments — Japan's embracing of quality and Western complacency — led to numerous "rude awakenings" in Western industry later. Perhaps the most profound realization was that quality had become inextricably linked with competitive strength in those industries that had at least one dominant quality player. Government intervention alone was not enough to enable Western industry to survive and flourish in this new age.

Industries in Western countries responded in a number of ways, many successful and some less so. The Malcolm Baldrige National Quality Award in the United States (like comparable awards of other nations) has focused attention on a select few firms who use quality tools to drive organizational excellence. A "mutual fund" of Baldrige winners has outperformed Standard & Poor's 500 by a factor of two or more since its inception. Success stories such as Motorola in the late 1980s, Allied Signal in the early 1990s, and General Electric vastly outnumber the alleged failures such as Florida Power & Light's.*

* In truth, Florida Power & Light (FP&L) reveals more about what happens when an organization dismantles its quality program than it does about such a program failing.

Sadly, however, there also have been some disappointments:

- During the SPC fad, control charts sprouted like proverbial weeds. Unfortunately, few managers bothered to interpret them, and fewer still permitted employees to invoke appropriate responses. As a result, the charts had minimal impact on outcomes.
- Dazzled by Japanese quality circles, representatives of warring factions were directed to convene and do likewise — without training, infrastructure support, or motivation for different outcomes. Although some successes can be reported, often the sole benefit was isolation of the war zone to a single theatre.
- Stubbornly refusing to recognize the crucial difference between awareness and what Deming called "profound knowledge," organizations slashed weeks of training to days and tried to achieve in months, or even weeks, what had taken years to germinate in Japan.
- ISO 9000 has been touted by some as a certification of world-class quality, spawning an entire industry of consultants and registrars. In reality, ISO 9000 represents a valid baseline of achievement, but falls well short of creating a Six Sigma organization. Thus, the number of ISO 9000 certifications vastly exceeds the number of truly world-class organizations in existence.

Western industry has had many practitioners who appreciate these shortcomings: the aforementioned Deming and Juran, along with Joiner, Peters, Feigenbaum, Shainin, and many others. Sadly, however, many managers chose to eschew the rigorous demands of these experts, opting instead to cast their lot with practitioners whose appreciation may have been less profound. The so-called failures of total quality management (TQM) (and a vast array of similar other quality approaches currently lumped under that appellation) are highly correlated with the decision to yield to the quick fix.

Six Sigma is not a new philosophy, a new set of problem-solving tools, or a new expert skill level. In fact, many highly effective Six Sigma practitioners appear to have repackaged prior offerings under this popular new title!

What *is* new is that industry leaders such as Lawrence Bossidy (formerly CEO of Allied Signal, now Honeywell International) and Jack Welch (formerly CEO of General Electric) accepted personal responsibility for making Six Sigma succeed. They finally heeded the sine qua non shared by TQM and Six Sigma: It starts at the top. A chief executive officer alone cannot make a Six Sigma organization, but surely Six Sigma stands no chance without the deep personal commitment of the top executive.

Some enthusiasts insist that Six Sigma differs from fads in its focus on customers, its integration across entire organizations, its strategic targeting of problems to attack, and in the degree of improvement achieved by the typical project. However, the best practitioners of TQM understood those issues every bit as well as today's Johnny-come-lately Six Sigma practitioners do. To reiterate: The sole difference is that,

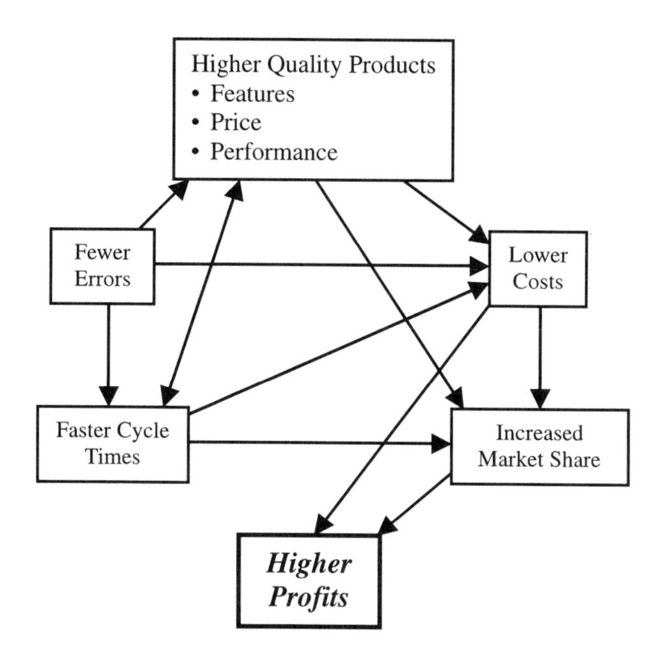

FIGURE 2.1 How Six Sigma drives the bottom line.

finally, business leaders have awakened to the mandate — and the benefits — of making this a personal commitment.

Quite frankly, impugning TQM practitioners is like blaming HMS *Titanic*'s shortage of lifeboats on the rowboat manufacturers. The goods were offered, but the decision-makers were not buying. Rather than berate the practitioners, let us rejoice that, at long last, decision-makers appreciate and accept their roles in making Six Sigma successful.

2.2 HOW SIX SIGMA AFFECTS THE BOTTOM LINE

There are many kinds of organizations. They could be classified by considering whether they exist to make a profit, or by whether their customers buy a manufactured or a service product. However, no matter the categorization, they all receive funding, which is expended to achieve organizational objectives. To the extent that Six Sigma reduces waste, even non-profit (e.g., governmental, educational, religious, or philanthropic) establishments can expend less of their budgets internally, thus freeing more funds for the benefit of their customers.

However, this book focuses on the manufacturer, presumably one who intends to turn a profit. Figure 2.1 uses a quality tool called an interrelationship diagraph to display how the benefits of Six Sigma contribute to one another and ultimately to the capitalistic success of a manufacturer — or of any business, for that matter.

Please note the comparative tone of the adjectives, *higher, lower,* etc. The meaning is that better performance is always possible, no matter how well an organization performs. In fact, if the reader's competition is reading and heeding this publication, continuous improvement well might be less a matter of domination and more one of survival.

Later, we will address how to undertake the transformation toward Figure 2.1. First, however, consider how the opposite condition comes about (after all, nobody sets out to create or operate an inefficient organization). When you have an appreciation of how a non-Six Sigma organization comes to be, the steps to rectify the situation may make more sense.

Organizations usually start small and grow (even spin-off businesses do this until they are rendered independent). As a result of this growth, tasks formerly done by one or two people eventually are performed so frequently that the job function(s) must be staffed. Unless a formal methodology is used, the ways various tasks — or processes — are done tend to propagate almost haphazardly.

Such organizational growth, along with the lack of formal process development or analysis, leads to a vast number of processes with shortcomings, which play havoc on the bottom line. Some examples are

- Unnecessary approval cycles, resulting in late deliveries, work lost in piles of paper, time wasted chasing down signatures, and decisions based on "How can I get this signed?" rather than "What best serves the customer?"
- Steps in the wrong sequence, increasing defects and rework — thus wasting resources
- Steps or subprocesses that benefit one part of the business at the expense of other parts
- Errors or defects in the delivered products that consume resources and drive away business

It is vital to recognize that these shortcomings also apply to processes off the factory floor: sales, order entry, accounting, etc. In fact, it is possible for a manufacturing defect to be due primarily to a "transactional" process. An example would be a perfectly designed and manufactured product that was not the one the customer wanted, reflecting an error in the process that converted customer orders into shop orders.

Even if an organization has yet to apply Six Sigma analysis to its processes, management is often acutely aware that things are going poorly. A common response is to determine who touched the process last and "counsel" that poor soul (such a benign-sounding euphemism!). Not only does this not solve the problem, but it also adds a brand-new category of loss: employee turnover.

What is the alternative? Six Sigma. Let us examine what a Six Sigma organization looks like. Afterward, we will review some roles and responsibilities associated with successful Six Sigma programs. Once the obligations and players are identified, it will be easier to see how implementation happens.

2.3 CHARACTERISTICS OF A SIX SIGMA ORGANIZATION

To start down the path toward Six Sigma, let us develop a vision of life "on the other side of the rainbow." A simple definition of a Six Sigma organization might be that the bulk of its decision-making supports and sustains the outcomes described in Figure 2.1. Of course, those outcomes depend on some day-to-day characteristics, listed and discussed below.

Please note: Although many organizations successfully display some of the following characteristics, becoming a true Six Sigma organization depends on being effective at all of them.

2.3.1 Customer Focus

The selection and execution of every project start with three critical questions about the process: (1) What are the deliverables of this process? (2) Who receives them? and (3) What are their requirements?

It is tempting to overestimate our understanding of these issues. Some common lapses include

- Excluding crucial customer communities. For a manufacturer of automobile components, the factory floor's customers (with deliverables indicated in parentheses) might include shipping, auto manufacturer, repair shop, driver of the car (the manufactured product), government (reports and data), engineering (prototypes), management, accounting, sales (data), and so on. Many departments erroneously believe they have but one customer and one deliverable.
- Favoring easy-to-measure over necessary-to-measure. For example, manufacturers frequently scrutinize the features and quality of the delivered product, while neglecting service products that might drive customers away. Manufacturers must understand all the products they provide and must know the truth about their ability to satisfy customers in every regard.
- Presuming full awareness of customers' priorities. Frequently, we can generate an accurate list of things about which customers might care. It is quite rare for us as suppliers to rank those requirements correctly.

Any one of these can lead to improvements that don't benefit customers, while ignoring major sore points. That's a substantial waste of organizational resources. The Six Sigma organization invests wisely in order to know the customers and requirements for every process. Throughout subsequent problem-solving activities, the ultimate test of any proposed change becomes "How will this benefit the customers?"

2.3.2 Everybody on the Same Page

Some managers avoid overemphasizing specific programs, customers, or product lines lest a change in the environment be interpreted as their failure. When pressed to identify priorities, they spout platitudes about there being no trivial tasks, followed by threats toward the underling who fails to deliver across the board.

Of course, when "everything is priority number one," the reality becomes that everybody is left to set his or her own priorities. With this approach, crucial competitive initiatives get no more priority than ones that could be delayed or even scrapped. Furthermore, since every effort is regarded as urgent, efforts to obtain budgets and personnel become monumental yet needless battles in which one department must lose so that another can win.

In the Six Sigma organization, top management owns up to its obligation to establish and communicate a fundamental direction and vision. Then the organization mobilizes to align priorities, resources, projects, metrics, and rewards. People don't have to wonder, "Why am I doing this?" because the reason is incorporated into the marching orders of the tasks.

2.3.3 Extensive and Effective Data Usage

The discussion on "Fanatical Customer Focus" mentioned the requirement to determine what our customers need and how to measure it. Objective, quantifiable measures — what Deming called "data-driven" management — replace opinions, power struggles, and politics as the dominant bases of decision-making. To paraphrase some Motorola pundits:

If we can't quantify it, we can't understand it.
If we can't understand it, we can't control it.
If we can't control it, it controls us.

Vince Lombardi put it even better: "If you aren't keeping score, it's only practice." Just as Six Sigma tasks and projects have a "food chain" up to the organization's top priorities, so do the things we measure. In the broadest sense, we measure the following:

- *Customer Satisfaction:* the core metrics of how a Six Sigma organization measures up against its competition
- *Process Performance:* the key internal indicators that drive customer satisfaction, determined near the outset of Six Sigma projects
- *Process Inputs:* those factors objectively demonstrated to control process performance upon completion of a Six Sigma project
- *Organizational Indicators:* metrics that track whether people's behaviors support the metrics listed above and are aligned with strategic objectives
- *Cost of Poor Quality:* the penalties that an organization pays for failing to meet customer requirements, for waste and rework — ultimately, the cost of bad decisions

Make no mistake about it, the task of determining what to measure and how, is far from trivial. Making a metric "bullet-proof," that is, robust against playing games with the numbers, takes a lot of work. On top of that, the organization and its environment are in a constant state of flux, so even the best of metrics must be scrutinized periodically.

Finally, the entire organization must follow some straightforward but uncompromising rules regarding how the data are interpreted. This does not demand awesome statistical prowess. In fact, a high schooler can learn the basics in a day.

It does imply, though, that everybody up and down the organizational chart must measure and interpret performance using criteria that are objective, shared, and understood by all.

2.3.4 Empowerment: Autonomy, Accountability, and Guidance

Just because something is factual does not mean it will be accepted. Columbus, Magellan, and many of their partners shouldered considerable personal risk before most people finally accepted the fact that the world is in fact round. In that spirit, here is a statement that riles highly traditional managers, but is absolutely ironclad in its certainty:

We cannot expect the best effort from people who don't feel trusted and respected.

This is a major personal obstacle against the transition to a Six Sigma organization. Not only must management behave in new ways, but also those being managed must respond differently than before. One should anticipate major resistance here.

Ultimately, empowerment is the recognition that routine process decisions are best left to those doing the work. Here's how to make empowerment a practical aspect of Six Sigma:

- Give people the *autonomy* to make appropriate "line-of-sight" decisions without supervisory approval. This may mean that appropriately trained operators might decide how to configure their workspace, when to perform maintenance, and so on. It does not confer the authority to approve a $250,000 expenditure.
- Build in *accountability* to ward off anarchy. Although employees at Ritz-Carlton Hotels have authority to spend $100 without prior approval, spending it on a drunken binge almost certainly would precipitate severe consequences. Likewise, management's obligation not to let abusers off the hook is often a challenge, because enforcement initially increases one's workload.
- Provide *guidance* so people know how far their authority goes. Once the organization is well into Six Sigma, management is consulted mainly when the boundaries warrant widening.

2.3.5 Reward Systems that Support Objectives

The surest way to derail a Six Sigma effort is to reward people for avoiding it, and to punish people for practicing it. Unfortunately, many traditional performance measurements do just that. Some examples:

- *Production Volume.* People rewarded solely for how much stuff they jam through the factory — or who inevitably face punishment for failing to do so — know that protecting the customer comes at great personal risk.

- *Sales Commission Structure.* If a product line carries a high commission, personal outcomes might conflict with the customer's best interests. The Six Sigma organization assumes responsibility for aligning sales incentives with customer needs.
- *Reporting and Correcting Defects.* Traditional supervisors insist that empowerment is like "putting inmates in charge of the asylum" — a clear message that those doing the work can't be trusted to make decisions. However, those "untrustworthy" workers are the first to bear the brunt when mistakes do occur. As a result, mistakes are often hidden and passed along to where they cost vastly more to rectify.
- *Shooting the Messenger.* Rather than resolving situations, management becomes defensive and retaliates against those who point out problems. The Six Sigma organization strives to reward people for behaviors that align with customer needs. A structure is established where pointing out problems constitutes neither attack nor suicide. Only in such an environment can breakthrough levels of improvement pervade the organization.

2.3.6 Relentless Improvement

Notice that the right side of Figure 2.1 — lower costs, increased market share, and profits — is driven by the left side: reductions in errors and cycle times along with higher quality products. Its workings are reminiscent of a bicycle: the front wheel (financial outcomes) steers and the rear wheel (process improvements) drives. The Six Sigma organization uses customer focus, a single vision, data, empowerment, and rewards to drive improvements where they are needed most.

The need for improvement never disappears. As targeted improvements are realized, previously low-priority issues emerge as new targets. Furthermore, priorities evolve along with technology, markets, and competitors' strengths. Thus, the Six Sigma organization remains in a constant state of identifying, prioritizing, and attacking opportunities for improvement.

2.4 DEPARTMENTAL ROLES AND RESPONSIBILITIES

The dominant challenge of becoming a Six Sigma organization is not in finding opportunities to improve, finding and developing talent, or applying problem-solving tools. These tasks have proven methodologies.

The hardest part is changing the way the people and departments in the organization work with one another. Everybody, starting with the person in charge, has to address the two themes of empowerment and data analysis. At the risk of redundancy, let us review the need to abandon Taylorism and to embody the teachings of Shewhart.

Traditional management unconsciously applies the model developed by Frederick Taylor near the beginning of the Industrial Revolution. It is based on two beliefs: (1) everything works when managers do the thinking and "worker bees" follow the instructions, and (2) things go wrong only when instructions aren't followed.

The Six Sigma philosophy, like TQM, recognizes that those who best understand a process are those who are immersed in its daily operation. Any executive who wishes to test this theory should try doing another's work for an hour or two. It inevitably is far tougher than it looks, frequently due to well-intended management directives. Empowerment is the antithesis of Taylorism.

Complementing Taylor-engendered distrust for workers is management's failure to distinguish whether their work reflects common cause or special cause variation. The concept is explained fully in Chapter 15 but let's consider a working example.

Suppose an automobile's average fuel efficiency is 20 miles per gallon (MPG). Readings of 22 MPG for one tankful, and 18 MPG for another, are expected. This reflects "common cause" variation. The only way to add, say, 5 MPG would be through major modifications of some sort. Thus, a value of 30 MPG might arouse the suspicion of a measurement error. Likewise, a value of 10 MPG could mean that repairs were warranted. Extreme readings represent "special cause" variation. One can deal with special cause incidents individually, but not common cause incidents. That's why it makes sense to visit the garage after noting a reading of 10 MPG, while a checkup following every 18-MPG reading would be pointless.

Applying special cause responses to common cause problems is a colossal waste of resources — and a cherished tradition in highly traditional environments. Probably 80 to 95% of the times that somebody is chastised for an unwanted outcome (punishment assuredly is a special cause "solution"), the underlying process actually reflects common cause variation. People are being penalized for no greater offense than being on the job while the process behaves normally!

Abandoning Taylorism and adopting Shewhart's teachings, and using these changes as the first steps toward being a Six Sigma organization, tend to represent radical departures from many organizations' approaches — even if top management is truly enlightened regarding Six Sigma! In the next section we will address how to anticipate and handle the inevitable resistance to such changes.

For now, however, let us examine what those changes look like, since Six Sigma impacts the *what* and *how* of nearly every job in an organization. Table 2.2 summarizes role differences between a traditional and a Six Sigma organization. Since the table's entries are cryptic, we will elaborate on specific roles.

2.4.1 Top Management

Whether he or she is called president, general manager, or grand high Pooh-Bah, the person with ultimate authority has some unique and specific tasks. To reiterate, Six Sigma starts at the top.

If the organization's leader truly expects employees to make Six Sigma decisions, his or her leadership had better be by example — it cannot be delegated. The top executive's actions must percolate through to his or her staff, thence to their staffs, and so on. Once again, we mean less Taylor and more Shewhart.

At the start of the Six Sigma journey, the top executive leads his or her staff in developing and communicating their vision with the guidance of an appropriate expert. As resistance is encountered, they must be steadfast in holding people

TABLE 2.2
Departmental Role Transitions

Who	Traditional Duties	Six Sigma Roles	Needs
Leadership	Impose will Take heat for decisions	Vision Model behaviors Enforce reward system Allocate resources	Courage Consistency Integrity
Cost accounting	Gatekeeper of expenditures and budgets	Drive COPQ[a] tracking Validate savings Ensure funding	Training Accurate data
Information technology	Screen requests Implement solutions	Implement COPQ[a] Revise data systems Collection Access and reporting	Priorities Resources
Human resources	Enforce policy	Reward system Legality Application Communications	Data Timing Expectations Outcomes
Factory management	Move product Develop processes Discipline workers	Empowerment Accountability	Training and resources Reward system
Sales and marketing	Close every sale	Customer advocacy Source of data Market Satisfaction Forecast	Reward system Data: specs/$/dates Training and resources
Engineering and design	Technical expertise Product designs technology driven	Technical resource Market driven designs Concurrent designs Manufacturable	Reward system Training resources Data Customer needs Product performance
Quality	Enforce compliance Sell Six Sigma	Training Consulting Facilitation	Resources Reward system

[a] COPQ = cost of poor quality.

accountable — few organizations complete the transition without some involuntary departures along the way. Additionally, the staff must force availability of people and funds before the infrastructure can drive such decisions objectively.

Once the organization achieves a sort of Six Sigma steady state, the staff continues to lead by example, using Six Sigma techniques to make crucial decisions. As the organization gradually acclimates to its new culture, executives spearhead positive reinforcement of desired behaviors throughout the organization. The ongoing obligation to set and communicate strategy becomes an integral part of how the staff functions.

2.4.2 Cost Accounting

Customary financial controls reflect perhaps the ultimate Tayloristic notion: that the only group motivated to preserve cash flow is the one responsible for reporting it. Six Sigma demands a new way of thinking. Cost accounting becomes the resource for a continually improved understanding of the cost of poor quality (COPQ). In turn, the rest of the organization must provide much more detailed and accurate data than ever before. This requires overcoming entrenched mutual distrust; once again, top management's clarity and consistency will be put to the test.

During the sustaining phases of Six Sigma, the cost accounting department becomes the reality check for claims of project savings and the advocate to allocate resources where potential benefits are greatest. Ultimately, they compile defensible summations of financial benefits attributable to Six Sigma.

2.4.3 Information Technology

Conventional information technology (IT) groups often establish priorities on behalf of the entire organization (after all, every request is "number one priority"), in order to restrict their workload within budgetary constraints. Some IT groups also favor technical elitism over customer focus.

Existing data systems almost always need modifications, if not outright replacement, in order to support the Six Sigma organization. The IT department must adopt a "fanatical customer focus" at the outset of a Six Sigma transition, since an internally focused group cannot even contemplate such an ambitious undertaking. In return, the remainder of the business must provide IT with resource support and clear priorities.

Once the new data system is operational, IT will be the resource for continuous improvement in gathering, understanding, and sharing information. Rather than fending off requests from the rest of the business, the transformed IT organization needs to be vigilant in identifying and proposing opportunities to drive such improvements.

2.4.4 Human Resources

Typical human resources (HR) departments have diverse obligations: some are conscripted as the official mouthpiece of the pre-Six Sigma status quo, while others espouse enlightened but unsupported ideals; occasionally, they must shoulder both duties. A Six Sigma HR group ensures that proposed reward system revisions conform to legal and regulatory requirements — not as obstructionist gatekeepers, but by using Six Sigma problem-solving methods to identify and deploy plausible alternatives.

They communicate organizational changes consistently and clearly. Since the credibility of the entire Six Sigma effort hinges on whether the words match management's actions, HR must channel feedback upward. Once the transition is well underway, it is the responsibility of the HR group to apply the reward system fairly and consistently.

2.4.5 Factory Management

Factory managers often bemoan their inability to handle anything other than getting product out the door. The Six Sigma organization must establish and enforce requirements to measure more. Implementing team findings will require customized training for operators and supervisors. Empowerment with accountability becomes indispensable for improvements to become permanent. Management must allocate resources for people to attend training and team meetings, gather data, and conduct DOE runs, all without crippling the very production that brings in revenue.

2.4.6 Sales and Marketing

Without a clear vision of which markets a business serves, and with which products, the people who close sales get their sole direction from a catalog and a commission structure. This puts the organization at risk of providing the customer with less-than-optimal solutions. The Six Sigma sales force has the tools to drive customer satisfaction, which in turn drives business success. If commissions remain, they should align markets, products, and customer needs. Ironically, the Six Sigma organization actually may refer some business to competitors, just to ensure customer satisfaction.

The other side of the coin comes into play, too; the people who interact most with customers become a resource for a customer-focused organization. They must obtain and relay crucial information about marketing opportunities, customer satisfaction issues, and sales forecasts, and they must do so with accurate and objective data. In order to bring about these skills, training will be needed — potentially as much training as the problem-solving experts get. In order to ensure compliance, accountability must be enforced consistently and fairly.

2.4.7 Engineering and Design

Traditional product design is yet another bastion of Taylorism. Inputs from Manufacturing or Quality are perceived as distracting; those from Sales are considered downright irrelevant. Design quality is measured strictly in terms of technical specifications whose connection to customer requirements may be tenuous at best. Failure to meet said specifications is attributed to factory deficiencies. Using DOE and SPC are said to detract from the designer's "art." Needless to say, a fully "traditional" design community is rife with potential for resistance against Six Sigma.

The Six Sigma business holds Design accountable for ensuring widespread participation throughout the design process, for validating and addressing the requirements of diverse customer communities, and for applying appropriate methodologies along the way. In return for such radical changes, the remainder of the organization must allocate resources to meet a design process that demands participants who have the available time, knowledge, and decision-making authority to represent their departments effectively. Naturally, this transition will not transpire without an appropriate blend of training, guidance, and accountability.

2.4.8 Quality

Just think: we have discussed Six Sigma all this time before finally bringing up the quality department! It goes a long way to indicate where the real responsibility for Six Sigma lies. The next section, as well as Chapter 14, should more than compensate for any perceived shortcomings in attention devoted to quality experts.

Quality departments in traditional businesses often provide one final vestige of Taylorism: the notions that only "independent" assessors can be trusted to acquire and report data honestly, and that only adversarial process audits can prevent people from shirking their duties. Despite these perceptions, such organizations frequently have enlightened practitioners striving vainly to bring another paradigm to the business. When Six Sigma comes to town, these people frequently enjoy dramatic transformations from pariahs to heroes.

Ideally, the quality people can serve as invaluable internal consultants: sources of guidance and feedback to executives, providers and coordinators of training, and experts to facilitate initial uses of the problem-solving methodologies.

2.4.9 Other Organizations

We could include an array of other departments. For example, groups responsible for facilities, maintenance, safety, and environmental compliance all represent opportunities to identify customers and requirements, to reduce waste and rework, and to develop efficient processes. For now, let us note that the departments listed in Table 2.2 represent the minimum participants in making Six Sigma work for a manufacturing business. Each individual organization will have specifics to address.

2.5 INDIVIDUAL ROLES AND RESPONSIBILITIES

In addition to modifying departmental missions and obligations, Six Sigma also affects the job of nearly every individual. Table 2.3 shows how individuals contribute to Six Sigma, no matter the department. The roles presented below are specialists in aspects of Six Sigma, with the exception of team members and executive staff.

2.5.1 Executive Staff

The tasks of the executive staff have been discussed, but not how they attain the knowledge necessary to do the job. Most organizations provide customized training to the staff, covering 5 days of contact time over 3 to 5 weeks. Topics usually include

TABLE 2.3
Individual Assignments in Six Sigma

Role	#	Prerequisites	Training (Days)	Six Sigma Roles
Executive staff	5–10	Member of staff	Executive Six Sigma (5)	See Table 2.2
Coordinator	1	Master Trainer Project manager	Attend all executive and champion sessions	Top-level coordination Planning and metrics Facilitation and training Progress tracking
Champion	5–10	6σ problem solver Project manager	Practitioner (5–10) Change management (5)	Project selection Project implementation Progress tracking
Middle managers		Manager of supervisors	Orientation (3–5) 6σ project management (2)	Enforce reward system Eliminate obstacles Gather improvement data
Master	1 per 1000	Recognized expert Facilitator Trainer	Master (10–15) Change management (5) 6σ project management (2)	Advanced problem solving Mentor to experts Train-the-trainer
Expert	1 per 100	Recognized practitioner	Expert (30–40) Facilitation (5) Train-the-trainer (5) 6σ project management (2)	Lead teams and projects Mentor to practitioners Trainer
Practitioner	1 per 12–25	6σ problem solver People skills	Practitioner (5–10) Understanding people (2–3)	Coordinate task work Data entry and analysis
Sponsor/ supervisor	1 per project	Authority over process being studied	Basic problem solving (1–2) Change management (2) 6σ project management (2)	Implement team findings Enforce reward system Track improvements
Team member	All	Current job assignments	Basic problem solving (1–2) Understanding people (1–2)	Attend team meetings Complete action items

- Benefits of Six Sigma
- Shortcomings of Taylorism
- Variation: common cause vs. special cause
- Change management
- Project management in a Six Sigma environment

This may not seem like a lot of training, considering the overwhelming personal changes required of the staff. That is where the coordinator and the implementation plan come in.

2.5.2 Coordinator

Consider the world-class athlete, blessed with natural gifts and an outstanding work ethic. Certainly the executive staffs of manufacturing businesses have an analogous combination of skill and will. However, unlike athletes, executives perceive a stigma against seeking out personal trainers. Many Six Sigma initiatives have been crippled by executives' steadfast refusal to acknowledge that a single topic might lie outside their realm of expertise.

For those who accept our shared human limitations, the Six Sigma Coordinator is akin to the personal trainer.* She or he maps out the game plan, facilitates executive sessions, provides feedback, develops and conducts a lot of just-in-time training, and generally ensures that executive actions and decisions are as constructive as possible. Clearly, this job demands consummate Six Sigma skills, to coordinate all aspects of organization-wide implementation and to facilitate applying the methods with the staff. This must be backed up with the credibility to reinforce assertions and the ability to balance when to take a stand and when to bide one's time.

2.5.3 Champions

Champions monitor and report the vital signs of the Six Sigma effort, as they strive to sustain an environment in which the new culture can thrive. As a rule of thumb, each major department needs access, and needs to provide access, to at least one designated champion.

Champions and the coordinator are a close team, sharing successes and working issues among departments. Just as the coordinator needs credibility at the highest level, champions must exert influence in departments. Since champions lack the expertise to serve as personal trainers, their contacts with the coordinator provide departments with access to the coordinator's Six Sigma skills on an as-needed basis.

Generally, champions initiate specific projects, as well as work to overcome obstacles the projects encounter such as funding, personnel support, resistance to changes, and so on. They compile progress reports on projects and high-level metrics.

2.5.4 Problem-Solving Practitioners, Experts, and Masters

Some organizations call them "green belts," "black belts," and "master black belts," respectively. Each level represents an increasing aptitude in solving problems and working with people and organizations. Masters and experts tend to be full-time positions, especially at the outset.

A major flaw propagated by many Six Sigma consultants is the elitist notion that every project needs an expert or a master — an insidious form of Taylorism. In reality, practitioners and line workers solve many of their own problems in a stable Six Sigma environment.

The projects that always call for an expert or master include (1) those whose priority and scope demand high-caliber leadership, such as to establish the Six Sigma infrastructure, (2) those crossing multiple departmental boundaries, and (3) those

---

* The author gratefully acknowledges Ms. Sandra Claudell for permission to use her idea.

involved when Six Sigma is new. As time goes on, the organization will develop the resources and experience to entrust teams led by practitioners.

2.5.5 Team Members and Supervisors

Those who do a task routinely should be on problem-solving teams. This isn't easy. Higher-ups fear loss of power and control, while workers wonder if this will cause more work or layoffs. All this resistance manifests itself as "lack of resources" or "no time." Later in the project, as the team proposes process changes, more resistance materializes based on the same fears: loss of power, control, or jobs.

There are two ways to address this: (1) consequences that encourage empowerment, with clear and truthful messages about power, control, and job security and (2) assigning experts or even masters to initial projects.

Those managers and supervisors destined to thrive in a Six Sigma environment will come to see how Six Sigma leads to the outcomes of Figure 2.1. They will manage an implementation project, getting from "as is" to "should be" in an aggressive yet feasible time frame. Tasks along the way include training and testing, revising procedures — and the who, what, when, where, how and how much of gathering, understanding, reporting, and responding to new kinds of process data.

Likewise, team members destined for Six Sigma success will start to appreciate the fact that empowerment works in their interest, and will initiate their own improvement projects. Within these enthusiastic workers and supervisors reside the seeds of future practitioners, experts, and maybe even a master or a champion. It has happened more than once.

2.6 SIX SIGMA IMPLEMENTATION STRATEGIES

Many organizations urgently need results in the first 6 to 12 months, even if short-term improvements are dwarfed by subsequent opportunities. A business that effectively handles the project management aspects of Six Sigma can enjoy both. The good news: handling task issues is the easy part. The bad news: handling task issues is the easy part.

We have said it before: Six Sigma starts at the top. The situation described in Figure 2.1 will neither start nor continue without leaders bringing to bear vast amounts of will and skill, along with a willingness to learn. Of course, one rarely ascends to leadership without those characteristics.

The difference with Six Sigma is subtle but crucial. Not only does it demand that executives learn new skills, but it also demands that they forget others.

Consider the implications. Becoming an executive is the culmination of years of behaviors that are a cherished and integral aspect of one's very success. And now Six Sigma requires executives to trade in those comfy old shoes for new ones that guarantee downright painful moments! Not only that, but just about everybody else will be issued new shoes somewhere along the way, with like implications. No wonder responses to cultural change resemble grief — we mourn the death of our beloved status quo.

Thus, rolling out Six Sigma presents two challenges: (1) the logistical aspects, along with (2) getting people to make personal changes — starting with the top person in the organization. It is imperative to recognize that addressing the second

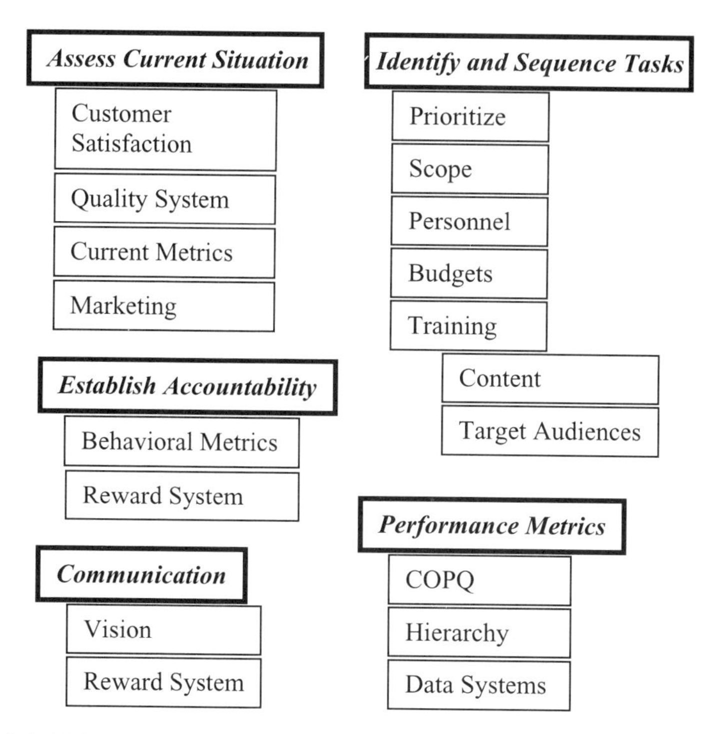

FIGURE 2.2 Major organizational tasks.

challenge is neither optional nor trivial. Without widespread personal changes, the outcomes of Six Sigma inevitably disappoint.

Exacerbating the challenge is the fact that every situation is unique, depending on the products, the competition, customer satisfaction, organizational culture, and so on. The reader is urged to avoid a one-size-fits-all approach, whether advocated by internal or external sources.

Fortunately, there are some overall guidelines, a set of questions we need to ask and answer in order to implement Six Sigma effectively. Figure 2.2 (an "affinity diagram") shows one way to organize the high-level issues that must be addressed.

In every case, the organization's Six Sigma coordinator is expected to play a major role in ensuring that the questions are asked and answered correctly, and the executive staff is expected to provide resources as well as their own time and effort.

2.6.1 Assess Current Situation

In order to customize Six Sigma to the situation, a clear picture of that situation is necessary. There are four components to that picture:

- *Customer Satisfaction.* To apply "fanatical customer focus" appropriately, we must make certain we know our customers' priorities and perceptions. Although formal surveys yield the best data, they are costly and slow, so interim approaches should be considered as well.

- *Quality System.* An objective assessment of the current quality system will reveal an organization's strengths and weaknesses. Many state governments and corporations offer effective assessment tools, based on the Malcolm Baldrige National Quality Award, which are more comprehensive than ones based on ISO 9000.
- *Current Metrics.* The organization should compile all metrics, including how they are gathered and for what they are used. Later, each will be scrutinized and retained, modified, or scrapped, and the compilation will become a basis for strategic planning through the years.
- *Marketing.* The firm's business plan contains most if not all pertinent information. A Six Sigma organization uses these data to help prioritize customer segments for surveying, and to help select the manufactured product areas where Six Sigma improvement projects are needed most.

2.6.2 Establish Accountability and Communication

For people to make behavioral changes, they must know (1) the desired behaviors, (2) why the changes are beneficial, (3) that behaviors will be tracked objectively, (4) the positive consequences associated with desired behaviors, (5) the negative consequences associated with unwanted behaviors, and (6) the certainty of both positive and negative consequences.

Items (1) and (2) derive their power from the executive vision of the Six Sigma organization. Item (3) comes from developing effective metrics to track people's behaviors. Items (4), (5), and (6) represent the reward system.

These six factors start with executive staff, but also link each individual's tasks with organizational needs. Throughout the effort, vigilance and scrutiny ensure that the system supports the correct behaviors, with minimal fudging. Measures and rewards will need to evolve — and be communicated — as the Six Sigma program matures.

2.6.3 Identify and Sequence Tasks

This activity establishes much of the Six Sigma infrastructure. It starts at the outset of the organization's commitment to Six Sigma, but also uses assessment results for fine-tuning. The Six Sigma coordinator facilitates numerous sessions with senior staff and their staffs to establish realistic priorities, sequence, personnel, and budgets.

Realistic means that mission-critical projects are assigned to masters and established experts, with time frames appropriate to the scope. Experts-in-training need projects to develop their skills, meaning that major payoffs will be the exception. All training should be as just-in-time as possible, so the new skills can be put to work right away.

2.6.4 Performance Metrics

Having too many metrics is as bad as having too few. The organization should track Six Sigma with five or six top-level metrics, each supported by five or six more. The coordinator and the executive staff develop the primary metrics. Once these are disseminated, department staffs and masters develop secondary metrics; subordinate metrics are developed in turn by line organizations. Thus, every metric has a linkage to at least one top-level metric.

Initiating metrics begins with gathering data to determine their starting performance levels, including the amounts of common cause variation associated with each. As Six Sigma progresses, charts clearly display improvements. Champions and the coordinator compile reports and work issues regarding primary and secondary metrics.

Existing systems rarely have the capability to provide data automatically. This means that some of the infrastructure work is to define, design, fund, and implement a new data system. Usually this must include revisions to the cost accounting systems, to support tracking cost of poor quality (COPQ). Until the revised system comes on line, resources must be allocated for gathering data manually.

2.7 CONCLUSION

Six Sigma can bring profound improvements to an organization. However, it is not easy. It demands profound changes of an organization: first, on the part of its leaders, and eventually, on the part of everybody else. All will be tested along the way.

So why do people do it? In this author's experience, the common thread seems to be this:

- Because it really works
- Because it makes things better
- Because it lets everyone make a positive difference

Or, as a mentor once said: "Happiness isn't a destination; it's the shoes one puts on in the morning." When taken with others, Six Sigma is a wonderfully rewarding journey. May it be so with you.

3 Design of Experiments

Jack B. ReVelle, Ph.D.

3.1 OVERVIEW

Design of experiments (DOE) does not sound like a production tool. Most people who are not familiar with the subject might think that DOE sounds more like something from research and development. The fact is that DOE is at the very heart of a process improvement flow that will help a manufacturing manager obtain what he or she most wants in production, a smooth and efficient operation. DOE can appear complicated at first, but many researchers, writers, and software engineers have turned this concept into a useful tool for application in every manufacturing operation. Don't let the concept of an experiment turn you away from the application of this most useful tool. DOEs can be structured to obtain useful information in the most efficient way possible.

3.2 BACKGROUND

DOEs grew out of the need to plan efficient experiments in agriculture in England during the early part of the 20th century. Agriculture poses unique problems for experimentation. The farmer has little control over the quality of soil and no control whatsoever over the weather. This means that a promising new hybrid seed in a field with poor soil could show a reduced yield when compared with a less effective hybrid planted in a better soil. Alternatively, weather or soil could cause a new seed to appear better, prompting a costly change for farmers when the results actually stemmed from more favorable growing conditions during the experiment. Although these considerations are more exaggerated for farmers, the same factors affect manufacturing. We strive to make our operations consistent, but there are slight differences from machine to machine, operator to operator, shift to shift, supplier to supplier, lot to lot, and plant to plant. These differences can affect results during experimentation with the introduction of a new material or even a small change in a process, thus leading to incorrect conclusions.

In addition, the long lead time necessary to obtain results in agriculture (the growing season) and to repeat an experiment if necessary require that experiments be efficient and well planned. After the experiment starts, it is too late to include another factor; it must wait till next season. This same discipline is useful in manufacturing. We want an experiment to give us the most useful information in the shortest time so our resources (personnel and equipment) can return to production.

One of the early pioneers in this field was Sir Ronald Fisher. He determined the initial methodology for separating the experimental variance between the factors and the underlying process and began his experimentation in biology and agriculture.

1-57444-300-3/02/$0.00+$1.50

The method he proposed we know today as ANalysis Of VAriance (ANOVA). There is more discussion on ANOVA later in this chapter. Other important researchers have been Box, Hunter, and Behnken. Each contributed to what are now known as classical DOE methods. Dr. Genichi Taguchi developed methods for experimentation that were adopted by many engineers. These methods and other related tools are now known as robust design, robust engineering, and Taguchi Methods™.

3.3 GLOSSARY OF TERMS AND ACRONYMS

TABLE 3.1
Glossary of Terms and Acronymns

Term	Definition
Confounding	When a design is used that does not explore all the factor level setting combinations, some interactions may be mixed with each other or with experimental factors such that the analysis cannot tell which factor contributes to or influences the magnitude of the response effect. When responses from interactions or factors are mixed, they are said to be *confounded.*
DOE	Design of experiments is also known as industrial experiments, experimental design, and design of industrial experiments.
Factor	A process setting or input to a process. For example, the temperature setting of an oven is a factor as is the type of raw material used.
Factor level settings	The combinations of factors and their settings for one or more runs of the experiment. For example, consider an experiment with three factors, each with two levels (H and L = high and low). The possible factor level settings are H-H-H, H-L-L, etc.
Factor space	The hypothetical space determined by the extremes of all the factors considered in the experiment. If there are *k* factors in the experiment, the factor space is *k*-dimensional.
Interaction	Factors are said to have an interaction when changes in one factor cause an increased or reduced response to changes in another factor or factors.
Randomization	After an experiment is planned, the order of the runs is randomized. This reduces the effect of uncontrolled changes in the environment such as tool wear, chemical depletion, warm-up, etc.
Replication	When each factor level setting combination is run more than one time, the experiment is *replicated.* Each run beyond the first one for a factor level setting combination is a *replicate.*
Response	The result to be measured and improved by the experiment. In most experiments there is one response, but it is certainly possible to be concerned about more than one response.

TABLE 3.1 (continued)
Glossary of Terms and Acronymns

Statistically significant	A factor or interaction is said to be statistically significant if its contribution to the variance of the experiment appears to be larger than would be expected from the normal variance of the process.

3.4 THEORY

This section approaches theory in two parts. The first part is a verbal, nontechnical discussion. The second part of the theory section covers a more technical, algebraic presentation that may be skipped if the reader desires to do so.

Here is the question facing a manager considering an experiment for a manufacturing line: What are my optimal process factors for the most efficient operation possible? There may be many factors to be considered in the typical process. One approach may be to choose a factor and change it to observe the result. Another approach might change two or three factors at the same time. It is possible that an experimenter will be lucky with either of these approaches and find an improvement. It is also possible that the real improvement is not discovered, is masked by other changes, or that a cheaper alternative is not discovered. In a true DOE, the most critical two, three, or four factors (although higher factors are certainly possible, most experiments are in this range) are identified and an experiment is designed to modify these factors in a planned, systematic way. The result can be not only knowledge about how the factors affect the process, but also how the factors interact with each other.

The following is a simple and more technical explanation of looking at the theory in an algebraic way. Let's consider the situation of a process with three factors: A, B, and C. For now we'll ignore interactions. The response of the system in algebraic form is given by

$$Y = \beta_0 + \beta_1 X_A + \beta_2 X_B + \beta_3 X_C + \varepsilon \tag{3.1}$$

where β_0 is the intercept, β_1, β_2, and β_3 are the coefficients for the factor levels represented by X_A, X_B, and X_C, and ε represents the inherent process variability. Setting aside ε for a while, we remember from basic algebra that we need four distinct experimental runs to obtain an estimate for β_0, β_1, β_2, and β_3 (note that ε and β_0 are both constants and cannot be separated in this example). This is based on the need for at least four different equations to solve for four unknowns.

The algebraic explanation in the previous paragraph is close to the underlying principles of experimentation but, like many explanations constructed for simplicity, it is incomplete. The point is that we need at least four pieces of information (four equations) to solve for four unknowns. However, an experiment is constructed to provide sufficient information to solve for the unknowns *and* to help the experimenter determine if the results are statistically significant. In most cases this requires that an experiment consist of more runs than would be required from the algebraic perspective.

3.5 EXAMPLE APPLICATIONS AND PRACTICAL TIPS

3.5.1 Using Structured DOEs to Optimize Process-Setting Targets

The most useful application for DOEs is to optimize a process. This is achieved by determining which factors in a process may have the greatest effect on the response. The target factors are placed in a DOE so the factors are adjusted in a planned way, and the output is analyzed with respect to the factor level setting combination.

An example that the author was involved in dealt with a UV-curing process for a medical product. This process used intense ultraviolet (UV) light to cure an adhesive applied to two plastic components. The process flow was for an operator to assemble the parts, apply the adhesive, and place the assembly on a conveyor belt that passed the assembly under a bank of UV lights. The responses of concern were the degree of cure as well as bond strength. An additional response involved color of the assembly since the UV light had a tendency to change the color of some components if the light was too intense. The team involved with developing this process determined that the critical factors were most likely conveyor speed, strength of the UV source (the bulb output diminishes over time), and the height of the UV source. Additionally, some thought that placement of the assembly on the belt (orientation with respect to the UV source bulbs), could have an effect, so this factor was added.

An experiment was planned and the results analyzed for this UV-curing process. The team learned that the orientation of the assemblies on the belt was significant and that one particular orientation led to a more consistent adhesive cure. This type of find is especially important in manufacturing because there is essentially no additional cost to this benefit. Occasionally, an experiment result indicates that the desired process improvement can be achieved, but only at a cost that must be balanced against the gain from improvement. Additional information acquired by the team: the assembly color was affected least when the UV source was farther from the assemblies (not surprising), and sufficient cure and bond strength were attainable when the assemblies were either quickly passed close to the source or dwelt longer at a greater distance from the source. What surprised the team was the penalty they would pay for process speed. When the assembly was passed close to the light, they could speed the conveyor up and obtain sufficient cure, but there were always a small number of discolored assemblies. In addition, the shorter time made the process more sensitive to degradation of the UV light, requiring more preventive maintenance to change the source bulbs. The team chose to set the process up with a slower conveyor speed and the light source farther from the belt. This created an optimal balance between assembly throughput, reduction in defective assemblies, and preventive line maintenance.

Another DOE with which the author was involved was aimed at improving a laser welding process. This process was an aerospace application wherein a laser welder was used to assemble a microwave wave guide and antenna assembly. The process was plagued with a significant amount of rework, ranging from 20 to 50% of the assemblies. The reworked assemblies required hand filing of nubs created on the back of the assembly if the weld beam had burned through the parts. The welder had gone through numerous adjustments and refurbishment over the years. Support engineering believed that the variation they were experiencing was due to attempted piecemeal improvements and that they must develop an optimum setting that would still probably result in rework, but the result would be steady performance. The experiment was conducted using focus depth, power level, and laser pulse width (the laser was not continuous, rather it fired at a given power level for a controlled time period or pulse). The team found that the power level and pulse width ranges they had been using over the years had an essentially negligible impact on the weld. The key parameter was the beam focus depth. What's more, upon further investigation, the team found that the method of setting the focus depth was imprecise and, thus, dependent on operator experience and visual acuity. To fix this process, the team had a small tool fabricated and installed in the process to help the operator consistently set the proper laser beam focus. This resulted in a reduction of rework to nearly zero!

3.5.2 Using Structured DOEs to Establish Process Limits

Manufacturers know it is difficult to maintain a process when the factor settings are not permitted any variation and the limits on the settings are quite small. Such a process, often called a "point" process, may be indicative of high sensitivity to input parameters. Alternatively, it may indicate a lack of knowledge of the effect of process settings and a desire to control the process tightly *just in case.*

To determine allowable process settings for key parameters, place these factors in a DOE and monitor the key process outputs. If the process outputs remain in specification and especially if the process outputs exhibit significant margin within the factor space, the process settings are certainly acceptable for manufacturing. To determine the output margin, an experimenter can run sufficient experimental replicates to assess process capability (C_{pk}) or process performance (P_{pk}). If the output is not acceptable in parts of the factor space, the experimenter can determine which portion of the factor space would yield acceptable results.

3.5.3 Using Structured DOEs to Guide New Design Features and Tolerances

As stated previously, DOE is often used in development work to assess the differences between two potential designs, materials, etc. This sounds like development work only, not manufacturing. Properly done, DOE can serve both purposes.

3.5.4 Planning for a DOE

Planning for a DOE is not particularly challenging, but there are some approaches to use that help to avoid pitfalls. The first and most important concept is to include many process stakeholders in the planning effort. Ideally, the planning group should include at least one representative each from design, production technical support, and production operators. It is not necessary to assemble a big group, but these functions should all be represented.

The rationale for their inclusion is to obtain their input in both the planning and the execution of the experiment. As you can imagine, experiments are not done every day, and communication is necessary to understand the objective, the plan, and the order of execution.

When the planning team is assembled, start by brainstorming the factors that may be included in the experiment. These may be tabulated (listed) and then prioritized. One tool that is frequently used for brainstorming factors is a cause-and-effect diagram, also known as a fishbone or Ishikawa diagram. This tool helps prompt the planning team on some elements to be considered as experimental factors.

Newcomers to DOE may be overly enthusiastic and want to include too many factors in the experiment. Although it is desirable to include as many factors as are considered significant, it must be remembered that each factor brings a cost. For example, consider an experiment with five factors, each at two levels. When all possible combinations are included in the experiment (this is called a full factorial design), the experiment will take $2^5 = 32$ runs to complete each factor level setting combination just once! As will be discussed later, replicating an experiment at least once is very desirable. For this experiment, one replication will take 64 runs. In general, if an experiment has k factors at two levels, l factors at three levels, and m factors at four levels, the number of runs to complete every experimental factor level setting is given by $2^k * 3^l * 4^m$. As you can see, the size of the experiment can grow quickly. It is important to prioritize the possible factors for the experiment and include what are thought to be the most significant ones with respect to the time and material that can be devoted to the DOE on the given process.

If it is desirable to experiment with a large number of factors, there are ways to reduce the size of the experiment. Some methods involve reducing the number of levels for the factors. It is not usually necessary to run factors at levels higher than three, and often three levels is unnecessary. In most cases, responses are linear over the range of experimental values and two levels are sufficient. As a rule of thumb, it is not necessary to experiment with factors at more than two levels unless the factors are qualitative (material types, suppliers, etc.) or the response is expected to be nonlinear (quadratic, exponential, or logarithmic) due to known physical phenomena.

Another method to reduce the size of the experiment is somewhat beyond the scope of this chapter, but is discussed in sufficient detail to provide some additional guidance. A full factorial design is generally desirable because it allows the experimenter to assess not only the significance of each factor, but *all* the interactions between the factors. For example, given factors T (temperature), P (pressure), and M (material) in an experiment, a full factorial design can detect the significance of T, P, and M as well as interactions TP, TM, PM, and TPM. There is a class of experiments wherein the experimenter deliberately reduces the size of the experiment and gives up some of the resulting potential information by a strategic reduction in factor level setting combinations. This class is generally called "fractional factorial" experiments because the result is a fraction of the full factorial design. For example, a half-fractional experiment would consist of 2^{n-1} factor level setting combinations. Many fractional factorial designs have been developed such that the design gives up information on some or all of the potential interactions (the formal term for this loss of information is *confounding* — the interaction is not lost, it is confounded or mixed with another interaction's or factor's result). To use one of these designs, the experimenter should consult one or more of the reference books listed at the end of this chapter or employ one of the enumerated software applications. These will have guidance tables or selection options to guide you to a design. In general, employ designs that confound higher level interactions (three-way, four-way, etc.). Avoid designs that confound individual factors with each other or two-way interactions (AB, AC, etc.) and, if possible, use a design that preserves two-way interactions. Most experimental practitioners will tell you that three-way or better interactions are not detected often and are not usually of engineering significance even if noted.

The next part of planning the experiment is to determine the factor levels. Factor levels fall into two general categories. Some factors are quantitative and cover a range of possible settings; temperature is one example. Often these factors are continuous. A subset of this type of factor is one with an ordered set of levels. An example of this is high-medium-low fan settings. Some experimental factors are known as attribute or qualitative factors. These include material types, suppliers, operators, etc. The distinction between these two types of factors really drives the experimental analysis and sometimes the experimental planning. For example, while experimenting with the temperatures 100, 125, and 150°C, a regression could be performed and it could identify the optimum temperature as something between the three experimental settings, say 133°C, for example. While experimenting with three materials, A, B, and C, one does not often have the option of selecting a material part way between A and B if such a material is not on the market!

Continuing our discussion of factor levels, the attribute factors are generally given. Quantitative factors pose the problem of selecting the levels for the experiment. Generally, the levels should be set wide enough apart to allow identification of differences, but not so wide as to ruin the experiment or cause misleading settings. Consider curing a material at ~100°C. If your oven maintains temperature ± 5°C, then an experiment of 95, 100, 105°C may be a waste of time. At the same time, an experiment of 50, 100, 150°C may be so broad that the lower temperature material doesn't cure and the higher temperature material burns. Experimental levels of 90, 100, and 110°C are likely to be more appropriate.

After the experiment is planned, it is important to randomize the order of the runs. Randomization is the key to preventing some environmental factor that changes over time from confounding with an experimental factor. For example, let's suppose you are experimenting with reducing chatter on a milling machine. You are experimenting with cutting speed and material from two suppliers, A and B. If you run all of A's samples first, would you expect tool wear to affect the output when B is run? Using randomization, the order would be mixed so that each material sample has an equal probability of the application of either a fresh or a dulled cutting edge.

Randomization can be accomplished by sorting on random numbers added to the rows in a spreadsheet. Another method is to add telephone numbers taken sequentially from the phone book to each run and sort the runs by these numbers. You can also draw the numbers from a hat or any other method that removes the human bias.

When you conduct an experiment that includes replicates, you may be tempted to randomize the factor level setting combinations and run the replicates back-to-back while at the combination setting. This is less desirable than full randomization for the reasons given previously. Sometimes, an experiment is difficult to fully randomize due to the nature of experimental elements. For example, an experiment on a heat-treat oven or furnace for ceramics may be difficult to fully randomize because of the time involved with changing the oven temperature. In this case, one can relax the randomization somewhat and randomize factor level combinations while allowing the replicates at each factor level setting combination to go back-to-back. Randomization can also be achieved by randomizing how material is assigned to the individual runs.

3.5.5 Executing the DOE Efficiently

The experimenter will find it important to bring all the personnel who may handle experimental material into the planning at some point for training. Every experimenter has had one or more experiments ruined by someone who didn't understand the objective or significance of the experimental steps. Errors of this sort include mixing the material (not maintaining traceability to the experimental runs), running all the material at the same setting (not changing process setting according to plan), and other instances of Murphy's Law that may enter the experiment. It is also advisable to train everyone involved with the experiment to write down times, settings, and variances that may be observed. The latter might include maintenance performed on a process during the experiment, erratic gauge readings, shift changes, power losses, etc. The astute experimenter must also recognize that when an operator makes errors, you can't berate the operator and expect cooperation on the next trial of the experiment. Everyone involved will know what happened and the next time there is a problem with your experiment, you'll be the last to know exactly what went wrong!

3.5.6 Interpreting the DOE Results

In the year 2000, DOEs were most often analyzed using a statistical software package that provided analysis capabilities such as ANalysis Of VAriance (ANOVA) and regression. ANOVA is a statistical analysis technique that decomposes the variation of experimental results into the variance from experimental factors (and their interactions if the experiment supported such analysis) and the underlying variation of the process. Using statistical tests, ANOVA designates which factors (and interactions) are statistically significant and which are not. In this context, if a factor is statistically significant, it means that the observed data are not likely to normally result from the process. Stated another way, the factor had a discernible effect on the process. If a factor or interaction is not determined to be statistically significant, the effect is not discernible from the background process variation under the experimental conditions. The way that most statistical software packages implementing ANOVA identify significance is by estimating a p-value for factors and interactions. A p-value indicates the probability that the resulting variance from the given factor or interaction would normally occur, given the underlying process. When the p-value is low, the variance shown by the factor or interaction is less likely to have normally occurred. Generally, experimenters use a p-value of 0.05 as a cut-off point. When a p-value is less than 0.05, that factor/interaction is said to be statistically significant.

Regression is an experimental technique that attempts to fit an equation to the data. For example, if the experiment involves two factors, A and B, the experimenter would be interested in fitting the following equation:

$$Y = \beta_0 + \beta_1 X_A + \beta_2 X_B + \beta_{12} X_{AB} + \varepsilon \tag{3.2}$$

Regression software packages develop estimates for the constant (β_0) as well as the coefficients (β_A, β_B, and β_{AB}) of the variable terms. If there are sufficient experimental runs, regression packages also provide an estimate for the process standard deviation (ε). As with ANOVA, regression identifies which factors and interactions are significant. The way regression packages do this is to identify a p-value for each coefficient. As with ANOVA, experimenters generally tend to use a p-value of 0.05 as a cut-off point. Any coefficient p-value that is less than 0.05 indicates that the corresponding factor or interaction is statistically significant.

These are powerful tools and are quite useful, but are a little beyond further detailed discussion in this chapter. See some of the references provided for a more detailed explanation of these tools. If you do not have a statistical package to support ANOVA or regression, there are two options available for your analysis. The first option is to use the built-in ANOVA and regression packages in an office spreadsheet such as Microsoft Excel. The regression package in Excel is quite good; however, the ANOVA package is somewhat limited. Another option is to analyze the data graphically. For example, suppose you conduct an experiment with two factors (A and B) at two levels (2^2) and you do three replicates (a total of 16 runs). Use a bar chart or a scatter plot of factor A at both of its levels (each of the two levels will have eight data points). Then use a bar chart or scatter plot of factor B at both of its levels (each of the two levels will have eight data points). Finally, to show interactions, create a line chart with one line representing factor A and one line for factor B. Each line will show the average at the corresponding factor's level. Although this approach will not have statistical support, it may give you a path to pursue.

3.5.7 Types of Experiments

As stated in previous paragraphs, there are two main types of experiments found in the existing literature. These are full factorial experiments and fractional factorial experiments. The pros and cons of these experiments have already been discussed and will not be covered again. However, there are other types of DOEs that are frequently mentioned in other writings.

Before discussing the details of these other types, let's look at Figure 3.1a.

We see a Venn Diagram with three overlapping circles. Each circle represents a specific school or approach to designed experiments: classical methods (one thinks

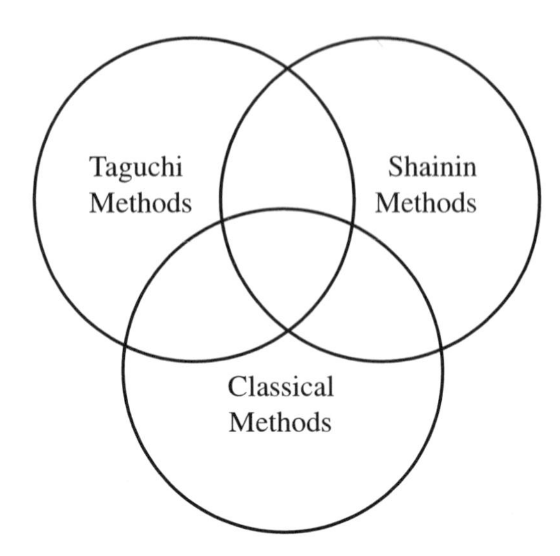

FIGURE 3.1a Design of experiments — I.

of Drs. George Box and Douglas Montgomery), Taguchi Methods (referring to Dr. Genichi Taguchi), and statistical engineering (established and taught by Dorian Shainin). In Figure 3.1b we see that all three approaches share a common focus, i.e., the factorial principle referred to earlier in this chapter. Figure 3.1c demonstrates that each pairing of approaches shares a common focus or orientation, one approach with another. Finally, in Figure 3.1d, it is clear that each individual approach possesses its own unique focus or orientation.

The predominant type of nonclassical experiment that is most often discussed is named after Dr. Genichi Taguchi and is usually referred to as Taguchi Methods or robust design, and occasionally as quality engineering. Taguchi experiments are fractional factorial experiments. In that regard, the experimental structures are not as significantly different as is Dr. Taguchi's presentation of the experimental arrays and his approach to the analysis of results. Some practicing statisticians do not promote Dr. Taguchi's experimental arrays due to opinions that other experimental approaches are superior. Despite this, many knowledgeable DOE professionals have noted that practicing engineers seem to grasp experimental methods as presented by Dr. Taguchi more readily than methods advocated by classical statisticians and quality engineers. It may be that Dr. Taguchi's use of graphical analysis is a help. Although ANOVA and regression have strong grounds in statistics and are very powerful, telling an engineer which factors and interactions are important is less effective than showing him or her the direction of effects using graphical analysis.

Despite the relatively small controversy regarding Taguchi Methods, Dr. Taguchi's contributions to DOE thinking remain. This influence runs from the promotion of his experimental tools such as the signal-to-noise ratio and orthogonal array and, perhaps more importantly, his promotion of using experiments designed to reduce the influence of process variation and uncontrollable factors. Dr. Taguchi would describe uncontrollable factors, often called noise factors, as elements in a process

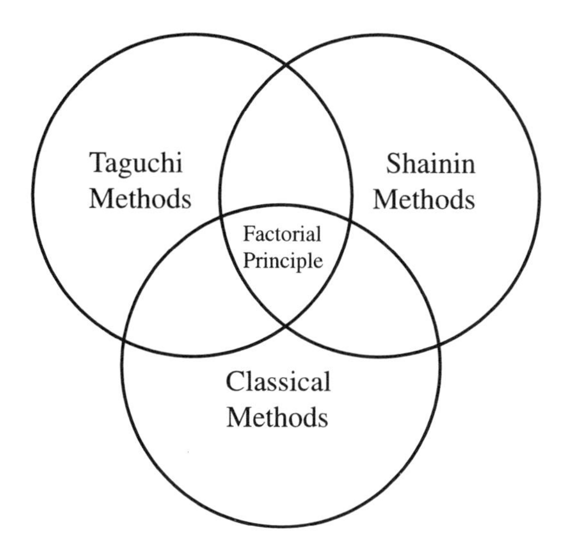

FIGURE 3.1b Design of experiments — II.

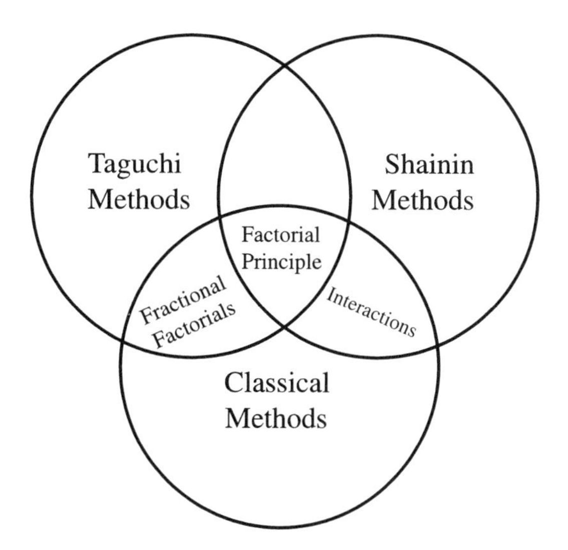

FIGURE 3.1c Design of experiments — III.

that are too costly, or difficult — if not impossible — to control. A classic example of an uncontrollable factor is copier paper. Despite our instructions and specifications, a copier customer will use whatever paper is available, especially as a deadline is near. If the wrong paper is used and a jam is created, the service personnel will be correct to point out the error of not following instructions. Unfortunately, the customer will still be dissatisfied. Dr. Taguchi recommends making the copier's internal processes more robust against paper variation, the uncontrollable factor.

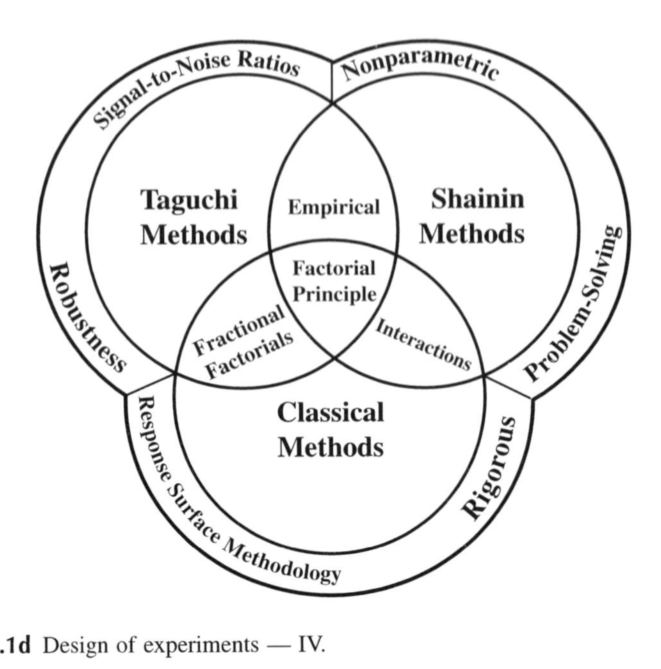

FIGURE 3.1d Design of experiments — IV.

Other types of experimental designs are specialized for instances where the results may be nonlinear, i.e., the response may be a polynomial or exponential form. Several of these designs attempt to implement the requirement for more factor levels in the most efficient way. One of these types is the Box-Behnken design. There are also classes of designs called central composite designs (CCDs).

Two specialized forms of experimentation are EVolutionary OPerations (EVOP) and mixture experiments. EVOP is especially useful in situations requiring complete optimization of a process. An EVOP approach would consist of two or more experiments. The first would be a specially constructed screening experiment around some starting point to identify how much to increase or decrease each factor to provide the desired improvement in the response(s). After determining the direction of movement, the process factors are adjusted and another experiment is conducted around the new point. These experiments are repeated until subsequent experiments show that a local maximum (or minimum, if the response is to be minimized) has been achieved. Mixture experiments are specialized to chemical processes where changes to a factor (for example, the addition of a constituent chemical) require a change in the overall process to maintain a fixed volume.

This discussion of designed experiments would not be complete without at least some mention of Dorian Shainin and his unique perspective on this topic. Although there may be some room for debate regarding Shainin's primary contributions to the field, most knowledgeable persons would probably agree that he is best known for his work with multi-vari charts (variable identification), significance testing (using rank order, pink x shuffle, and b[etter] vs. c[urrent]), and techniques for large experiments (variable search and component search).

Some important terms that are considered to be unique to Shainin's work are the red x variable, contrast, capping run, and endcount.

3.6 BEFORE THE STATISTICIAN ARRIVES

Most organizations that have not yet instituted the use of Six Sigma have few, if any, persons with much knowledge of applied statistics. To support this type of organization, it is suggested that process improvement teams make use of the following process to help them to define, measure, analyze, improve, and control (DMAIC).

Create Organization

- Designator

Column 1	Column 2	Column 3
Process	Improvement	Team
Product	Action	Group
Project	Enhancement	Task Force
Problem	Solution	Pack

- Appoint cross-functional representation
- Appoint leader/facilitator
- Agree on team logistics
  - Identify meeting place and time
  - Extent of resource availability
  - Scope of responsibility and authority
- Identify who the team reports to and when report is expected

Definitions and Descriptions

- Fully describe problem
  - Source
  - Duration (frequency and length)
  - Impact (who and how much)
- Completely define performance or quality characteristic to be used to measure problem
  - Prioritize if more than one metric is available
  - State objective (bigger is better, smaller is better, nominal is best)
  - Determine data collection method (automated vs. manual, attribute vs. variable, real time vs. delayed)

Controllable Factors and Factor Interactions

- Identify all controllable factors and prioritize
- Identify all significant interactions and prioritize

- Select factors and interactions to be tested
- Select number of factor levels
  - Two for linear relationships
  - Three or more for nonlinear relationships
  - Include present levels

Uncontrollable Factors

- Identify uncontrollable (noise) factors and prioritize
- Select factors to be tested
- Select number of factor levels
  - Use extremes (outer limits) with intermediate levels if range is broad

Orthogonal Array Tables (OATs)

- Assign controllable factors to inner OAT
- Assign uncontrollable factors to outer OAT
- Assignment considerations:
  - Interactions (if inner OAT only)
  - Degree of difficulty in changing factor levels (use linear graphs or triangular interaction table)

Consulting Statistician

- Request and arrange assistance
- Inform statistician of what has already been recommended for experimentation
- Work, as needed, with statistician to complete design, conduct experiment, collect and validate data, perform data analysis, and prepare conclusions/recommendations

Taguchi Approach to Experimental Design

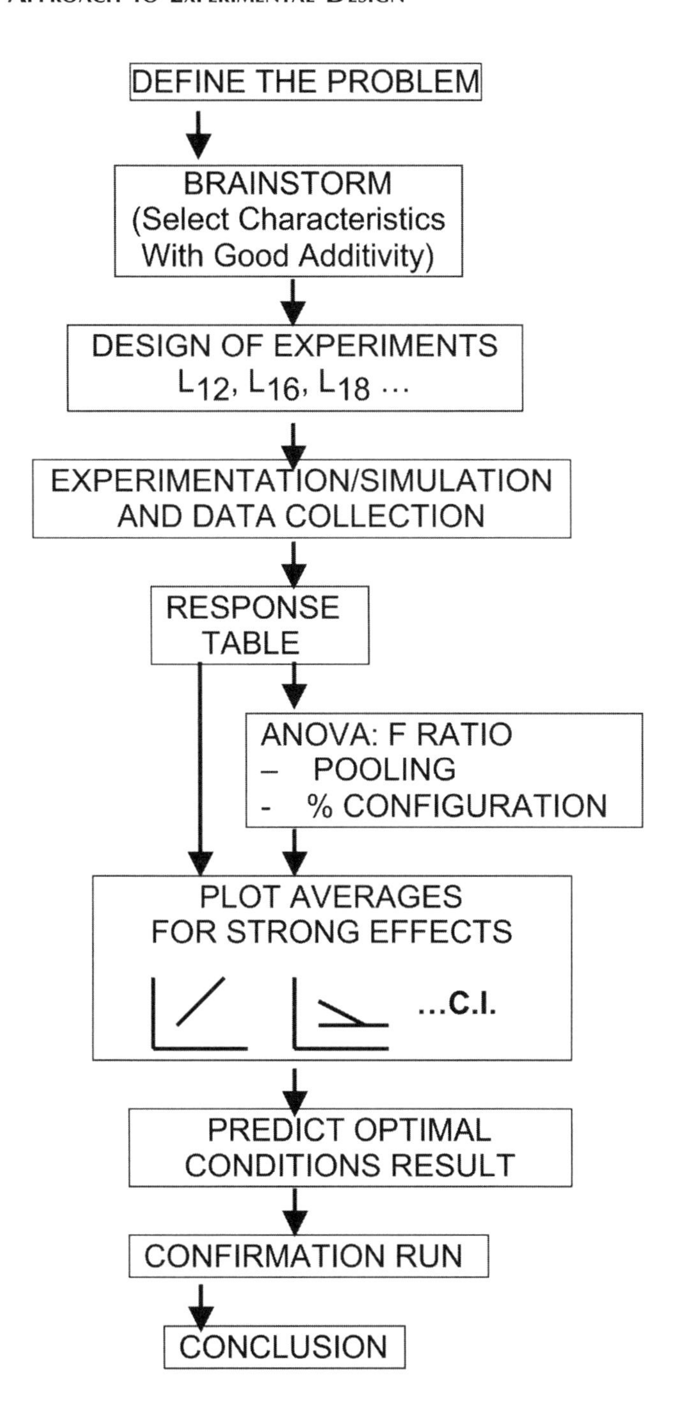

3.7 CHECKLISTS FOR INDUSTRIAL EXPERIMENTATION

In this final section a series of checklists is provided for use by DOE novices. The reader is encouraged to review and apply these checklists to assure that their DOEs are conducted efficiently and effectively.

CHECKLIST — INDUSTRIAL EXPERIMENTATION

1. DEFINE THE PROBLEM
   - A clear statement of the problem to be solved.
2. DETERMINE THE OBJECTIVE
   - Identify output characteristics (preferably measurable and with good additivity).
3. BRAINSTORM
   - Identify factors. It is desirable (but not vital) that inputs be measurable.
   - Group factors into control factors and noise factors.
   - Determine levels and values for factors.
   - Discuss what characteristics should be used as outputs.
4. DESIGN THE EXPERIMENT
   - Select the appropriate orthogonal arrays for control factors.
   - Assign control factors (and interaction) to orthogonal array columns.
   - Select an outer array for noise factors and assign factors to columns.
5. CONDUCT THE EXPERIMENT OR SIMULATION AND COLLECT DATA
6. ANALYZE THE DATA BY:

Regular Analysis	**Signal to Noise Ratio (S/N) Analysis**
Avg. response tables	Avg. response tables
Avg. response graphs	Avg. response graphs
Avg. interaction graphs	S/N ANOVA
ANOVA	

7. INTERPRET RESULTS
   - Select optimum levels of control factors.
     - For nominal-the-best use mean response analysis in conjunction with S/N analysis.
   - Predict results for the optimal condition.
8. ALWAYS, ALWAYS, ALWAYS RUN A CONFIRMATION EXPERIMENT TO VERIFY PREDICTED RESULTS
   - If results are not confirmed or are otherwise unsatisfactory, additional experiments may be required.

DOE — GENERAL STEPS — I

Step	Activity
• Clearly define the problem.	Identify which input variables (parameters or factors) may significantly affect specific output variables (performance characteristics or factors). Also, identify which input factor interactions may be significant.
• Select input factors to be investigated and their sets of levels (values).	Apply Pareto analysis to focus on the "vital few" factors to be examined in initial experiment.

DOE — GENERAL STEPS — II

Step	Activity
• Decide number of observations required.	Determine how many observations are needed to ensure, at predetermined risk levels, that correct conclusions are drawn from the experiment.
• Choose experimental design.	Design should provide an easy way to measure the effect of changing each factor and separate it from effects of changing other factors and from experimental error. Orthogonal (symmetrical/balanced) designs simplify calculations and interpretation of results.

DOE PROJECT PHASES

Phase	Activity
• Process characterization experiments.	Identify significant variables that determine output performance characteristics and optimum level for each variable.
• Process control.	Determine if process variables can be maintained at optimum levels. Upgrade process if it cannot. Provide for training and documentation.

PROCESS CHARACTERIZATION EXPERIMENTS

Objective	Activity
• Screening	Separate "vital few" variables from "trivial many."
• Refining	Identify interactions between variables and set optimum ranges for each variable.
• Confirmation	Verify ideal values and optimum ranges for key variables.

SCREENING EXPERIMENT

Step	Activity
1	Identify desired responses.
2	Identify variables.
3	Calculate sample size and trial combinations.
4	Run tests.
5	Evaluate results.

REFINING EXPERIMENT

Step	Activity
1	Select, modify, and construct experimental matrix design.
2	Determine optimum ranges for key variables.
3	Identify meaningful interactions between variables.

CONFIRMATION EXPERIMENT

Step	Activity
1	Conduct additional testing to verify ideal values of significant factors.
2	Determine extent to which these factors influence the process output.

PROCESS CONTROL

Step	Activity
1	Determine capability to maintain process within new upper and lower operating limits, i.e., evaluate systems used to monitor and control significant factors.
2	Initiate statistical quality control (SQC) to establish upper and lower control limits.
3	Put systems into place to monitor and control equipment.
4	Develop and provide training materials for use by manufacturing.
5	Document process, control system, and SQC.

POTENTIAL PITFALLS

It is possible to

- Overlook significant variables when creating experiment.
- Miss unexpected factors initially invisible to experimenters. The significance of unknown factors and process random variations will be apparent by the degree to which outcomes are explained by input variables.
- Fail to control all variables during experiment. With tighter ranges, it is harder to hold process at one end or other of range during experiment.
- Neglect to simultaneously consider multiple performances. Ideally, significant variables affect all responses at same end of process window.

PROCESS OPTIMIZATION

- OBJECTIVE
  Find best overall level (setting) for each of a number of input parameters (variables) such that process output(s), i.e., performance characteristics, are optimized.
- APPROACHES
  - One-dimensional search: all parameters except one are fixed.
  - Multidimensional search: uses selected subsets of level setting combinations (for controllable parameters). Fractional factorial design.
  - Full-dimensional search: uses all combinations of level settings for controllable parameters. Full factorial design.

DIMENSIONAL SEARCH SCALE

LEVEL SETTING CRITERIA

- Level settings for input parameters should be carefully chosen.
  - If settings are too wide, process minimum or maximum could occur between them and thus be missed.
  - If settings are too narrow, effect of that input parameter could be too small to appear significant.
  - Settings should be selected so that process fluctuations are greater than sampling error.
  - For insensitive input parameters, i.e., robust factors, large differences in settings are required to bring parameter effect above noise level.

WHY REPLICATION?

- Experimental results contain information on
  - Random fluctuations in process.
  - Process drift.
  - Effect of varying levels of input parameters.
- Thus, it is important to replicate (repeat) at least one experimental run one or more times to estimate extent of variability.

REFERENCES

Barker, T. R., *Quality by Experimental Design,* 2nd ed., Marcel Dekker, New York, 1994.

Barker, T. R., *Engineering Quality by Design,* Marcel Dekker, New York, 1986.

Bhote, K. R., *World Class Quality: Using Design of Experiments to Make it Happen,* ASQ Quality Press, Milwaukee, WI, 1991.

Box, G. E. P., Hunter, W. G., and Hunter, J. S., *Statistics for Experimenters,* John Wiley, New York, 1978.

Dehnad, K., *Quality Control, Robust Design, and the Taguchi Method,* Wadsworth & Brooks/Cole, Pacific Grove, CA, 1989.

Hicks, C. H., *Fundamental Concepts in the Design of Experiments,* 3rd ed., Holt, Rinehart & Winston, New York, 1982.

Lochner, R. H. and Matar, J. E. *Designing for Quality: An Introduction to the Best of Taguchi and Western Methods of Statistical Experimental Design,* Quality Resources, White Plains, NY, 1990.

Montgomery, D. C., *Design and Analysis of Experiments,* John Wiley, New York, 1976.

Phadke, M. S., *Quality Engineering Using Robust Design,* Prentice Hall, Englewood Cliffs, NJ, 1989.

ReVelle, J. B., Frigon, N. L. Sr., and Jackson, H. K., Jr., *From Concept to Customer: The Practical Guide to Integrated Product and Process Development and Business Process Reengineering*, Van Nostrand Reinhold, New York, 1995.

Ross, P. J., *Taguchi Techniques for Quality Engineering,* McGraw-Hill, New York, 1988.

Roy, R., *A Primer on the Taguchi Method,* Van Nostrand Reinhold, New York, 1990.

Schmidt, S. R. and Launsby, R. G., *Understanding Industrial Designed Experiment,* 2nd ed., CQG Printing, Longmont,CO, 1989.

Taguchi, G., *Introduction to Quality Engineering: Designing Quality into Products and Processes*, Quality Resources, White Plains, NY, 1986.

4 DFMA/DFSS

John W. Hidahl

Design for manufacture and assembly (DFMA) and design for Six Sigma (DFSS) are complementary approaches to achieving a superior product line that maximizes quality while minimizing cost and cycle time in a manufacturing environment. DFMA is a methodology that stresses evolving a design concept to its absolute simplest configuration. It embodies ten simple rules, which can have an incredible impact on minimizing design complexity and maximizing the use of cost-effective standards. DFSS applies a statistical approach to achieving nearly defect-free products. It uses a scorecard format to quantify the parts, process, performance, and software (if applicable) capabilities or sigma level. It facilitates the effective design of a product by aiding the selection of (1) suppliers (parts), (2) manufacturing and assembly processes (process), (3) a system architecture and design (performance), and (4) a software process (software) that minimizes defects and thus produces a high-quality product in a short cycle time.

4.1 DESIGN FOR MANUFACTURE AND ASSEMBLY (DFMA)

The DFMA methodology consists of six basic considerations and ten related rules, as shown in Table 4.1.

DFMA is intended to increase the awareness of the engineering design staff to the need for concurrent product and process development. Several studies have proven that the design process is where approximately 80% of a product's total costs are determined. Stated differently, the cost of making changes to a product as it progresses through the product development process increases by orders of magnitude at various stages. For instance, if the cost of making a change to a product during its conceptual design phase is $1000, then the cost of making the same change after the drawings are released and the initial prototype is fabricated is approximately $10,000. If this same change is not applied until the production run has started, the cost impact will be approximately $100,000. If the need for the design change is not recognized until after the product has been purchased by the consumer or delivered to the end user, the total cost for the change will be approximately 1000 times as great as if it had been implemented during the conceptual design review. In addition to driving product cost, design is also a major driver of product quality, reliability, and time to market. In today's marketplace, customers are seeking the best value for their investment, and the most effective way to incorporate maximum value into a product's design disclosure is through the use of DFMA.

1-57444-300-3/02/$0.00+$1.50

TABLE 4.1
DFMA Considerations and Commandments

Considerations
1. Simplicity
2. Standard materials and components
3. Standardized design of the product itself
4. Specify tolerances based on process capability
5. Use of the materials most processed
6. Collaboration with manufacturing personnel

The Ten Commandments
1. Minimize the number of parts
2. Minimize the use of fasteners
3. Minimize reorientations
4. Use multifunctional parts
5. Use modular subassemblies
6. Standardize
7. Avoid difficult components
8. Use self-locating features
9. Avoid special tooling
10. Provide accessibility

4.1.1 Simplicity

Simplicity is the first design consideration, and it bridges the first five DFMA commandments, namely, (1) minimize the number of parts, (2) minimize the use of fasteners, (3) minimize reorientations, (4) use multifunctional parts, and (5) use modular assemblies. There are several approaches that can be used to minimize the part count in a design, and specific workbook and software techniques have been developed on this, but the driving principles revolve around three questions: (1) Does the part move? (2) Does the part have to be made from a different material than the other parts? and (3) Is the part required for assembly or disassembly? If the answer to all three is no, then that part's function can be combined with another existing part. Using this approach progressively, existing assemblies that were not based upon DFMA principles can often be redesigned to eliminate 50% or more of their existing parts count. Reduced part counts yield (1) higher reliability; (2) lower configuration management, manufacturing, assembly, and inventory costs; (3) fewer opportunities for defects; and (4) reduced cycle times. Minimizing the use of fasteners has several obvious advantages, and yet it is the most frequently disregarded principle of DFMA. Excessive fasteners in a design are often the result of engineering design uncertainty, and are often justified as offering flexibility, adjustment, quick component replacement, or modularity. The reality is that excessive fasteners increase the cost of assembly, increase inventory costs, reduce automation opportunities, reduce product reliability, and contribute to employee health risks such as carpal tunnel syndrome. Prototype designs may require additional fasteners and interfaces to test various design or component options, but the production design should be stripped of any excessive fasteners. The five *why*'s approach as used commonly in root cause analysis is recommended for testing the minimal requirements for fasteners. Unless one of the sequential answers to, "*Why* do we need this fastener?" can be traced directly to a stated operational requirement, the fastener(s) should be eliminated from the production design disclosure. With respect to minimizing reorientations during assembly, the guiding principles are to create a design that can be easily assembled (with a minimum amount of special tooling) and to always use gravity to aid you in assembly. Minimizing the number of fasteners will obviously contribute toward minimizing the number of reorientations necessary. The use of multifunctional parts is a primary method of reducing the total parts count, thus enhancing design simplicity. Similarly, the use of modular subassemblies is a good design method to predesign for continuous product improvement through block upgrades and similar product line enhancements over time. As new technology moves into practice and becomes cost effective, modular subassemblies can be easily replaced to provide expanded capabilities, higher processing speeds, or more economical (market competitive) modular substitutions. Although modular subassemblies may increase the total part count of the original product, the added ease and speed of implementing improvements are a positive trade-off for many products or product families.

4.1.2 Use of Standard Materials Components and Designs

The second and third design considerations, standard material and components and standardized design of the product, are described by the sixth commandment: standardize. Design reuse is one of the most cost-effective methods used in the design process. By defining company- or product family-related standard materials, standard parts, and specific design process standards, the product cost and time to market will be reduced, while reliability and customer value will be maximized. The key element in standardization is establishing the discipline within the organization to keep the standards current and readily available to the product development team, and enforcing their effective and consistent use.

4.1.3 Specify Tolerances

The fourth design consideration is specifying or establishing design tolerances based upon process capability rather than the typical design engineer's affinity for closely toleranced parts. This approach is embodied in the seventh design commandment: avoid difficult components. The most effective way to apply this consideration is through the concurrent product development team environment where the design engineer and the manufacturing (producibility) engineer work collaboratively to ensure that the designed parts can be efficiently manufactured without excessive costs or scrapped material. This imposes the requirement that the manufacturing engineer have full knowledge of the process capabilities of in-house equipment and processes, as well as supplier equipment and processes.

4.1.4 Use of Common Materials

The fifth design consideration is use of the materials most processed. This simply means that materials that are commonly machined or processed in some manner within the company or within the company's supplier base should be the first materials of choice for the various components. Exotic or state-of-the-art processes and materials should be avoided whenever possible to preclude extended process development activities associated with low process capability, which typically increase cost and cycle time while reducing quality and reliability.

4.1.5 Concurrent Engineering Collaboration

The sixth and final design consideration is collaboration with manufacturing personnel. As identified previously, it is essential that the design team include cross-functional personnel such as manufacturing engineers, quality engineers, and procurement specialists to ensure that all the appropriate design trade-offs are properly analyzed and selected throughout the product development process by the experts in the respective disciplines involved. The traditional "Throw the design over the wall to manufacturing when engineering is done with it" approach is guaranteed to produce product attributes that contribute to higher production costs and extended time to market.

The other three design commandments that remain to be described are (8) to use self-locating features, (9) to avoid special tooling, and (10) to provide accessibility. The use of self-locating features is an assembly aid that can dramatically reduce assembly costs and cycle time. Parts that naturally nest together or contain self-centering geometries reduce the handling, alignment, reorientation, and inspection costs of assembly. Automated assembly processes in particular benefit tremendously from self-locating features to minimize the tooling and fixturing often required to ensure proper part alignment during assembly. Similarly, the avoidance of special tooling is a key consideration in complex assembly processes. Special tooling should be used only when other design elements or part geometries cannot incorporate self-locating features. Special tooling harbors an extensive array of hidden costs when fully analyzed. In addition to the cost of designing, fabricating, checkout, inventory, maintenance, spares, and planned replacement of special tooling, it can also add substantial cycle time to the assembly process. The added cycle time can accrue from issuing it from stores, moving it, installing it, and then verifying its proper placement, alignment, attachment, and operation over its intended design life. The final commandment is to provide accessibility, which implies the need for maintenance, inspection, part adjustment, part replacement, or other product access requirements over its design life. The key here is to define the requirements for accessibility based on the customers' (end-users') needs and the product development team's comprehensive vision of the product's possible applications, as well as its growth or evolution in the future. This requires a balance between satisfying current minimum needs and anticipating the most likely future needs, while still keeping the design simplicity DFMA consideration in mind.

All the aforementioned DFMA considerations and commandments should be applied as an integrated and balanced approach in the design process. A well-documented product development process, in combination with clearly defined team member roles and responsibilities, will greatly improve the application of DFMA in most organizations.

4.2 DESIGN FOR SIX SIGMA (DFSS)

DFSS methodology encompasses all the DFMA principles and adds proven statistical techniques to drive the design process, and thus the product, to lower defect counts. The typical DFSS statistical applications in design include (1) tolerance analysis, (2) process mapping, (3) use of a product scorecard, (4) design to unit production costs, and (5) design of experiments.

4.2.1 Statistical Tolerance Analysis

Statistical tolerance analysis employs a root-sum-squared approach to evaluating tolerancing requirements in lieu of the more traditional "worst-case analysis." Its methodology is based on the statistical fact that the probabilities of encountering the worst-case scenario are extremely remote. For instance, if an assembly involves the interfacing of four different parts, and each part is known to have a ±3 sigma dimensional capability, then the defect probability can be calculated to be 2.7 in 1000, or 0.0027. By applying statistics, the probability of encountering the worst-case situation can be calculated to be 5 in 100 billion or 0.0000000000534. This clearly demonstrates the ultraconservatism of this approach and the consequent extremely tightly toleranced part call-outs required to achieve it. Tightly toleranced parts have inherent hidden manufacturing costs associated with them, because they dictate detailed inspection requirements and often require scrap or rework of a significant percentage of the manufactured parts. Most of these scrapped or reworked parts would have, in fact, worked perfectly well, but were rejected due to excessively demanding part tolerancing.

A product generally consists of both parts and processes. This relationship means that to be successful you should seek to understand both the upstream and downstream capabilities of the various processes that will be used to produce the product. A product must be designed to not only meet the customer's requirements, but must also complement the process capabilities of the manufacturing company and its supplier base. It is unlikely that a company will ever reach a goal of Six Sigma quality without understanding the capability of the entire supply (or value) chain. Design teams must understand and properly apply the process capabilities of their manufacturing facilities and those of their suppliers in order to repeatedly produce near zero-defect products. Process capability data are the enabling links needed to create robust designs. The preferred graphical method of describing the key process capabilities and how they relate to the overall product manufacturing activity is through the process map.

4.2.2 Process Mapping

Six Sigma process-mapping techniques encompass several statistical measures of process performance and capabilities in addition to the typical process flows and related process operation information. As you will see, this information is extremely useful when a team of individuals has been assigned to improve a process. Let's

TABLE 4.2
Process Mapping Vocabulary

Process map: a graphical representation of the flow of a process. A detailed process map contains information that is beneficial to improving the process, i.e., cycle times, quality, costs, inputs, and outputs.
Y: key process output variable; any item or feature on a product that is deemed to be "customer" critical, referred to as "y1, y2, y3."
X: key process input variable; any item which has an impact on Y, referred to as "x1, x2, x3."
Controllable X: knob variable; an input that can be easily changed to measure the effect on a Y.
Noise X: inputs that are very difficult to control.
S.O.P. X: standard operating procedure; clearly defined and implemented work instructions used at each process step.
XY matrix: a simple spreadsheet used to relate and prioritize X's and Y's through numerical ranking.

start with some of the common vocabulary used in process mapping to become familiar with the terminology (Table 4.2).

Now that the basic terms have been defined, why do you suppose a process map is important when improving an existing process or implementing a new one? There are several visual features that a process map provides to aid a team's understanding of the operations involved in a given process:

1. A process map allows everyone involved in improving a process to agree on the steps it takes to produce a good product or service.
2. A map will create a sound starting block for team breakthrough activities.
3. It can identify areas where process improvements are needed most, such as the identification and elimination of non-value-added steps, the potential for combining operations, and the ability to assist with root-cause analysis of defects.
4. It will identify areas where data collection exists and ascertain its appropriateness.
5. The map will identify potential X's and Y's, leading to determining the extent to which various x's affect the y's through the use of designed experiments.
6. The map serves as a visual living document used to monitor and update changes in the process.
7. It acts as the baseline for an XY matrix and a process failure modes and effects analysis (PFMEA).

A Six Sigma process map for a manufacturing operation is shown in Figure 4.1. The map was created by a focused team working on a product-enabling process. The team consisted of operators, maintenance technicians, design engineers, material and process engineers, shop floor supervisors, and operations managers. The basic elements of this process map include (1) the process boundaries, (2) the major operations involved, (3) process inputs, (4) process outputs, and (5) the process metrics. There are several steps that must be followed to create a valid process map, as outlined in Table 4.3.

TABLE 4.3
Steps to Creating a Process Map

Step 1: Define the scope of the process you need to work on (actionable level).
Step 2: Identify all operations needed in the production of a "good" product or service (include cycle time and quality levels at each step).
Step 3: Identify each operation above as a value-added or non-value-added activity. A value added operation "transforms the product in a way that is meaningful to the customer."
Step 4: List both internal and external Y's at each process step.
Step 5: List both internal and external X's at each process step.
Step 6: Classify all X's as one or more of the following:
- Controllable (C)
- Standard operating procedures
- Noise

Step 7: Document any known operating specifications for each input and output.
Step 8: Clearly identify all process data-collection points.

The key statistical information often described on a Six Sigma process map includes the defects per unit (DPU) at each operation step, rolled throughput yield (RTY), and key process capability (CPk) values. The design team needs to analyze these process parameters and understand their influence on RTY in order to design quality into the product rather than attempting to inspect quality into the product.

4.2.3 Six Sigma Product Scorecard

The Six Sigma product scorecard is an excellent method for applying process capability information to the conceptual phase as well as subsequent phases of the design evolution. The scorecard is derived from the Six Sigma requirements for process definition, measurement, analysis, improvement, and control. By individually analyzing four elements of a design (parts, process, performance, and software), scorecard sigma levels can be identified. Initial scorecard values can be used to evaluate conceptual design alternatives and to influence the downselect criteria; refined scorecards can be used to aid trade studies to optimize the baseline design configurations. In these design studies, product sigma levels can be evaluated as independent variables that drive cost, schedule, and other critical parameters. Baseline design selection at an overall 3 Sigma level, for instance, would yield 66,807 parts per million (ppm) defective, whereas achievement of a 6 Sigma design level would yield only 3.4 ppm defective, or a ratio of approximately 20,000 to 1 in improved quality!

An example of a Six Sigma product scorecard is shown in Figure 4.2. This summary-level scorecard includes the four assembly level evaluation elements: parts, process, performance, and software, with the software element being nonapplicable for this simple mechanical configuration. Note that for each of the elements, the DPU estimate and the opportunity counts are described for each major subassembly. These are then totaled near the bottom of the table, and first time sigma, DPU/opportunity, sigma/opportunity long term and short term are all calculated through algorithms built into the Excel spreadsheet. Each element results in a separate short-term sigma

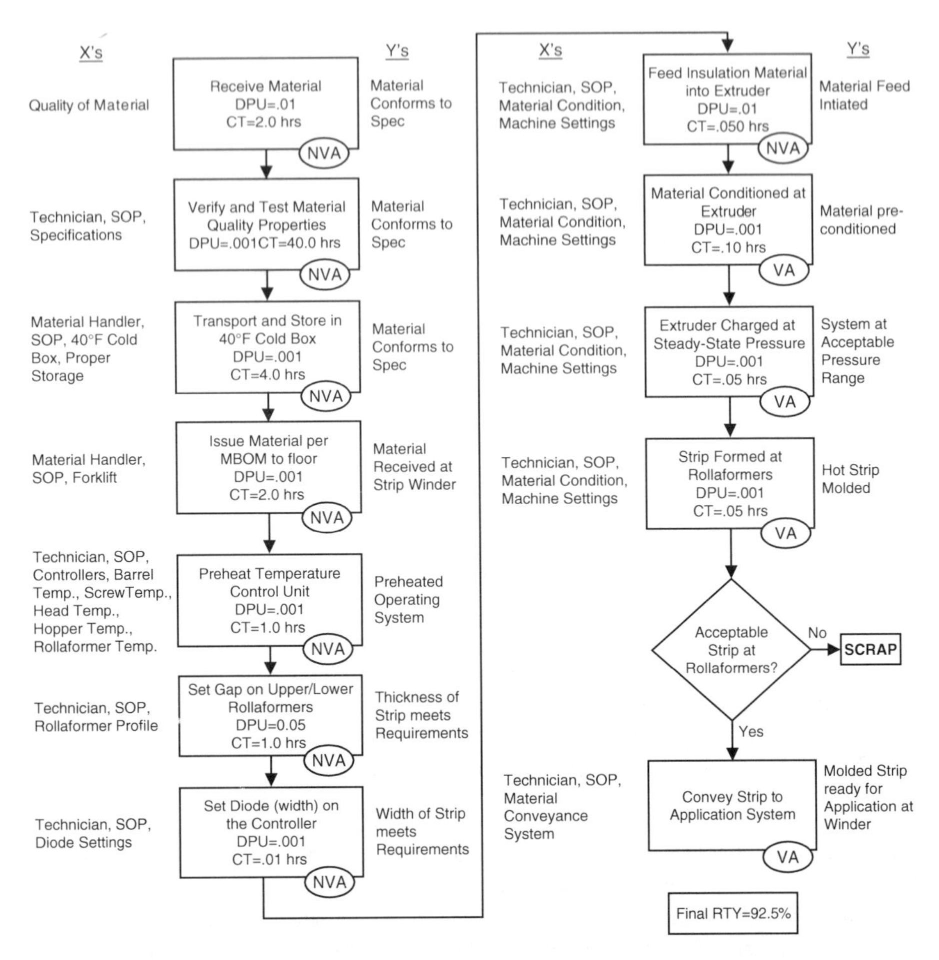

FIGURE 4.1 Solid rocket motor strip winding process map. CT = cycle time, DPU = defects per unit, MBOM = manufacturing bill of materials, NVA = non-value added, RTY = rolled throughput yield, SOP = standard operating procedures, VA = value addeed, X = input variables, Y = output variables.

that is used as the design basis for most applications. The minimum sigma value for any of the elements constitutes the design sigma limitation. Unless all the elements are fairly equivalent in value, the overall sigma score will be heavily influenced by the lowest element sigma value. Each of the elements uses a separate worksheet accessible through the Excel worksheet tabs at the bottom of the spreadsheet layout.

The parts worksheet shown in Figure 4.3 is completed by defining all the major purchased or manufactured individual parts that will make up the assembly or subassembly. This is most easily accomplished through the use of a bill of materials, or parts listing. The supplier, part number, part description, quantity, part defect rate in ppm defective, and the total DPU, an alternate description for ppm, are all defined. A separate worksheet is completed for each major subassembly to be built by manufacturing. The overall intent of this methodology is to drive the previously

Date	04/04/00
Part Number	xxxxxxxx
Name	ACME
Period of Data	4/1/00 -

AI&T Cost	$2,599
Critical Path Cycle Time	0
Raw Process Multiplier	8.29

DPU	38.2127
DPMO	7843.3
Sigma	2.42

	Part (σ)			Process (σ)							Performance (σ)	
Assembly	DPU	Opp. Count	Parts Cost	DPU	Opp. Count	Labor Cost	Cycle Time (min.)	Total Time (min.	VA Time (min.)	RPM	DPU	Opp. Count
Scan Drive	0.1649	1	$2,500	37.9711	2680	$99	250	290	35	8.29	0.07667732	2191
Antenna												
Receiver												
Electronics												
System												
Totals	0.1649	1	$2,500	37.9711	2680	99	250	290	35	8.29	0.0767	2191
First Time Sigma	1.03			<-6							1.45	
RTY	84.8%			0.0%							92.6%	
DPU/Opp	0.1649			0.0142							0.0000	
Sigma/Opp	1.03			2.20							3.98	

FIGURE 4.2 Six Sigma product scorecard.

described DFMA principles of fewer parts and part types into the design and to ultimately select quality suppliers and processes to manufacture the individual parts.

The process worksheet portrayed in Figure 4.4 describes the assembly process information, much of which is taken directly from the process map previously considered. Here again, one worksheet per major assembly or subassembly is compiled for each assembly level built by manufacturing. The process worksheet identifies all the major internal processes used to build the product. The DFMA intent here is to use high quality processes and simplify the build process to the greatest practical extent. For each process step, the load center, cycle time, labor hours and cost, process target, specification or tolerance, upper specification limit (USL), lower specification limit (LSL), process mean value, standard deviation, process capability (CPk), number of applications, process opportunities, and product opportunities are all defined. From this information the spreadsheet algorithms are used to calculate the total number of product opportunities, average defects per opportunity, average yield per opportunity, average process sigma long term (LT), average process sigma short term (ST), as well as the total defects per unit, the rolled throughput yield, and the sigma (z) score. As evidenced by the amount of statistical process data required, this methodology involves extensive process capability data collection and knowledge to be used successfully. It requires taking the operator "black magic" out of the process capability equation, and replacing it with parametrically driven process knowledge and control features, which can be derived from design of experiments, and other Six Sigma methodologies.

An example of the performance worksheet is presented in Figure 4.5. It is used to identify all the customer-focused, top-level system performance parameters, and to quantify the probability that the design configuration will successfully achieve them. Its intent is to quantifiably assess the design's capability against the defined system-level requirements. It also provides insight into the production acceptance testing requirements and needed measurement system accuracy (MSA). The worksheet lists the key customer-based performance parameters that can be obtained from a customer's specification, a technical requirements document, or from a quality function deployment (QFD) process. It defines target values, units, upper specification limit (USL), lower specification limit (LSL), performance mean value, standard deviation, z score USL, z score LSL, rolled throughput yield, and DPU.

A software worksheet is presented in Figure 4.6. It identifies the entire software build process, tracks defects found during each phase of the software development, and calculates the efficiency of each software phase in detecting and eliminating defects. It also provides a future extrapolation of overall delivered software quality, based on defect rates demonstrated during the build process.

The top-level product scorecard results are calculated by algorithms internal to the spreadsheet using all the individual worksheet inputs. As previously identified, Figure 4.2 illustrates the combined results from this Six Sigma tool, and its influence on designing quality into the product. This methodology provides a powerful method of positively influencing the design process through the use of data and removes the mystery (or mystique) that surrounds many modern-day manufacturing facilities about their ability to produce high-quality products on a consistently repetitive basis.

Part Number	
Name	Scan Drive
Period of Data	4/1/00 -
Total Part Count	1
Avg Defects/Part	0.1649
Avg Yield/Part	84.8%
Avg. Part Sigma	1.03

TDU	0.1649
Yield	84.8%
Sigma	1.03

COQ	$500
Part Cost	$2,500
Variance	($500)

Part Description			Measured Feature						Defect Data					Cost Data				
Supplier	Part No.	Description	Feature	Qty.	LSL	USL	Mean	St.Dev.	Units	Defects	PPM	DPU	Sigma	Unit Cost	COQ	Total	Planned	Variance
Ace	1349594-1	Printed Wiring Board		1					291	48	164948	0.1649	1.03	$2,000	$500	$2,500	$2,000	($500)

FIGURE 4.3 Six Sigma product scorecard — parts worksheet.

Part Number	xxxxxxxx	TDU	37.971	Total Unit Cost	$99	Cycle Time - Mean (min.) 250
Name	Scan Drive	Yield	0.0%	Total COQ	$20	Cycle Time - Std. Dev. (min.) 36
Period of Data	4/1/00 -	Sigma (Z)	< -6	Total Cost	$119	Total Process Time - Mean (min.) 290
Total # of Product Opps.	2680			Total Variance	$81	Total Process Time - Std. Dev. (min.) 36
Average Defects/Opp.	0.01417					Value Added Time (min.) 35
Average Yield/Opp.	98.59%					Raw Process Multiplier 7.14
Avg. Process Sigma	2.20					

Process Step	Operation Number	Defect Identification Method	LSL	USL	Mean	Std. Dev.	Cpk	Number of Units	Number of Defects	# of Times Used	Operation Opportunities	Product Opportunities	Defects per Unit	DPMO	First Time Sigma	Sigma/Opportunity	Number of Times Process Implemented	Std Hours/Unit	Unit Labor Rate ($)	Extended Cost ($)	COQ ($)	Total Cost ($)	Planned Cost ($)	Variance (Plan-Actual) ($)	Critical Path Process	Cycle Time - Mean (minutes)	Cycle Time - Std. Dev. (minutes)	Value Added Time - Mean (minutes)	Raw Process Multiplier
Form & Tin	2306	Insp. 3324					< 0	382	3647	1	908	908	9.547	10514	-3.80	2.31	1	0.3	$37	$11	$2	$13	$200	$187	1	120	30	10	12.00
Identification	3044	Insp. 3445					0.56	332	16	1	0	0	0.048	48193	1.67	1.67	1	0.3	$37	$11	$2	$13		($13)	0	0	4	5	58.00
Stencil Print	3196	Insp. 3324					< 0	382	3475	1	886	886	9.097	10267	-3.69	2.32	1	0.1	$37	$4	$1	$4		($4)	1	100	20	0	inf
Pick & Place	3196	TOTAL					0.67			1	886	886	19.279	21760	< -6	2.02	1	2	$37	$73	$15	$88		($88)	1	0	3	120	1.50
		Insp. 3196					< 0	361	405				1.122	1266	-0.45	3.02													
		Insp. 3324					0.68	382	6936				18.157	20493	< -6	2.05													

FIGURE 4.4 Six Sigma product scorecard — process worksheet.

Part Number	xxxxxx
Name	Scan Drive
Period of Data	4/04/00 -
# of Parameters	7
Avg. Defects/Parameter	0.0110
Avg. Yield/Parameter	98.91%
Avg. Parameter Sigma	2.29

Units Tested	313
Units Failed	24
TDU	0.0767
Yield	0.93
Sigma (Z)	1.45

Performance Parameter	Process Step at Which Measurement is Made	Operation Number	Target	Units	Failures	LSL	USL	μ, mean	Std. Dev.	Z, LSL	Z, USL	Cpk	Calc yield	Actual Yield	Calc. DPU	Actual DPU
UUT CRNT				313	0	0.95	1.5	1.13855017	0.03	7.0	13.4	2.3	100.0%	100.0%	0.0000	0.0000
V1				313	2	5.38	5.565	5.44587197	0.19	0.4	0.6	0.1	53.6%	99.4%	0.6230	0.0064
V12				313	0	11.5	11.96	11.8114671	0.03	9.9	4.7	1.6	100.0%	100.0%	0.0000	0.0000
ACT DELAY				313	1	575	745	666.189273	17.60	5.2	4.5	1.5	100.0%	99.7%	0.0000	0.0032
ACT AMP F1				313	0	1.4	3.25	2.10889273	0.20	3.6	5.8	1.2	100.0%	100.0%	0.0001	0.0000
ACT AMP 196				313	0			2.11858131	0.22					100.0%		0.0000
ACT AMP F3				313	0			1.85948097	0.20					100.0%		0.0000

FIGURE 4.5 Six Sigma product scorecard — performance worksheet.

4.2.4 Design to Unit Production Cost (DTUPC)

The design to unit production cost (DTUPC) methodology is yet another opportunity to apply statistical methods to a design optimization process. In this case, the critical dependent variable is cost, and the design must be evolved to meet this driving requirement. DTUPC offers a method of determining how much it costs to build a product, what each DPU costs the company, how much work-in-progress (WIP) the factory or shop has, and how much WIP your suppliers are holding that you are ultimately paying for. Many companies do not find out until the end of their accounting cycle, whether annually, monthly, or weekly, what profit they have made. DTUPC offers the opportunity to know the true cost of every unit produced. The cost of defects is typically ignored in most factory operations, but in reality, the additional labor, inventory, overhead, inspection, and other hidden costs, including warranty coverage can completely undermine the product profit margin. Excessive WIP, whether in your factory or at the supplier's, is yet another indication of carrying costs that limit profitability and cash flow. Six Sigma DTUPC includes seven basic manufacturing cost elements: (1) setup and assembly labor costs, (2) applicable overhead and general and administrative costs (G&A), (3) bill of material (BOM) cost of parts, (4) inspection costs, (5) DPU, (6) rework cost to correct defects, and (7) warranty costs for escaping defects. Most organizations have cost estimating or collection methods for determining the contributions of cost elements (1), (2), (3), and (4), but the "hidden-factory" or Cost-of-Poor-Quality elements (5), (6), and (7) are often overlooked or ignored, and yet can contribute substantially to the cost of the product. For instance, if supplier A prices a part at $35/unit that has a DPU of 1.0, and your labor (hidden) to repair the part is

$$\tfrac{1}{4} \text{ hour} \times \$60/\text{hour} = \$15$$

then the total cost is $50/unit. If supplier B offers the same part for $42/unit, but has a DPU of 0.05, and your hidden repair costs are, therefore, reduced to

$$5 \text{ defects}/100 \text{ units} \times \tfrac{1}{4} \text{ hour} \times \$60/\text{hr} = \$0.75/\text{unit average}$$

then the total cost is $42.75/unit, or a savings of $7.25/unit (roughly 17% of supplier B's total cost). This simple illustration points out the importance of knowing your supplier's part defect rates and avoiding merely selecting the apparent low-cost supplier in the source selection process. Detailed statistical analysis of DTUPC can be applied as an extension of the product scorecard to ascertain true unit production costs using various suppliers, in-house processes, and materials. This type of Six Sigma analysis facilitates cost trades and the ultimate approach to achieving the minimum production cost of any given product.

4.2.5 Designed Experiments for Design Optimization

The use of design of experiments (DOE) to solve design problems is yet another method of applying Six Sigma principles to the engineering design process. Similar

Intro. At	Detected At																
	System Design	Analysis	Preliminary Design	Detailed Design	Coding & Unit Test	Integration & Test	Formal Test	System Integration & Test	Flight Test/Post release	Grand Total	Leakage	Leaked	Opportunities	DPO	Yield	PPM	Process Sigma
System Design				5	4	29	19	4	24	85	100%	85	320	0.266	77%	233273	0.7
Analysis				2	16	128	18	6	70	240	100%	240	1230	0.195	82%	177266	0.9
Preliminary Design				6	31	23	3	7	0	70	100%	70	4330	0.016	98%	16036	2.1
Detailed Design				489	182	86	32	31	0	820	40%	331	8660	0.038	96%	37500	1.8
Coding &Unit Test					1921	490	107	28	25	2571	25%	650	109000	0.006	99%	5946	2.5
Integration & Test						177	5	3	0	185	4%	8	285	0.028	97%	27680	1.9
Formal Test							36	10	0	46	22%	10	302	0.033	97%	32570	1.8
System Integration & Test								2	0	2	0%	0	433	0.000	100%	0	Infinite
Grand Total	0	0	0	502	2154	933	220	91	119	4019	35%	1394	124560	0.011	99%	11129	2.3

	DPO	Yield	PPM	Process Sigma
Product Development Sigma*	0.032	0.968	31751	1.9
Delivered Product Sigma**	0.001	0.999	1091	3.1

FIGURE 4.6 Six Sigma product scorecard — software worksheet.

in context to a manufacturing DOE, engineering DOEs can be used to aid in downselecting design concepts and in defining the sensitivity of a design alternative to various parameters or environmental exposures. As an example, suppose a materials engineer recommends the use of an adhesive to bond two dissimilar materials together, which see a shear load and a temperature gradient during system start-up in a reusable application. We want to verify that the adhesive will meet the design requirements and identify which recommended application process produces the best bonds when exposed to the operating environmental loads, duty-cycle duration, and repeated cycling. We start by fully defining the engineering requirements. Let's assume that the shear load is 250 lb and that the temperature of the bond changes during the start-up transient from 70 to 180°F at a rate of 5°/second. By preparing a process map and a process FMEA, the critical few variables that are influencing the bond strength can be isolated. Let's assume that the five variables suspected of influencing the bond strength of the adhesive are (1) material surface preparation, (2) adhesive cure temperature, (3) adhesive pot life, (4) curing pressure, and (5) application area humidity. By running a 2^{5-1} order factorial experiment wherein each of the five variables has two values at which several test coupons were prepared and evaluated, the sensitivity of each of the tested variables can be ascertained, the bond strength requirement can be verified, and a margin of safety calculated. A design DOE of this type was run, which produced the results shown in Table 4.4.

From these DOE results we can conclude that (1) surface preparation makes a small difference in the bond strength, but both the low and high test point produce acceptable results; (2) cure temperature likewise has a small effect on the bond strength, but both the low and high test points produce acceptable results; (3) the adhesive pot life had almost no discernable effect on the bond strength over the range of values tested, and therefore if a pot life of one-shift (or 8 hours) is optimum from an operations standpoint, then an 8-hour pot-life test should be evaluated to determine its effect on bond strength; (4) curing pressure, like pot life, had almost no discernable effect on bond strength over the range of values tested; but (5) the local humidity had a great influence on the bond strength over the ranges tested. At 20% humidity, the bond strength is acceptable with about a 28% margin, but at 95% humidity, approximately 99% of the bonded parts failed at a shear load of 250 lb. This example demonstrates the importance and value of conducting design-based DOEs during the design process. By completing this DOE, the design engineer was

TABLE 4.4

Variable	Low Value	Sheer Strength	High Value	Sheer Strength
Surface preparation	Isopropyl alcohol wipe	319.8 ± 2.5 lb	Grit blast	325.2 ± 3.6 lb
Cure temperature	50°F	314.3 ± 2.9 lb	100°F	320.2 ± 3.0 lb
Adhesive pot life	1 hour	321.8 ± 2.4 lb	4 hours	320.6 ± 2.7 lb
Curing pressure	0.1 psi	320.9 ± 2.5 lb	1.0 psi	321.5 ± 2.9 lb
Local humidity	20%	325.6 ± 3.5 lb	95%	230.6 ± 20.2 lb

able to specify the desired adhesive for bonding the two parts. He could allow a wide range of process variables (as defined by the DOE), as long as the local humidity was maintained consistent with a humidity-controlled (air conditioned) environment as is found in most laboratories and clean rooms.

DFMA and DFSS are both effective methods for aiding the design engineer in conceptualizing and detailing the design disclosure package for a wide variety of parts, components, assemblies, subsystems, and systems. Proper application of the various tools described within this chapter will yield tremendous dividends to the company or organization that fosters a "near-zero" defect mindset into its design functions.

5 Integrated Product and Process Development

Robert Hughes

5.1 OVERVIEW

WANTED: Generalist with extrovert tendencies. Communicates openly and often. Willing to integrate many areas of expertise at a moment's notice. Accepts responsibility for goals that may seem unachievable. Will work side by side, physically or electronically, with designers and engineers, manufacturers, material management, marketing, and logistics. Reports to a team leader.

Look like a typical job ad? If not, why are so many organizations looking for just this person? Help-wanted ads for integrated product and process development (IPPD) team members usually read a little more conventionally. However, the positions and challenges are unconventional — so why don't we see this ad?

This chapter first defines how integrated product teams (IPTs) work and how to prepare a team and organization for success, then mentions some of the pitfalls. Second, it offers methodologies for designing the product and process simultaneously.

5.2 BACKGROUND

5.2.1 Design–Build–Test

Before the manufacturing revolution, an entrepreneur would design–build–test a product from the first to the last unit. As large-scale manufacturing evolved, a new product went to market by slugging its way through a serial process of concept to the design–build–test of a prototype, to marketing buy-in, and finally to design–build–test the manufacturing process. This often resulted in

- Delayed market entry and revenue streams
- Additional costs from losses in efficiency (serial vs. parallel development)
- Mental silos that created functional tunnel vision and minimal lateral movement of personnel

This disconnected, hands-off approach of leadership, intent, and knowledge almost always played like a version of that game where a whispered message is passed from person to person until the original message (what the customer asked for) is lost.

1-57444-300-3/02/$0.00+$1.50

With IPPD, each process is still needed. It's the sequencing, the sharing of knowledge, and the elimination of hand-offs that now offer a product that's made better and quicker and has fewer costs. The success of IPPD methodology equals proactive planning and participation.

5.2.2 Teams Outperform Individuals

As industry grew, one benefit was the gathering of multiple skills in one organization. This encouraged specialization. The specialists were then organized into departments or by function and the exchange of information happened through reports. As specialists got more specialized, unique lingo and tools further confused (divided) organizations, and meetings were held to facilitate better communication. Alas, darkness fell upon the land.

If you've been in business for any period of time, certainly you've had more than one moderator conduct an exercise demonstrating that a team yields better scores that any one individual. You can choose your favorite clichés, but mine are "Two heads are better than one," and "The more the merrier." If skeptics still exist, consider the analogy of basketball played five against one, even if the one is Michael Jordan at his prime.

5.2.3 Types of Teams

Four types of teams are successful in business: functional, lightweight, heavyweight, and autonomous. All are great, but choosing which is needed for each situation takes some thought. To aid a manager in selection, knowing the general characteristics of each is helpful, as shown in Table 5.1.

Using these characteristics, assess the type of team needed to complete the project. A limited focus, such as upgrades in interface software, may be effectively managed with a functional team. On the other extreme, a new product platform that involves significant investment and development is more likely to be successful with a heavyweight or autonomous team.

The differences between a heavyweight and an autonomous team may not be readily apparent to the casual observer. As its name implies, the autonomous team depends entirely on its own resources for success. However, the culture shift required for this can be too much for many organizations. Functional silos within the business must accept this loss of turf, while the team's members must accept that they still are expected to comply with the systems of the organization.

5.2.4 Fad of the Early 1990s

Teams and *concurrent engineering* became corporate buzzwords in the early 1990s. Most organizations, seeking to accelerate product development, attempted IPPD with either success or failure.

However, combining product and process development together, and challenging project teams to achieve a "stretch" goal didn't yield extraordinary returns without the business also experiencing the necessary culture changes. Therefore, as with

TABLE 5.1
Identifying Team Characteristics

Type of Team	Focus	Team Roster
Functional	One specialty (such as software)	High degree of competency within the specialty Members report to a functional manager Led by a functional manager
Lightweight	Multiple specialties	Moderate to high degree of competency within each specialty Members still report to their functional manager but are assigned to a project Led by a project manager (often one of the functional managers)
Heavyweight	Multiple specialties, usually lacking all resources needed to complete the project	Generalists with a focused specialty in which they have a high degree of competency Report to the project manager Project manager is highly visible within the organization
Autonomous	All specialties required to complete the project (equivalent to a business unit)	Generalists constitute the "core team," and "bit" players fill competency gaps Report to the project manager Project manager is highly visible within the organization

most business fads, organizations expecting a magic elixir without the hard work required for success became disillusioned and resigned to only the improvements experienced with lightweight teams.

Organizations, or leaders, that recognized the big payoffs sparked success stories and won professional accolades. Recognition opportunities, such as *Machine Design*'s annual Concurrent Engineering Award, offered ongoing evidence that IPPD was more than a fad.

5.2.5 DoD Directive 5000.2-R (Mandatory Procedures for Major Defense Acquisition Programs)

In the 1990s, the U.S. Department of Defense (DoD) recognized the value gained by projects utilizing IPPD and IPTs. This document is available to the public, and it offers a manual for organizations to follow. But be aware that it doesn't offer insight into the issues of culture change, internal turf wars, and career development challenges.

5.2.5.1 Benefits of IPPD

To a manager selling the implementation of IPPD to upper management, peers, or subordinates, the benefits must outweigh the costs. Benefits of IPPD are maximized when the three segments of a business — customers, employees, and the organization — profit from its use.

5.2.5.2 Why IPPD Benefits Employees

Enhanced communication is on everyone's list of things to improve. This could be listed under all three segments, but the employee is the largest benefactor. Creating an environment where product and process development feed each other keeps everyone in the communication loop.

So what's the benefit of all this communication? It's expanding the knowledge of each discipline by injecting the experience of others (such as manufacturing's assembly concerns being voiced during concept development, allowing the designer to add features for gripping and handling). In effect, free horizontal growth occurs in skill sets as team members learn from the experience of their peers.

Example: While developing a product, a manufacturer shared samples (components from recently completed production tooling) with several customers, who then indicated that significantly greater rigidity was desired. Product designers preferred a switch from polymer to steel shafts, but this meant radical changes and new tooling, causing a product launch-date delay of 6 months. Therefore, the prototyping of exotic and expensive polymer substitutes was initiated. However, manufacturing team members who were involved in the brainstorming quickly demonstrated that the existing tooling could be used to insert mold through a steel tube, reducing the delay to less than 1 month and providing the product improvement.

In your current organization, would the product designers have thought to insert molding? Would the manufacturing (molding) engineers have challenged the switch to a different polymer by offering this solution? Possibly, but why did this occur? Because they communicated!

Involving customers in pilot evaluations facilitated the proactive insight. In turn, this prevented a rushed response, at significantly higher costs to the employee (their time and stress) and to the business (dissatisfied customers, increased tooling costs, expediting fees).

Employees also benefit by being given *Focus*. IPPD won't be successful when the team members have to choose between responsibilities of the project and day-to-day priorities. The most common problem I have encountered is the daily fire fighting required to keep current products going. This reduces the resources assigned to a project. However, focus can be achieved with the recognition that the project has

- Limited duration — doesn't mean 100% dedication (although I discuss the work environment advantage later). It's the commitment to complete the project within a defined time frame.
- Measurable outputs — (for the employee) lets employees know how they will be measured as contributors to the project.

5.2.5.3 Why IPPD Benefits the Customer

Because customer satisfaction is the only assurance of continuing business, what does IPPD do for the customer? IPPD successes give the customer

- Alternatives sooner by reducing the time from concept to market launch
- Value by creating designs with reduced production costs and improved features

In our earlier example, our customers received the product earlier (1 month's delay vs. 6 months in product launch) and they helped the company improve the product features (increased rigidity). This is part of the common belief that 80% of a product's cost is determined during the first 20% of the project.

The early design phase (the 20%) chooses materials and the industrial design, thus establishing manufacturing costs (the 80%) for methods of assembly, and material sourcing. The effort to change materials before launch was successful, and it was cost effective because it incorporated the customer's review before the design was "frozen."

A word of warning: IPPD doesn't substitute for good market research or direct communication with the customer. In fact, we discuss later the importance of including a customer voice when organizing the team.

5.2.5.4 Why IPPD Benefits an Organization

"Show me the money!" is a requirement if upper management is going to champion IPPD and support a project or team when difficulties arise. Fortunately, every business has internal examples of product designs wherein product costs and quality have created challenges for manufacturing. Using these illustrations in a tactful manner will reinforce the benefits of maximizing manufacturing efficiencies earlier in the design.

From DoD's IPPD handbook, implementation principles that provide organizational benefit are

- Life-cycle planning that delivers a product with affordable production and servicing throughout its life
- Proactive risk management organized to contain project and product costs, reduce technical risks, and maintain completion dates

5.3 ORGANIZING AN IPT

5.3.1 Initial Challenges — What Are We Doing (Goals)? Why Change? How Are We Going to Do It (Roles)?

Sit down and write answers to the three questions above in a way that everyone can understand, and then you'll be a champion for IPPD/IPT's. Upper management, the team, and affected organizations need and deserve these answers. To aid your efforts, consider:

5.3.1.1 Goals

Don't limit these to the typical project milestones of budgets, launch dates, and quality expectations. They must include system or process measures, such as reducing the number of product engineering change notices to address manufacturing issues or reducing the budget for equipment modifications. The key is measured business results from the interaction of product and process.

Note: Consider a "metrics dashboard," wherein the project schedule or work breakdown is scored by the team. If a key team metric is reducing the product weight (let's say from 1.8 to 1.5 lb), a chart could be maintained in an area where the team often gathers. As each design exchange is processed, the weight would be revised even if the change does not relate to weight. The emphasis is reminding the team of a commitment. Similarly, the metric can be a systems objective, such as reducing approval times by off-site reviewers or customers. Tracking each approval time reminds team members to go the extra mile in expediting the process. Collectively displayed, these become a "dashboard" — a snapshot of how well the team is meeting performance expectations.

5.3.1.2 Why Change?

Effectively outlining the goals will go a long way in making this an easy answer. Goals define challenges that couldn't be met by the current practice. IPPD encourages their completion by

- Establishing a new environment that minimizes hand-offs by having design and manufacturing jointly develop the product and process
- Cutting down on rework or having to go backward in the development cycle
- Changing sequential activities to parallel so that the time to market is reduced
- Identifying where new approaches are required to meet new metrics

5.3.1.3 Roles

Creating a new job description for each team member involves defining the daily tasks, information flow, and interactions with other areas within the business. After the team starts to jell (within the first 2 weeks), everyone should meet to assure that all project areas have been captured, using the descriptions as a starting point, and to define who is the best member to lead that item.

5.3.2 Core Members (Generalists) vs. Specialists (Bit Players)

IPTs (excluding functional teams) will require members from a number of disciplines to complete the project. The risk of this is shown in the proverb, "Too many cooks spoil the broth." To help compensate for this, it is necessary to define who is accountable — the core team, and which disciplines are support roles — the specialists. Core teams of five to eight will usually offer sufficient technical expertise without creating a "committee" environment.

Construct the core team with generalists. They should be veteran members of the organization who are respected within their disciplines and have demonstrated knowledge (and experience) in other areas as well. Generalists often recognize the interactions between product and process, are aware of the formal and informal systems within the business, and often wear more than one hat.

However, the core team shouldn't be a collection of managers! They need to be individual contributors. If those who do the work aren't *working* side by side, then the opportunity for codevelopment of the product and process is lost. The team leader may not have tasks, but everyone else should. This isn't saying that the core members can't be supervisors as well. If a discipline, such as product design, requires the work of several individuals, simply assure that the core member is a hands-on contributor, in addition to being a supervisor. Otherwise, the details of day-to-day decisions are delayed.

A critical role in the team is the voice of the customer. Assigning a marketing representative is only part of the answer. Core members also have the responsibility for converting the input to technical specifications (a role usually assigned to the design engineer) and for validating the design with the customer (usually handled by the marketing and quality representatives).

5.3.3 Collocation and Communication Links

Multiply face-to-face contact with continuous interaction and you have collocation. It has been my experience that this simple action has the biggest payback, regardless of the type of team (refer to earlier descriptions). It isn't enough that team members are in the same city or even the same building — the key is to have them share their work area. Despite their drawbacks, open-air office cubes benefit IPTs by having all team members able to hear — and jump into — discussions on all facets of the product and process. You would be surprised how often mistakes have been avoided, shortcuts discovered, and a commitment to goals has occurred because physical proximity encourages continual participation.

The core team should be collocated at almost any cost. The specialty members can be collocated or accessible through communication links. However, when the specialty team member is off-site or not collocated, it's beneficial if one of the core team members is responsible for his or her activities. When feasible, include daily contacts and encourage video- or teleconferencing with as many teams members as possible.

5.3.4 Team Culture

Teams aren't functional departments. Therefore, most of us aren't trained or experienced in the culture of corporate teams. Numerous training sources exist in every city, but as a start, find a copy of Price Pritchett's 65-page *Team Member Handbook for Teamwork*, which I try to review at the start of each project. Sport teams and volunteer organizations offer some team-oriented reference points for many people. Using these experiences as examples can be helpful, but collectively, the group must create an environment that focuses on teamwork, not individual success.

Teams that spend downtime with each other often create a culture of work and fun, uniting the members and helping to break the functional ties that cause division within a team. However, the team will often be dependent on others in the functional organization for success, so don't burn bridges to those outside the team!

5.3.5 Picking the Right Team Leader

The team leader is the focus, inspiration, and manager for the IPT. Libraries are filled with literature defining good leaders. Some of their characteristics can also create good team leaders:

- More than an administrator. The leader should have experience in the traditional roles of management (planning, organizing, controlling), but these are support skills for this role. It is more important to be excellent at communicating, counseling, and consulting.
- A risk taker. The individual must be able to react and lead in an uncertain environment. If this is a new role within the organization, the risks and challenges are multiplied by the probable scrutiny of upper management.
- Motivational. This may be the most difficult to define, but my gauge is the team leader's ability to get more out of others than was thought possible.
- Knows the business. The leader must be able to access what the team needs from all levels of the organization. This requires knowledge of formal and informal business systems. In other words, don't bring in an outsider; the challenge is more than enough without the burdens of not having internal networks, mentors, and champions when the going gets tough.
- Knows the technology. The leader must be able to understand and communicate in the language of the technical experts involved in order to lead joint product and process development as well as be a contributor. If the members don't respect the leader's ability to grasp issues, the opportunity to lead will be sidelined by a lack of credibility.
- Puts the team first. Beyond this list of superhuman qualifications, everything must take a backseat to the leader's dedication and support of the team itself. While dealing with challenges such as turf wars with the functional areas, naysayers against the project, limited resources, and conflicts and frustrations among the members, the leader must remember that the ultimate goal is the success of the team.

5.4 BUILDING THE ENVIRONMENT (CULTURE) FOR SUCCESSFUL IPPD

Organizational change is traumatic for all employees. Preparing them and the organization for IPTs will support quicker identification of the challenges and implementation of solutions. Managing this includes the organizational issues (such as structure, information logistics, responsibilities, and authority) as well as effective change management.

5.4.1 Effective Change Management

Anticipate the initial drop in performance; change rarely results in immediate improvement. The team — and the rest of the organization — must accept this.

Helping them both understand the phases and typical attitudes for the change process will offer a roadmap to monitor and support change. Change occurs in a serial path of performance vs. time:

- First, a denial, voiced by comments such as "Why do we need change?" or "We can't do that."
- Then, an uncertain leap of faith, expressed in "I knew this wouldn't work" or "I don't know why I'm doing this."
- Followed by reaching rock bottom, sometimes explained by "Yes, it's working but I don't know why."
- Then a sense of gaining control, represented by "This is not as difficult as I feared" or, preferably, "I think I found a way to do this better."
- Finally, — a new cultural norm develops.

5.4.1.1 Fear and Jealousy of Change (From the Functional Manager's View)

Fear of loss of turf is the best way to explain the response from most functional managers. To understand this, consider Maslow's hierarchy of needs. In the short term, loss of security. The functional manager may see the change as a threat to job security. Why else would management have taken this from me? Therefore, reluctance to support the change should be expected.

Longer term, I have perceived considerable jealousy of IPTs from nonteam members of functional organizations. First, IPTs usually have higher visibility (not surprising if the team has multiple functions and upper management has multiple interests). Next, resources come from project budgets and headcounts and, regardless of reality, functional managers always feel that their area is under-resourced. Finally, assuming that the IPTs are successful and continuing with future projects, the functional managers may see themselves as "stuck in the past."

5.4.1.2 Organizational Issues Created by Change

Building different relationships with other departments	Product and process development silos must expand to communicate with the other's new contacts. These relationships, created by multidiscipline teams, often expand to upper management. Building a continuous communication loop among the team and these others requires a commitment of resources that wasn't necessary with the individual function approach.
Matching the team's goals to the organization's	Team goals need to be championed from above. To support this happening, each of the team's goals should clearly support one or more of the organization's goals. With good correlation among the goals, the organization's top management team is more likely to recognize how the team's success will benefit them.
Addressing the authority to make decisions	A final consideration is the amount of authority given to the team. If a seasoned team leader is assigned and the team is composed of senior members, the organization is usually comfortable entrusting the team with the authority to meet the goals successfully. But what do you do when the team is less seasoned?

You want an autonomous decision-making group. However, unproven team leaders or members will require a counseling process to help ensure that their decisions align with the organization's expectations. A common approach is assigning the unproven team leader or specialist a project mentor. To maximize the payback, this could be a member of upper management who can also provide additional support for the project.

5.5 THE TOOLS THAT AN IPT WILL REQUIRE

5.5.1 Technical Tools

Beyond the organizational environment and change management challenges, there is the work itself. For a specific project, management — in conjunction with the team leader — should define which skills are required from core team members. (Some specifics on how to pick team members were previously presented.)

5.5.2 Communication and Decision-Making Tools

The new types of business relationships may require different skills than the members previously employed. Communication will be broader both vertically (such as designers to manufacturing supervisors) and horizontally (engineers to marketing directors). Preparing meetings and reports to meet the needs of this varied audience will be critical for effective communication.

Decision-making will change! Well-organized teams quickly find that the autocratic process of proposals-presentations-buy-off is gone. Teams authorized to make decisions will still need to present status reports to confirm that the team is working to the correct end; however, the members will need to master nondirective leadership. Effective with strong core members and long-term projects, this approach demands equal responsibility among the members for decisions.

As a side note, involvement in the team doesn't relieve the group of the responsibilities of the business systems — they still apply! An IPT is also NOT responsible for reinventing the business. This must be clear to the team and the rest of the organization.

5.6 PROBABLE PROBLEM AREAS, AND MITIGATIONS

5.6.1 Reduced Development Time = Less Time for Corrections and Customer Review and Feedback

When you reduce development time utilizing IPTs, the risk of missing the customer's voice increases. Simply, less time in development equals less time to hear from your customers or detect an internal error in specifications. As the product and process rocket forward, corrections become more costly if you need to go back. Therefore, you will need to integrate processes to reduce the risk for both.

5.6.1.1 Customer Inputs

Successful quality function deployment (QFD) minimizes the risk in this area. If your organization isn't accustomed to this methodology, assure that first, customer input is documented (defined) and measurable and that the engineering specifications directly relate to these inputs.

Building the discipline to document the customer's voice encourages review and gains team concurrence. These yield invaluable payback through internal critique and a commitment for quicker execution.

5.6.1.2 Specification Errors

They always happen. How does your organization currently avoid them? A successful approach includes a change-approval system for affected disciplines. This can be a change-approval form detailing what changed and why it was necessary and demonstrating that this fixes the issue, or there is a review board within the team. However, limit the approval requirement of those not assigned to the team — or the team will see itself as lacking authority.

A common pitfall is either not adequately including the input of all necessary disciplines or not giving them sufficient veto power. Examples of opportunities to include all affected parties are

- Suppliers for changes to components
- Marketing, industrial design, and packaging for aesthetics
- Legal for user manuals and advertisements

5.6.2 "Silo" and "Group-Think" Mentality

Build good relationships, but include safeguards against developing a "silo" within the team — "group-think." As the team creates its own identity, a tendency to revert to the functional silo mentality is natural. Preventive action or systems will aid by discouraging this tendency. I have seen technical (engineers from other teams and the functional organizations) or business (upper management from multiple functions) reviews serve this purpose. However, these must be administered as proactive counseling, not "got-cha" sessions that leave the team circling the wagons.

Other options include scheduling customer reviews and focus groups, thesis-type presentations to an internal expert, and appointing a team member to be a devil's advocate during decision-making sessions.

5.6.3 Self-Sufficient vs. Too Large a Team

Who hasn't been in a situation where the group is too large to make a decision? Ideally, involve the necessary disciplines and notify the others. If the team is too limited, however, it lacks the resources or the empowerment to complete a project.

In general, keep the team as small as possible. The advantages of minimizing the team size include easier and quicker communication among members (fewer to reach), increased flexibility because of the need to cover multiple disciplines, and an increased likelihood of keeping the team leader involved, not just managing.

Deciding which disciplines should have members on the team draws on logical group dynamics. These guidelines are a good start:

- No more than 12 actual team members (some may be player-coaches for larger, function-focused work groups).
- Between three and five core members who lead meetings and approve specifications (only one core member per discipline).
- If a discipline is required to deliver work, include it on the team. However, all specialties aren't required to be on the core team. A discipline requiring fewer than 20 hours a week (such as a documentation administrator) or with an involvement of less than half the project's duration (such as industrial design at the earliest phases of design) shouldn't be part of the core team.

5.6.4 Recruiting — Internal (Why Were They Chosen?) vs. External

Filling the team roster requires an evaluation of attributes. Go back and read the "help-wanted" paragraph starting this chapter. Does it make more sense now? When you make *your* list and recruit, consider these factors:

- Past team experience
- Knowledge or expertise in the project's technology
- Level of past involvement and experience with the organization's business systems
- Problem-solving skills
- Ability to have a candidate assigned to the team
- Drive, and the ability to get it from others

As I indicated, a number of these attributes require seasoned veterans from within the business. You must be able to answer "Why me?" to those you want and "Why not me?" to those you don't. Having your selection criteria defined should prepare you to address both.

5.6.5 Retention and Career Paths Following Project Completion

The externally recruited members are the most vulnerable upon completion of a project. As teams dissolve, the member must either move to a new team or be in organizational limbo. Sound scary? Imagine being a recently hired employee (less than 2 years with your organization), and not having a functional area to call home. Would it be a threatening situation? This situation can be addressed — preferably within a month of hiring — by pairing new recruits with mentors and involving them in the functional area aligned to their career paths.

Members can be culled from within the organization by the traditional enticements of promotions, salary increases, and bonuses. However, these are maintenance needs. To encourage team retention, emphasize the motivators created by an IPT environment: lateral growth, achievement of project goals, increased responsibility and visibility, and recognition at a level generally unavailable within a functional setting.

Internally recruited team members will need a career path at the project's completion. Returning to the same position within a functional organization will almost certainly result in a lost employee. If the IPT is a rotational assignment, there are two options:

- Horizontal growth through a different discipline (with the employee's concurrence) to gain a higher level of expertise in that area. With the team environment, the employee can orient and experience that new function as part of the team process.
- Vertical promotion within their functional expertise. This is increasingly rare. Organizations shouldn't promise future promotions. However, when opportunities do arise, team members should know that they are eligible. Identifying multidiscipline management or interaction as a prerequisite for many opportunities may encourage IPPD teams as part of succession planning within an organization.

5.6.6 Costs Associated with IPTs

An argument of increased cost is often offered as a downside to using IPTs. Accepting the conservative position that the same number of hours or personnel, individually or as part of a team, is required to complete a defined assignment — the labor content remains unchanged. Performance enhancement is still gained through multiple individuals applying separate areas of expertise, and an accelerated completion date earned by dividing the work.

Performance improvement = quality

There will be fewer resources expended going backward in the development cycle to correct earlier mistakes, fewer design changes downstream to incorporate features for manufacturing. The product is also more likely to meet customer expectations.

Performance improvement = $

Reducing the costs of changes, rework, and improvements increases the project's value.

Performance improvement = time

Getting it to market earlier grows the revenue stream and frees the resources to work on other efforts sooner.

5.7 METHODOLOGIES OF SIMULTANEOUS PRODUCT AND PROCESS DEVELOPMENT

After an IPT is formed the work begins. This section describes the tasks involved in effective IPPD deployment day-to-day. The manager or team leader who monitors the product and process development metrics will assure that the parallel efforts are on schedule for equal completion.

Methods will be covered chronologically, following the standard development phases of *concept*, *design*, and *qualification*.

5.7.1 Concept and Prototyping

"The earlier the better" should be the motto for implementing IPPD. Often, the integrated development opportunity is missed at the early stages. By using the advantages of IPPD, the team can quickly experience better planning for schedules, risks, and budgets. Although you won't be able to demonstrate improved results until the project's completion, the advantage of better inputs will be recognized immediately.

Inasmuch as most project managers seek approval at this phase, using IPPD will reduce the unknowns, significantly improving the quality of the decision process. You can minimize risk in the early stages through the following:

- Don't "promise the moon" by presenting concepts that haven't included all the functions in the planning. You can end up with a product that can't be manufactured (or at least one that can't be cost effective). In addition to costs, using different designs, especially materials, in prototypes may give performance indications that result in dangerous assumptions by the team and customer. Potential disasters are created when all functions aren't included in defining what is possible in the final product.
- "Can it be done?" is a standard question raised by manufacturing personnel when viewing prototypes. Involving them in prototype construction allows for early trial and error. Although the results will be different in the final manufacturing process, this experience is beneficial in developing schedules and cost estimates.
- Identify sourcing options as you develop the prototypes. Where new processes or materials are anticipated, the first steps of feeling out the supply chain can be tenuous. Developing the contacts and potentially identifying the suppliers will have this effort already completed when you start the design phase.

5.7.2 Design and Development

Your project has passed the first gauntlet, and you are now executing to a plan and budget. In other words, you are starting to be measured for efficiency and effectiveness. These methods improve one or both.

5.7.2.1 CAD Databases

CAD databases are the bricks and mortar of project configuration control. With the plethora of design, analysis, and tooling software available, it's a good idea to have an internal expert from each area (including suppliers and prototyping) define what will be used, and where and when.

Start with product model databases that can be transferred for efficient creation of tooling paths and mathematical modeling or analysis. Second, determine the level of detail required on the paper version, such as not defining the complex geometry of external surfaces if the tooling is generated directly from the product database.

5.7.2.2 Codevelopment

Codevelopment with your suppliers can greatly reduce errors in cost and timing by increasing the level of shared knowledge and the success rate of early production units. Where possible, let the supplier do the detailed design with your review and approval. Consider how to use your suppliers to help minimize potential costs (external charges vs. internal headcount charges), database compatibility challenges, and timing miscues (Can the supplier commit the necessary amount of resources?).

Example: A product was failing transit tests (impact and vibration). A coil spring, with extended legs to latch it into position, was moving, resulting in lower applied forces. The original design was codeveloped between a young design engineer in our firm and the spring manufacturer. After the failure was explained, the young engineer proposed a longer leg to increase the engagement. Instead of a quote, the supplier offered an alternative two-coil design that virtually assured that the spring wouldn't move after assembly.

The additional cost (less than 2 cents per spring) was more than offset by eliminating potential customer dissatisfaction.

Internally, we may have arrived at a similar design, but the active involvement of the supplier provided quick prototypes, as well as a personal commitment to delivery dates.

5.7.2.3 Tooling (Molds and Dies)

Tooling often accounts for the largest chunk of project budgets. Unfortunately, piece price is an inverse.

Example: Consider that the cost of an eight-cavity injection mold (warranted for over 1 million cycles) for engineered components, easily exceeds $60,000, while a similar single-cavity mold (100,000-cycle warranty) can be had for about $35,000. However, the difference in component costs isn't fractionally different. Expect a two- to three-times difference in piece price because the expense (overhead) of the molding machine and labor are shared by eight times more parts per hour.

The issue that IPPD can address is which part is right for the project. I have been a strong proponent that new designs (especially platforms) equal a high probability of changes. Starting multicavity tools and expecting rapid turnaround for design changes are contradictory. Better to let the process development (high-cost tools with efficient volumes) focus on a low-priced tool, after a product development version (single-cavity, low-cost efficiency) has been utilized. Your total costs will actually be less! Consider the incremental costs of several changes to multiple cavities; they will quickly exceed the cost of the smaller tool.

Furthermore, experience with the process development tool will improve the design and manufacture of the high-efficiency tooling. It can identify where inserts vs. permanent tooling features should be incorporated for further refinement, development, and derivatives.

5.7.2.4 Passive Assurance in Production

As the process development occurs, assembly line layout and equipment concepts offer the opportunity to incorporate "foolproof" devices or features. The advantage of developing these early in the product design allows assembly features to be included in the design and tooling.

Use of proximity sensors, weights, optical scanners, and electrical resistance are often incorporated to assess if features (or components) are present and properly assembled.

Example: A cone-shaped component needed to be attached to a tube using adhesives. Features on the cone needed to be aligned with latches on the tube until the adhesive dried. As prototypes were built, the alignment created a challenge to the model-makers. The team's manufacturing representative recognized the potential assembly issues that this created. Together with the design engineer, they added notches to the cone that allowed passive visual assurance of the required alignment. The result was zero scrap for lack of alignment in over 1 million units.

5.7.3 Qualification

5.7.3.1 Tooling Qualification

The best qualifications start with an agreed-upon plan. The IPT design, manufacturing, and quality representatives, meeting with their counterparts from the supplier's organization, generate the best plans. The objective is to cover part by part and, if at all possible, specification by specification:

- *What* will be measured by documenting the agreement.
- *How* it will be measured by specifying who is responsible for the metrology or testing, including any gauge studies.
- *How many* will be measured for each specification by defining if a control chart is required vs. a single piece and how to address multiple cavities and tools.

  Example: A new polymer component was having a 64-cavity tool designed. During the team's discussions with the supplier, the hot topic was capability analysis. It was determined that an inordinate amount of inspection time would be required for control charting for every cavity (a total of

16,000 data points). Rather, it was determined that six cavities, one from each corner and two in the middle, would satisfactorily identify process control and stability. The result was 14,500 fewer data points!
- "*When*" the measurements are conducted, by defining the difference in first shot snapshots and when a formal submission occurs.

5.7.3.2 Design Verification First

The tricky part of this effort is determining where the design stops, and the process takes over. As products are developed, you eventually need to assemble and test against the design specifications. Inherently, we introduce variation from how they were constructed. So when processed-induced variation raises issues with the verification, it's reactive and often quite emotional.

An advantage of IPPD is the opportunity to incorporate "Band-Aids" around the initial samples to minimize process variation. This doesn't mean that the final product, including the impact of the process, isn't tested (see below). Rather, the focus is on proving that the design is adequate, with the process introduced later. This gives the team an opportunity to address the origin of undesirable performance. Gone is the potential rift between the design and processing camps. Without the baggage, quicker resolution occurs.

5.7.3.3 Assembly Qualification = Product Qualification

The customer doesn't care if it was the design or the process that broke in their hands. Therefore, you need to test samples from production in a manner most representative of actual use. I won't address reliability testing and modeling techniques. However, maximizing your ability to detect flaws by using these modern tools is fantastic. Instead, go back to the basics. Consider how well processes are actually tested. Use the following checklist:

- Are the gauges (inspection and equipment) calibrated to prevent measurement errors?
- Have the processing parameters been bracketed by technical data (such as a designed experiment — DOE) or previous experience?
- Are the parameters, work instructions, and qualification sequence documented to assure the desired controls and outputs?
- Are multiple operators, raw material lots, and cavities (or tools) intentionally introduced during the qualification? If so, are samples marked such that subset analysis can be conducted to determine if they introduced meaningful variation?
- Will the product be run at best- and worst-case parameter combinations? Are there significant differences in product performance?
- Is all scrap analyzed to determine if there are potential issues for subsequent production?
- Did customers (or marketing) review product testing to assure that the technical methods simulate actual (or adverse) use?

- Does the analysis follow accepted statistical protocol (such as confirmation of a normal distribution)?
- Are the results reviewed by design, manufacturing, and quality team members?

Expect the process to vary. How well you assess this can be greatly improved through IPPD. As you went through the checklist, how many items are traditionally assigned to manufacturing representatives on new projects? Calibration and statistical analysis are usually quality assurance (QA) turf. Instructions and scrap are concerns left to the production supervisors. Several often fall between functional disciplines. The IPT, however, brings all of these together and minimizes the opportunity for error.

5.7.4 Conclusion

Integrated product and process development is successful when individuals are encouraged to work as a team. Replacing serial and silo approaches to product development with an integrated method assures that the product and process are developed more quickly, with fewer errors (reducing costs), and will meet more of the customer's expectations (increased revenue).

The challenge to management is developing the environment (culture), systems, and team members. The culture needs to emphasize new metrics (focusing on what will make the product development process successful), and has to accept that the team has both the responsibility and the authority to make decisions. Business systems need to flex, allowing nontraditional reporting structures, collocation of team members, and enhanced communication tools. However, the most important change is the conversion of technical and functional experts to team members. It starts with the team leader! Culture and system changes require a risk taker who can motivate team (and nonteam) members, while demonstrating business and technology knowledge (but they DON'T have to be the "expert"). Then, surrounding the individual with similar personnel, who are good at listening and improving their skills horizontally (such as a supplier manager becoming adept at statistical analysis), makes the team complete.

In short, have you prepared your business for response to the "help-wanted" ad at the start of this chapter?

5.8 INTERNET SITES

- *IPPD/IPT.* <http://www.acq-ref.navy.mil/turbo2/topics/bk.cfm>

REFERENCES

Integrated Product and Process Development Handbook, Office of the Under Secretary of Defense, Washington, D.C.

Katzenbach, J. and Smith, D. K., *The Wisdom of Teams,* Harvard Business School Press, Boston, 1993.

Lientz, B. P. and Rea, K. P., *Project Management for the 21st Century*, Academic Press, San Diego, 1995.
Meyer, C., How the right measures help excel teams, *Harvard Business Review*, 72(3), 99–103, May–June, 1994.
Muirhead, B. K. and Simon, W. L., *HIGH Velocity Leadership — The Mars Pathfinder Approach to Faster, Better, Cheaper*, HarperCollins, New York, 1999.
Pritchett, P., *The Team Member Handbook for Teamwork*, Pritchett Publishing, Dallas, 1993.
Snee, R. D. et al., Improving team effectiveness, *Quality Progress*, 31(5), 43–48, May, 1998.
Teschler, L., CE's best practitioners, *Machine Design*, July 8, 1999.
Wheelwright, S. C. and Clark, K. B., *Revolutionizing Product Development*, Free Press, New York, 1992.

6 ISO 9001:2000 Initiatives

Syed Imtiaz Haider, Ph.D.

6.1 INTRODUCTION

In most commercial organizations, products or services are intended to satisfy a user's need or requirement. Such requirements are often incorporated in specifications. However, technical specifications may not in themselves guarantee that a customer's requirements will be consistently met if there happen to be deficiencies in the specifications or in the organizational system that designs and produces the product or services. Consequently, system standards and guidelines that complement the relevant product or services requirement are given in the technical specifications. The series of International Standards ISO 9001:2000 includes a rationalization of the many and various national approaches.

ISO 9001:2000 has been prepared by the ISO/TC 178 Quality Management and Quality Assurance, subcommittee 2, "Quality System." Transposition into a European Standard has been managed by the CEN Central Secretariat with the assistance of CEN/BT WG 107.

EN 29001:EN 29002 and EN 29003 were superseded by EN ISO 9001:1994, EN ISO 9002:1994, and EN ISO 9003:1994, respectively, which in turn are now superseded by EN ISO 9001:2000.

The title of ISO 9001 has been revised in the latest edition and no longer includes the term, "quality assurance." This reflects the fact that quality management system requirements specified in this edition of ISO 9001, in addition to quality assurance of product, also aim to enhance customer satisfaction.

The concept of end product is a global concern for every business, profession, occupation, or enterprise regardless of whether it is a consultancy, a service, or a manufacturing or processing activity.

Well-planned design is one of the major contributing factors to achieving quality within a product or service. It is necessary, therefore, to carefully control critical elements and a company's documentation system with particular reference to each phase of design, to ensure that effects on end product or services are not disastrous and are based on customer requirements.

To be successful, an organization must offer services or products that meet well-defined purposes or needs. User satisfaction includes delivery timelines. It should be ensured that applicable standards, statutory requirements, and specifications are in compliance. The offered products or services should be within budget and provide value for the cost, ensuring that the firm's total operations are carefully reviewed before implementation to reduce the occurrence of error, and to apply quick corrections for inadvertent problems. Control should be demonstrated on changes through

1-57444-300-3/02/$0.00+$1.50

proactive communication and feedback loops between the company and the external interfaces. Trainings needs should be identified to promote efficiency and cost effectiveness.

Over the past decade, we have seen drastic changes in the world around us. There has been more competition in terms of quality; organizations are bound to do more work, accomplish higher targets, and be proactive rather than reactive, all with fewer people. And the quality must be the better than ever, due to competitive forces and consumer demands. From the beginning of the new millennium, it is evident that we are working in a global village where survival without offering quality in products and services is impossible.

The quest to achieve excellence in products and services should be the organization's mission, and that can only be achieved with a vision to implement a documented quality system based on a standard that is globally acceptable to overcome the economic barriers. Compliance with global International Standard ISO 9001:2000 will lead a company to

- Supply products that are totally fit for use
- Satisfy its customers' expectations and contractual requirements
- Yield profits on the invested resources
- Achieve and sustain a defined level of quality
- Comply with applicable standards and specifications
- Focus on prevention of defect, rather than detection
- Provide value for the cost
- Supply products within delivery timelines agreed on or required by the customer
- Comply with environmental requirements
- Work in harmony within the organization without productivity losses

6.2 THE BASIC CHANGES

The third edition of ISO 9001 cancels and replaces the second edition, ISO 9001:1994, together with ISO 9002:1994 and ISO 9003:1994. Those organizations that have used ISO 9002:1994 and ISO 9003:1994 in the past may use this international standard by excluding certain requirements in accordance with clause 1.2.

The revised and adopted module ISO 9001:2000 makes the quality system management requirements extremely clear. The quality system requirements in ISO 9001:2000 are comprehensively grouped under clause 7, with additional emphasis on customer satisfaction and internal communication, where exclusions are made. Claims of conformity to this international standard are not acceptable unless these exclusions are limited to requirements within clause 7, and such exclusions do not affect the organization's ability or responsibility to provide product that fulfills customer and applicable regulatory requirements.

The quality management principles stated in ISO 9004 have been taken into consideration during the development of this international standard. This international standard promotes the adoption of a process appropriate when developing, implementing, and improving the effectiveness of a quality management system to enhance customer satisfaction by meeting customer requirements. The following eight management principles are integrated into the ISO 9001:1994 standard and revised as the third edition ISO 9001:2000.

- Customer satisfaction
- The role of leadership
- The involvement of employees
- The business process approach
- A systematic approach to management
- Continual improvement
- A factual approach to decision making
- Mutual beneficial relationship with supplier

It is important for the organization to understand current and future needs and expectations to quantify customer satisfaction and act on it.

Leadership should take the lead in deploying policies and verifiable objectives, establishing vision, and giving direction to shared values. Management should set challenging goals and implement strategies to achieve them and empower qualified employees within the documented system.

The standard emphasizes creating personal ownership of an organization's goals by using the experience and knowledge of qualified and trained personnel with identified authority and responsibility to continuously work toward process improvement.

A system should be established for identifying internal and external customers and suppliers of processes. Process activities should be focused to adequately use the people, materials, machines, and methods to demonstrate control.

All processes within a system should be identified and aligned for their interdependencies with measurable organization goals and objectives. Continuous improvement should be ensured with realistic and challenging goals; resources should be provided as adequate to the company needs in terms of the people and equipment needed to accomplish customer requirements. Continual improvement should be ensured through management services, internal and external audits, and corrective and preventive actions. Analysis of the data and information such as customer complaints, nonconforming products, and audits should be conducted on a continuous basis to enhance productivity and to minimize waste, rework, and rejections. Cost improvement projects should be implemented.

Efforts should be made to establish strategic alliances to ensure joint development and continuous improvement of products, processes, and systems. The company should ensure that subcontractors are meeting the customer requirements through review and evaluation of their performance.

ISO 9000 defines "system processes" as activities that use resources to transform inputs into output. Inputs to a process are often outputs from other processes. Any activity or operation that receives inputs and converts them to outputs can be considered a process. Almost all product and service activities and operations are processes. For organizations to function, they have to define and manage numerous linked processes. Often the output from one process will directly form the input into

FIGURE 6.1 Input–process–output sequence. (From *Quality Management Systems Requirements,* 3rd ed., ISO 9001:2000, BSI, London, 2000. With permission.)

the next process. The systematic identification and management of the various processes employed within an organization, particularly the interactions between such processes, may be referred to as the "process approach" to management.

The proposed new version of the international standard encourages the adoption of the process approach for the management of an organization and its processes, and as a means of readily identifying and managing opportunities for meeting customer satisfaction needs with continuous improvement.

6.3 QUALITY MANAGEMENT SYSTEM

Every business, profession, occupation, and enterprise has an end product, regardless of whether it is a manufacturing, service, consulting, or processing activity. Probably there is no business that does not follow the input–process–output pattern. Even departments or teams within an organization work according to this concept. If we go down to the micro level of an effectively managed company, we find the individual members working with the input–process–output approach. Each one keeps his or her customer in mind and performs to delight him or her. This is the philosophy of Quality Delivery.

Any performance or operating process can be characterized as a sequence of input–process–output. Implementing a quality management system has the goal of ensuring that the inputs are according to a specific standard, that the process itself is controlled, and that the output is monitored for conformity with the requirements of the customer (see Figure 6.1).

The purposes of the quality delivery process are to

1. Ensure that everyone works on those activities that are most important for the success of the business. This is done by fulfilling work group missions.
2. Improve the quality of deliveries (outputs) to the internal customers (the next person down the line) who receive the work.
3. Eliminate work that is wasted because it has not been done right the first time.
4. Harness the combined skills, ideas, and experience of the work group members to improve the business continuously through teamwork.
5. Ultimately satisfy the external customers.

A standard work group in a company consists of a manager and his or her team members. These are the people who can significantly affect the quality of the work they do. The work group manager leads the team. The work group produces outputs that are delivered to either external or internal customers. These outputs must be identified and measured for quality.

There are ten steps in the quality delivery approach. They are as follows:

1. Create a mission statement that defines the work group's activities. It is focused on the end objective rather than the means of achieving it.
2. Determine the outputs of the work group and ensure that they are according to the mission.
3. Identify the internal and external customers who receive the outputs.
4. Define the agreed-upon customer requirements for each output that must be met in order to achieve customer satisfaction.
5. Develop the work group's output specifications for each output.
6. Determine the work group's processes and identify the inputs that will deliver the outputs to the customers in a cost-effective manner.
7. Identify the measurements of each output that will compare the actual quality level delivered with the output specification.
8. Identify
   - Any problem caused by a measured shortfall to target.
   - Any opportunity to exceed target at no additional cost.
   - Any opportunity to meet customer requirements at a lower internal cost.
9. Establish a project team to solve the identified problem(s) that will improve the actual quality level delivered to the customer or capture the opportunity.
10. Measure customer satisfaction against the agreed-upon customer requirements.

In addition to the cost of quality, there are seven generic ways in which the quality of outputs can be measured.

1. Defects: work that has not been done according to specifications
2. Rework: work that requires correction
3. Scrap: work that has to be thrown away
4. Lost items: work that needs to be done again
5. Backlogs: work that is behind schedule
6. Late deliveries: work done after the agreed time
7. Surplus items or the work not required

The above measurements apply equally to office outputs such as paper work, electronic data, telephone calls, etc., as well as to the outputs of manufacturing units, laboratories, warehouses, workshops, hotels, professional service, etc., such as finished products, test results, parts, materials, tools, and so on.

There are five measurements for each output.

1. Targets: the budget or target level of performance to be achieved.
2. Forecast: the forecast level of performance that may be better or worse than the target, depending on the current business situation. The forecast also shows when the target will be reached.
3. Actual: the actual level of performance achieved to date.

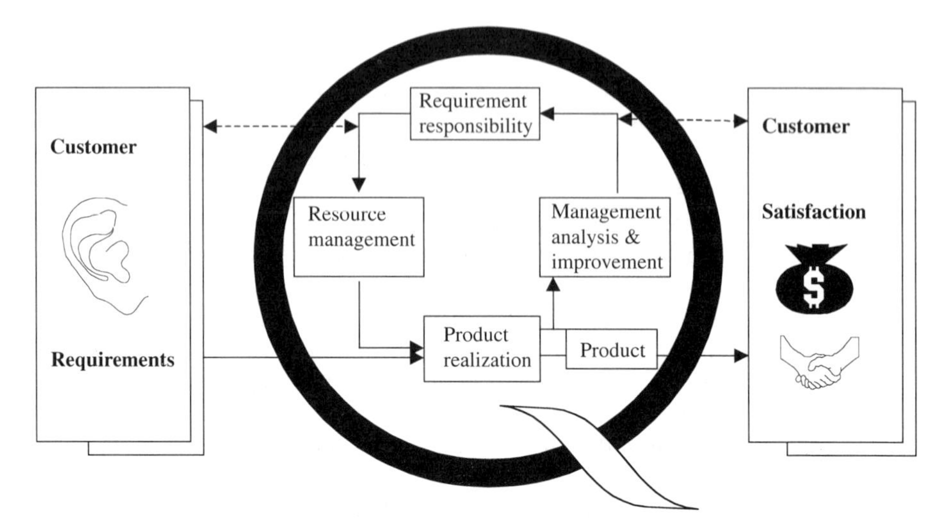

FIGURE 6.2 Information flow = ----->; value-adding activities ⟶.

4. Problem: the difference between the actual and target level of performance where "actual" is worse than "target."
5. Opportunity: the opportunity to improve quality over target at no extra cost.

The objective of the quality management system is to ensure that the product offered meets the expectations and needs of the user and that the product is continuously maintained to the prescribed standard/specification/contract/order agreement.

The Quality Management System ISO 9001:2000 indicates the adoption of a process approach when developing, implementing, and improving to enhance customer satisfaction by meeting customer requirements. The model of a process-based quality management system as shown in Figure 6.2 illustrates the process linkages presented in clauses 4 to 8. It is evident from the illustration that the role of customer is significant in defining the requirements as inputs. The evaluation of information by the organization plays a key role in determining whether the customer perception is met in the end product.

The documented quality management system should be composed of but not limited to the following key documents:

- Quality policy and quality objectives
- Quality System Manual
- Procedures required by this international standard
- Documents needed by the organization to ensure the effective planning, operation, and control of its processes
- Quality records required by this international standard

The documentation is maintained both in hardware and software form as appropriate. The organization should establish and maintain a quality system manual to describe the scope of the quality management system, including details of and justification for any exclusions, reference to the procedures established, and a description of the interaction between the processes of the quality management system. Procedures, documents, and quality records required by the quality management system should be reviewed to ensure their adequacy prior to approval and issue, and updated and reapproved as necessary, to ensure that changes and the current revision status of documents are identified. The relevant versions of applicable documents should be available at points of use as appropriate and should remain legible and readily identifiable. The documents of external origin should be identified, and their distribution should be controlled to prevent the unintended use of obsolete documents. A suitable identification system should be established to identify the documents if they are withheld for any purpose.

Very few publications are available on the market to explain exactly how to transcribe the elements of the existing ISO 9001:2000 standard into practically applicable language. For the easy understanding of readers, the text describes briefly the following elements of ISO 9001:2000 standards starting from clause 4 (Quality Management System) to clause 8 (Measurement, Analysis and Improvement) and provides audit checklists with the hope that they will serve as a valuable tool to ensure that the application of principles and procedures will result in quality, with a particular reference to customer satisfaction.

6.3.1 Quality Management System Audit Checklist Based on ISO 9001:2000 Clause 4

The Quality Management System Audit Checklist covers the following clauses: 4, 4.1, 4.2, 4.2.1, 4.2.2, 4.2.3, and 4.2.4.

The questions in the checklist in Appendix 6.1 are based on International Standards ISO 9001:2000, clause 4. Because quality management system audits can be performed by more than one person, signature spaces are provided at the end in order to provide a record of each individual auditor's activities.

6.4 MANAGEMENT RESPONSIBILITY

Management responsibility is one of the most critical element of the International Standard ISO 9001:2000, to ensure that quality aims are achieved and the firm's reputation is promoted in the market through customer satisfaction.

The organization should provide objective evidence of its commitment to the development and implementation of the quality management system (QMS). The QMS continually improves its effectiveness through communicating to the organization the importance of meeting customer specifications, as well as statutory and regulatory requirements, establishing a quality policy, and ensuring that quality objectives are established. Adequate resources should be provided and management reviews should be conducted at specified frequencies. A documented quality system should be maintained to ensure that the products conform to the specified requirements and meet the expectations and needs of the user and that the products are continuously maintained to the prescribed standards, specifications, contracts, and order agreements. The quality policy should be appropriate to the purpose of the organization, with a commitment to comply with requirements and continually improve the effectiveness of the quality management system. The policy should be communicated and understood within the organization and be reviewed on a periodic basis for its suitability. To meet customer (satisfaction) and regulatory requirements, the organization should ensure effective planning at all levels. Quantifiable quality objectives consistent with the quality policy should be established, including those needed to meet requirements for product-relevant functions at all levels within the company. The integrity of the quality management system should be maintained when changes to the system are planned and implemented. Communication should be conducted through identification of job responsibilities, training, and awareness. The responsibilities and authorities including interrelations and overlapping responsibilities should be defined and communicated within the organization. The company should hire or identify a management representative who, irrespective of other responsibilities, will have the responsibility and authority to

- Identify processes and implement and maintain the quality management system needed for quality management
- Evaluate and report the performance of the quality management system and any need for further improvement
- Promote the awareness of customer requirements throughout the organization

Top management should review the organization's quality management system at specified intervals to ensure its continuing suitability, adequacy, and effectiveness. This review should include assessing opportunities for improvement and the need for changes in the quality management system, including the quality policy and quality objectives. Records of the management reviews should be maintained.

The input to management review should include information, which may not be limited to

- Recommendations for improvement
- Planned changes that could affect the quality management system
- Status of preventive and corrective actions
- Follow-up actions from previous management reviews
- Process performance and product conformity
- Customer feedback
- Results of audits

Output from the management reviews should include decisions and actions related to improvement of the effectiveness of the quality management system and its processes, improvement of product related to customer requirements, and resource needs. The checklist in Appendix 6.2 is prepared to serve as an auditing tool to evaluate the management responsibilities based on International Standard ISO 9001:2000.

6.4.1 Management Responsibility Audit Checklist Based on ISO 9001:2000 Clause 5

The Management Responsibility Audit Checklist covers the following clauses: 5, 5.1, 5.2, 5.3, 5.4, 5.4.1, 5.4.2, 5.5, 5.5.1, 5.5.2, 5.5.3, 5.6, 5.6.1, 5.6.2, and 5.6.3.

The questions in the checklist shown in Appendix 6.2 are based on International Standard ISO 9001:2000, clause 5. Because management responsibility audits can be performed by more than one person, signature spaces are provided at the end in order to provide a record of each individual auditor's activities.

6.5 RESOURCE MANAGEMENT

The organization should determine, establish, and maintain the infrastructure needed to achieve conformity to product requirements, including human resources, materials and methods, machines, buildings, work environment, workspace, and associated utilities. Process equipment, both hardware and software, and supporting services such as transport or communication should also be reviewed and maintained on a continual basis.

Adequate resources should be established to achieve management objectives, customer satisfaction, and product compliance. The organization should determine and provide the resources needed to implement and maintain the quality management system and continually improve its effectiveness and to enhance customer satisfaction by meeting customer requirements. Human resources play an important role in the effectiveness of the quality system. Records of human resource development should be documented and maintained. The competency of the employees to fulfill their job functions on the basis of appropriate education, training, skills, and background experience should be ensured.

The checklist in Appendix 6.3 is provided to evaluate the resource management requirements based on an International Standard ISO 9001:2000.

6.5.1 Resources Management Audit Checklist Based on ISO 9001:2000 Clause 6

The Resource Management Audit Checklist covers the following clauses: 6, 6.1, 6.2, 6.2.1, 6.2.2, 6.3, and 6.4.

The questions in the checklist shown in Appendix 6.3 are based on International Standards ISO 9001:2000, clause 6. Because resources management system audits can be performed by more than one person, signature spaces are provided at the end section in order to provide a record of each individual auditor's activities.

6.6 PRODUCT REALIZATION

Top management should plan and develop the processes needed for product realization. Planning of product realization should be consistent with the requirements of the other processes of the quality management system. The following elements should be determined as appropriate:

- Quality objectives and product requirements
- Processes establishment, documents, and provision of resources specific to the product
- Verification, validation, monitoring, inspection, and test activities specific to the product to fulfill requirements
- Records needed to provide evidence that the realization processes and resulting product fulfill requirements
- Documentation of the output of the planning

The organization should ensure that it understands current and future customer needs and expectations through the role of leadership, involvement of people, continuous improvement, and mutually beneficial relationships with the suppliers. The organization should determine:

- Customer-specified requirements, including delivery and post-delivery activities
- Requirements necessary for specified use or known and intended use, even though not stated by the customer
- Regulatory and statutory requirements related to the product
- Additional requirements determined by the organization, if any

Organization reviews should address the requirements related to the product prior to the organization's commitment to supply a product to the customer (e.g., submission of tenders, acceptance of contracts of orders, acceptance of changes to contracts or orders) to ensure that

- Product requirements are defined
- Contract or order requirements differing from those previously expressed are resolved
- The organization has the ability to meet the defined requirements

If product requirements are changed, the organization should ensure that relevant documents are amended and that relevant personnel are made aware of the changed requirements.

The organization should establish and implement effective arrangements for communicating with customers in relation to

- Product information
- Inquiries, contracts, or order handling, including amendments
- Customer feedback, including customer complaints

Design and development should be carried out through effective quality planning. Both product and service should be considered with particular reference to customer focus. The design and development of products should include the review, verifications, and validations that are appropriate for each design and development stage, and the responsibilities and authorities for the design and development team.

The interfaces between different groups involved in design and development should be managed to ensure effective communication and clear assignment of responsibility. Inputs related to product requirements should be determined and documented, including

- Functional and performance requirements
- Applicable statutory and regulatory requirements
- Information derived from previous similar designs, where applicable
- Other requirements essential for design and development

These inputs should be reviewed for adequacy and completeness of requirements, to ensure that they are unambiguous and not in conflict with each other. The outputs of design development should be provided in a form that enables verification against the design and development input and approved prior to release. Design and development outputs should

- Meet the input requirements for design and development
- Provide appropriate information for purchasing, production, and service provisions
- Contain or reference product acceptance criteria
- Specify the characteristics of the product that are essential for its safe and proper use

Systematic reviews of design and development should be conducted at suitable stages. The ability of design and development to meet customer needs should be evaluated to fulfill requirements and to identify any problems and propose necessary actions. The representatives of functions concerned with the design and development stage should participate in the review, and a record of the necessary actions and review participation should be maintained.

Verification should be performed to ensure that the design and development outputs have satisfied the design and development input requirements. Design and development validation should be performed in accordance with the planned arrangement to ensure that the resulting product is capable of fulfilling the requirements for the specified or known use or application. Wherever practical, validation should be completed prior to the delivery or implementation of the product.

Design and development changes should be identified and documented. The changes should be reviewed, verified, and validated, as appropriate, and approved before implementation. The review of design and development changes should include an evaluation of the effect of the changes on constituent parts and delivered product.

Quality purchasing should be achieved through procurement from approved vendor sources. The company should ensure that the purchased product conforms to specified purchase requirements. The type and extent of control applied to the supplier and the purchased product are dependent on the effect of the purchased product on subsequent product realization or the final product. Purchasing information should be established to describe the product to be purchased, including, where appropriate

- Requirements for approval of product, procedures, processes, and equipment
- Requirements for qualification of personnel and for a quality management system

The organization should ensure the adequacy of specified purchase requirements prior to their communication to the supplier. The organization should establish and implement the inspection or other activities necessary for ensuring that purchased product meets specified purchase requirements. Where the organization or its customer intends to perform verification at the supplier premises, the organization should state the intended verification arrangements and method of product release in the purchasing information. The company should plan and carry out production and service provisions under controlled conditions as applicable, but not limited to the availability of information that describes the characteristics of the product, the availability of work instructions, the use of suitable equipment, the availability and use of monitoring and measuring devices, the implementation of monitoring and measurement, and the implementation of release, delivery, and post-delivery activities.

Any process for production and service provision wherein the resulting output cannot be verified by subsequent monitoring or measurement should be validated. Specific consideration should be given to the processes where deficiencies become apparent only after the product is in use or the service has been delivered. The planned results should be achieved through validation and where applicable through

- Defined criteria for review and approval of the processes
- Approved equipment and qualifications of personnel
- The use of specific methods and procedures
- Record maintenance (see Section 6.4.2.4)
- Planned and conducted revalidation

The organization should identify the product by suitable means throughout product realization.

The company should identify, verify, protect, and safeguard customer property provided for use or incorporation into the product, including intellectual property. If any customer property is lost, damaged, or otherwise found to be unsuitable for use, that information should be reported to the customer.

The conformity of product and its constituents should be preserved during internal processing and delivery to the intended destination. This preservation should include identification, handling, packaging, storage, and protection.

The organization should establish processes to ensure that monitoring and measurements undertaken to provide evidence of product conformity to its requirements are carried out in a manner that is consistent with monitoring and measurement requirements. As appropriate, measuring equipment should be

- Calibrated or verified at specified intervals, or prior to use, against measurement standards traceable to international or national measurement standards; where no such standards exist, the basis used for calibration or verification is recorded

- Adjusted or readjusted where necessary
- Identified to enable the calibration status to be determined
- Protected from adjustments that would invalidate the measurement result
- Protected to avoid damage and deterioration during handling, maintenance, and storage

In addition, when equipment is found not to conform to requirements, the organization should take appropriate action on the equipment and any product affected. Records of the results of calibration and verification should be maintained. The organization should assess and record the validity of previous measuring results.

The ability of computer software to satisfy the intended application should be confirmed prior to initial use when used in the monitoring and measurement of specified requirements.

The checklist in Appendix 6.4 is prepared to achieve consistency in evaluating compliance of the product realization process within the organization in accordance with the request of ISO 9001:2000.

6.6.1 Product Realization Audit Checklist Based on ISO 9001:2000 Clause 7

The Product Realization Audit Checklist covers the following clauses: 7, 7.1,7.2, 7.2.1, 7.2.2, 7.2.3, 7.3, 7.3.1, 7.3.2, 7.3.3, 7.3.4, 7.3.5, 7.3.6, 7.3.7, 7.4, 7.4.1, 7.4.2, 7.4.3, 7.5, 7.5.1, 7.5.2, 7.5.3, 7.5.4, 7.5.5, and 7.6.

The questions in this checklist, shown in Appendix 6.4, are based on International Standards ISO 9001:2000, clause 7. Because product realization audits can be performed by more than one person, signature spaces are provided at the end in order to provide a record of each individual auditor's activities.

6.7 MEASUREMENT, ANALYSIS, AND IMPROVEMENT

The revised standard ISO 9001:2000 places great emphasis on measurement, analysis, and continual improvement in the effectiveness of the quality management system through management review (measurement and analysis), internal and external audits, and corrective actions. Adequate monitoring and measurements should be conducted to ensure customer satisfaction and product specification compliance. It is recommended that suitable methods be used for obtaining information regarding customer perception about whether the company has fulfilled customer requirements through internal audits, the monitoring and measurement of process, and the monitoring and measurement of product. Internal audits at specified intervals should be conducted to determine whether the quality management system

- Is effectively implemented and maintained; and
- Conforms to the planned arrangements, the requirements of these international standards, and the quality management system requirements established by the organization.

Program audits should be conducted that consider the status and importance of the processes and areas to be audited and the results of previous audits. The audit criteria, scope, frequency, and methods should be defined. The selection of auditors and conduct of audits should ensure the objectivity and impartiality of the audit process. The responsibilities and requirements for planning and conducting audits as well as for reporting results and maintaining records should be defined in a documented procedure. The management responsible for the area being audited should ensure that actions are taken without undue delay to eliminate detected nonconformities and their causes. Follow-up activities should include the verification of actions taken and the reporting of verification results. Suitable methods for monitoring and, where applicable, measurement of the quality management system processes should be applied to demonstrate the ability of the processes to achieve planned results. When planned results are not achieved, correction and corrective action should be taken, as appropriate, to ensure the conformity of the product.

The product requirements should be fulfilled through monitoring and measuring the characteristics of the product. Evidence of conformity with the acceptance criteria should be maintained. Records should indicate the person(s) authorizing release of product. Product release and service delivery should not proceed until all the planned arrangements have been satisfactorily completed, unless otherwise approved by a relevant authority, and where applicable by the customer.

Any nonconforming product should be identified and controlled to prevent its unintended use or delivery. The controls and related responsibilities and authorities for dealing with nonconforming product should be defined in a documented procedure. Nonconforming products that are corrected should be subjected to reverification to demonstrate conformity to the requirements. When a nonconforming product is detected after delivery or use has started, the organization should take action appropriate to the effects of the nonconformity.

The data generated as a result of monitoring and measurement and from other relevant sources should be analyzed to demonstrate the suitability and effectiveness of the quality management system and continual improvement. The data should broadly based but not limited to

- Customer satisfaction
- Conformance to product requirements
- Characteristics and trends of processes and products including opportunities for preventive action

The organization's top management should ensure continuous improvement through creating personal ownership of the organization's goals by using its people's knowledge and experience, and through education achieved as a result of involvement in operational decisions and process improvement. It should be management's intention to continually improve the effectiveness of the quality management system through the use of the quality policy, quality objectives, audit results, analysis of data, corrective and preventive actions, and management reviews. Actions should be taken to eliminate the cause of nonconformities in order to prevent recurrence. Suitable corrective actions should be initiated to address the effects of the nonconformities encountered. A documented procedure should be established to define requirements for measurement, analysis, and continuous improvement. The audit checklist in Appendix 6.5 was developed to ensure that the critical elements of measurement, analysis, and improvement are in compliance with the International Standard ISO 9001:2000.

6.7.1 Measurement Analysis and Improvement Audit Checklist Based on ISO 9001:2000 Clause 8

The Measurement Analysis and Improvement Audit Checklist covers the following clauses: 8, 8.1, 8.2, 8.2.1, 8.2.2, 8.2.3, 8.2.4, 8.3, 8.4, 8.5, 8.5.1, 8.5.2, and 8.5.3.

The questions in this checklist, shown in Appendix 6.5, are based on International Standards ISO 9001:2000, clause 8. Because measurement and improvement audits can be performed by more than one person, signature spaces are provided at the end in order to provide a record of each individual auditor's activities.

6.8 DISCLAIMER

Although every effort has been made to ensure that the audit checklist is in accordance with the requirement of ISO 9001:2000 standard, the author accepts no responsibility for any occurrences or actions taken by the company subsequent to following this checklist.

APPENDIX 6.1

ISO 9001:2000 AUDIT CHECKLIST

Facility: ______________________________
Address: ______________________________
Audit date: ______________________________
Auditor (head): ______________________________
Audit Team Members (member 1): ______________________________
(member 2): ______________________________
(member 3): ______________________________
Purpose of audit: ______________________________

General Requirements	Rating				Ref. Clause for Int'l. Standard ISO 9001:2000
	(EX)	(AD)	(PO)	(U.S.)	
3.1.1 QUALITY MANAGEMENT SYSTEM					**4**
3.1.1.1 GENERAL					**4.1**
• Is there an adequate documented quality management system?	□	□	□	□	**4.1**
• Is the documented quality management system implemented and maintained?	□	□	□	□	**4.1**
• Does the quality management system indicate continuous improvement?	□	□	□	□	**4.1**
• Are the processes needed for quality management system application identified throughout the organization?	□	□	□	□	**4.1**
• Are all the sequences and interaction of these processes identified?	□	□	□	□	**4.1**
• Are the criteria and methods to ensure effectiveness of the operation and control of these processes established?	□	□	□	□	**4.1**
• Is the availability of resources and information necessary to support the operation and monitoring of these processed identified and established?	□	□	□	□	**4.1**
• Is there a procedure to monitor, measure, and analyze these processes?	□	□	□	□	**4.1**
• Are implementation actions monitored, measured, and analyzed to achieve planned results and continual improvement of these processes?	□	□	□	□	**4.1**
• If an organization chooses to outsource any processes that affect product quality, is control over such processes identified and demonstrated?	□	□	□	□	**4.1**
3.1.1.2 DOCUMENTATION REQUIREMENTS					
3.1.1.2.1 GENERAL					**4.2.1**
• Are there documented quality policy and quality objectives?	□	□	□	□	**4.2.1**
• Is there a documented quality manual addressing the elements described in the International Standard ISO 9001:2000?	□	□	□	□	**4.2.1**

• Are required written standard operating procedures available in accordance with International Standard ISO 9001:2000?	☐	☐	☐	☐	**4.2.1**
• Are effective planning operation and control in effect supported with adequate documents?	☐	☐	☐	☐	**4.2.1**
• Are the quality records required by this International Standard identified, established, documented, implemented, and maintained?	☐	☐	☐	☐	**4.2.1**
• Are the documented procedures established based on the size of the organization and type of activities?	☐	☐	☐	☐	**4.2.1**
• When writing documented procedures, are the complexity of processes and their interaction considered?	☐	☐	☐	☐	**4.2.1**
• Is the documented system adequate to support the competence of the personnel?	☐	☐	☐	☐	**4.2.1**
3.1.1.2.2 QUALITY MANUAL					**4.2.2**
• Does the Quality Manual describe its scope?	☐	☐	☐	☐	**4.2.2**
• Are the exclusions described in the Quality Manual are justified?	☐	☐	☐	☐	**4.2.2**
• Are the documented procedures established for the quality management system referred to in the Quality Manual?	☐	☐	☐	☐	**4.2.2**
3.1.1.2.3 CONTROL OF DOCUMENTS	☐	☐	☐	☐	**4.2.3**
• Are the documents required by the quality management system controlled?	☐	☐	☐	☐	**4.2.3**
• Is there a procedure to describe the control needed over the special type of documents?	☐	☐	☐	☐	**4.2.3**
• Are documents approved for adequacy prior to use?	☐	☐	☐	☐	**4.2.3**
• Is there a procedure to review and update and reapprove documents as necessary?	☐	☐	☐	☐	**4.2.3**
• Are changes and the current revision status of documents identified?	☐	☐	☐	☐	**4.2.3**
• Are relevant versions of applicable documents available at the point of use?	☐	☐	☐	☐	**4.2.3**
• Does the system ensure that the documents in use are legible and readily identifiable?	☐	☐	☐	☐	**4.2.3**
• Are obsolete documents retained for any purpose marked with suitable identification to prevent unintended use?	☐	☐	☐	☐	**4.2.3**
3.1.1.2.4 CONTROL OF QUALITY RECORDS	☐	☐	☐	☐	**4.2.4**
• Are quality records established and maintained to provide evidence of conformity with requirements and of the objective operation of the quality management system?	☐	☐	☐	☐	**4.2.4**
• Is there a system to ensure that quality records remain legible, readily identifiable, and retrievable?	☐	☐	☐	☐	**4.2.4**
• Is there a procedure describing the types of quality records, their identification, storage, protection, retrieval, retention time, and disposition?	☐	☐	☐	☐	**4.2.4**

Wrap-up session date: ____________________

Attendees: ____________________

____

____

____

Observations/Comments:

____

____

____

____

Audited by:____________________

Key: **EX = Excellent, AD = Adequate, PO = Poor, U.S. = Unsatisfactory**

(From *Quality Management Systems Requirements*, 3rd ed., ISO 9001:2000, BSI, London, 2000. With permission.)

APPENDIX 6.2

ISO 9001:2000 AUDIT CHECKLIST

Facility: ____________________

Address: ____________________

Audit date: ____________________

Auditor (head): ____________________

Audit Team Members (member 1): ____________________

(member 2): ____________________

(member 3): ____________________

Purpose of audit: ____________________

General Requirements	Rating (EX)	(AD)	(PO)	(U.S.)	Ref. Clause for Int'l. Standard ISO 9001:2000
4.1.1 MANAGEMENT RESPONSIBILITY					
4.1.1.1 MANAGEMENT COMMITMENT					**5.0**
• Is there evidence that top management is committed to the development and implementation of the quality management system and continually improving its effectiveness?	□	□	□	□	**5.1**
• Is there a procedure describing how the organization communicates internally the importance of meeting customer statutory and regulatory requirements?	□	□	□	□	**5.1**
• Is the quality policy established, documented, and understood?	□	□	□	□	**5.1**
• Is there a procedure to ensure that quality objectives are established?	□	□	□	□	**5.1**
• Is there a procedure to conduct management reviews?	□	□	□	□	**5.1**
• Is there a procedure to ensure that the resources required are identified and provided?	□	□	□	□	**5.1**

4.1.1.2 CUSTOMER FOCUS

• Do the procedures ensure that customer requirements are determined and fulfilled with the aim of enhancing customer satisfaction by the top management?	☐	☐	☐	☐	**5.2**

4.1.1.3 QUALITY POLICY

• Is there a written quality policy from top management?	☐	☐	☐	☐	**5.3**
• Is the quality policy appropriate for the purpose of the organization?	☐	☐	☐	☐	**5.3**
• Does the quality policy reflect a commitment to comply with requirements and continually improve the effectiveness of the quality management system?	☐	☐	☐	☐	**5.3**
• Does the quality policy provide a framework for establishing and reviewing quality objectives?	☐	☐	☐	☐	**5.3**
• Is the quality policy communicated and understood within the organization?	☐	☐	☐	☐	**5.3**
• Is there a system to review quality policy on a timely basis for continuing suitability?	☐	☐	☐	☐	**5.3**

4.1.1.4 PLANNING

4.1.1.4.1 QUALITY OBJECTIVES — 5.4

• Is a system to establish plans to ensure quality objectives identified?	☐	☐	☐	☐	**5.4.1**
• Are the quality objectives measurable and consistent with the quality policy?	☐	☐	☐	☐	**5.4.1**

4.1.1.4.2 QUALITY MANAGEMENT SYSTEM PLANNING

• Is quality management system planning carried out in accordance with the requirements given in 4.1 as well as the quality objectives?	☐	☐	☐	☐	**5.4.2**
• Is there a procedure to ensure that quality management system integrity is maintained when changes to the quality management system are planned and implemented?	☐	☐	☐	☐	**5.4.2**

4.1.1.5 RESPONSIBILITY, AUTHORITY, AND COMMUNICATION

4.1.1.5.1 RESPONSIBILITY AND AUTHORITY

• Is there a procedure defining responsibilities, authorities, and their interrelation?	☐	☐	☐	☐	**5.5.2**

4.1.1.5.2 MANAGEMENT REPRESENTATIVE

• Is there a person appointed or designated by top management, irrespective of other responsibilities, to be responsible for the overall quality management system?	☐	☐	☐	☐	**5.5.2**

- Is the management representative responsible for ensuring that processes needed for the quality management system are established, implemented, and maintained? ☐ ☐ ☐ ☐ **5.5.2**
- Is the management representative authorized and responsible to report to top management on the performance of the quality management system and any needs for improvement? ☐ ☐ ☐ ☐ **5.5.2**
- Is the management representative ensuring that awareness of customer requirements is communicated through the organization? ☐ ☐ ☐ ☐ **5.5.2**
- Is it the responsibility of the management representative to be the liaison with external parties on matters relating to the quality management system? ☐ ☐ ☐ ☐ **5.5.2**

4.1.1.5.3 INTERNAL COMMUNICATION

- Is there a documented procedure to ensure the adequacy of the communication processes within the organization? ☐ ☐ ☐ ☐ **5.5.3**
- Is there a procedure to ensure that communication is carried out regarding the effectiveness of the quality management system? ☐ ☐ ☐ ☐ **5.5.3**

4.1.1.6 MANAGEMENT REVIEW

4.1.1.6.1 GENERAL

- Is there a procedure to review the organization quality management system by top management at planned intervals? ☐ ☐ ☐ ☐ **5.6.1**
- Does the management review ensure the quality management system's continuing suitability, adequacy, and effectiveness? ☐ ☐ ☐ ☐ **5.6.1**
- Does the management review include assessing opportunities for improvement and the need for changes to the quality management system, including the quality policy and quality objectives? ☐ ☐ ☐ ☐ **5.6.1**
- Are the records of management reviews maintained? ☐ ☐ ☐ ☐ **5.6.1**

4.1.1.6.2 REVIEW INPUT

- Does the input to the management review include results of audits? ☐ ☐ ☐ ☐ **5.6.2**
- Is customer feedback included in each management review's input? ☐ ☐ ☐ ☐ **5.6.2**
- Are the preventive and corrective actions included in management reviews? ☐ ☐ ☐ ☐ **5.6.2**
- Are the follow-up actions from previous management reviews included in each management review's agenda? ☐ ☐ ☐ ☐ **5.6.2**
- Does the management review input consider planned changes that could affect the quality management system and recommendations for improvement? ☐ ☐ ☐ ☐ **5.6.2**

4.1.1.6.3 REVIEW OUTPUT

• Does the output from the management review include decisions and actions related to improvement of the effectiveness of the quality management system and its processes?	□	□	□	□	**5.6.3**
• Does the management review output include decisions ensuring improvement of product based on customer requirements and the resources needed?	□	□	□	□	**5.6.3**

Wrap-up session date: ____________________

Attendees: ____________________

____

____

____

Observations/Comments:

____

____

____

____

Audited by:__________

Key: **EX = Excellent, AD = Adequate, PO = Poor, U.S. = Unsatisfactory**

(From *Quality Management Systems Requirements*, 3rd ed., ISO 9001:2000, BSI, London, 2000. With permission.)

APPENDIX 6.3

ISO 9001:2000 AUDIT CHECKLIST

Facility: ____________________

Address: ____________________

Audit date: ____________________

Auditor (head): ____________________

Audit Team Members (member 1): ____________________

(member 2): ____________________

(member 3): ____________________

Purpose of audit: ____________________

General Requirements	Rating (EX)	(AD)	(PO)	(U.S.)	Ref. Clause for Int'l. Standard ISO 9001:2000
5.1.1 RESOURCE MANAGEMENT					
5.1.1.1 PROVISION OF RESOURCES					
• Is there a procedure to determine the resources needed?	□	□	□	□	**6.1**
• Is there a system in place to ensure adequacy of resources necessary to maintain the quality management system and continually improved its effectiveness?	□	□	□	□	**6.1**
• Does the resourcing procedure consider customer requirements to ensure customer satisfaction?	□	□	□	□	**6.1**

5.1.1.2 HUMAN RESOURCES

5.1.1.2.1 GENERAL

Question					Clause
• Is there a procedure available to evaluate competence and performance of the employees adequate to perform work affecting product quality?	☐	☐	☐	☐	**6.2.1**
• Does a system exist to provide on-the-job training to personnel relevant to their work?	☐	☐	☐	☐	**6.2.1**
• Is there a procedure to evaluate the results of training for personnel performing work affecting product quality?	☐	☐	☐	☐	**6.2.1**
• Is there a documented system to keep appropriate records of education, training, skills, and experience for all personnel?	☐	☐	☐	☐	**6.2.1**
• Is there a procedure to ensure that personnel are aware of the relevance and importance of their activities and how they contribute to achieving quality objectives?	☐	☐	☐	☐	**6.2.1**

5.1.1.2.2 COMPETENCE, AWARENESS, AND TRAINING

Question					Clause
• Are there procedures available to determine the competence of the personnel performing work affecting product quality?	☐	☐	☐	☐	**6.2.2**
• Is there a system in place to take necessary actions to satisfy competency requirements?	☐	☐	☐	☐	**6.2.2**
• Is there a procedure to review the effectiveness of the system?	☐	☐	☐	☐	**6.2.2**
• Are records of education, training, skills, and experience kept adequately?	☐	☐	☐	☐	**6.2.2**

5.1.1.3 INFRASTRUCTURE

Question					Clause
• Is there a system available to determine the infrastructure requirements such as buildings, workspace and associated utilities, process equipment, both hardware- and software-supporting services such as transport, to achieve conformity with product requirement?	☐	☐	☐	☐	**6.3**

5.1.1.4 WORK ENVIRONMENT

Question					Clause
• Is there a procedure available to evaluate and determine the work environment needed to achieve conformity with product requirement?	☐	☐	☐	☐	**6.4**

Wrap-up session date: ____________________

Attendees: ____________________

____

Observations/Comments:

____

____

Audited by:____________________

Key: **EX = Excellent, AD = Adequate, PO = Poor, U.S. = Unsatisfactory**

(From *Quality Management Systems Requirements*, 3rd ed., ISO 9001:2000, BSI, London, 2000. With permission.)

APPENDIX 6.4

ISO 9001:2000 AUDIT CHECKLIST

Facility: ______________________________
Address: ______________________________
Audit date: ______________________________
Auditor (head): ______________________________
Audit Team Members (member 1): ______________________________
(member 2): ______________________________
(member 3): ______________________________
Purpose of audit: ______________________________

General Requirements	Rating (EX)	(AD)	(PO)	(U.S.)	Ref. Clause for Int'l. Standard ISO 9001:2000
6.1.1 PRODUCT REALIZATION					
6.1.1.1 PLANNING OF PRODUCT REALIZATION					
• Are plans being made for product realization?	☐	☐	☐	☐	**7.1**
• Does this planning include the determination and review of the requirements of the product?	☐	☐	☐	☐	**7.1**
• Are SOPs available for it?	☐	☐	☐	☐	**7.1**
6.1.1.2 CUSTOMER-RELATED PROCESS					
6.1.1.2.1 DETERMINATION OF REQUIREMENTS RELATED TO PRODUCT					
• Do these SOPs cover the review of requirements that are not specified by the customer but are necessary for the intended use?	☐	☐	☐	☐	**7.2.1**
• Do these reviews include the requirements for delivery and post-delivery activities?	☐	☐	☐	☐	**7.2.1**
• Does the review include the regulatory requirements related to the product?	☐	☐	☐	☐	**7.2.1**
6.1.1.2.2 REVIEW OF REQUIREMENTS RELATED TO PRODUCT					
• Are these reviews being made prior to the organization's commitment to supply a product to the customer, e.g., at the time of submission of tenders?	☐	☐	☐	☐	**7.2.2**
• Do these reviews also include the organization's ability to fulfill the orders or tenders, with clarity and unambiguity in the customer requirements?	☐	☐	☐	☐	**7.2.2**
• Is there a procedure to review and to deal with situations where order or contract requirements differ from what was tendered or previously expressed?	☐	☐	☐	☐	**7.2.2**
• Are records being kept and maintained of all these reviews and actions arising as a results of these reviews?	☐	☐	☐	☐	**7.2.2**

• Is there a formal procedure to ensure that relevant documents are amended and personnel (relevant) are made aware when there is a change in requirements?	☐	☐	☐	☐	**7.2.2**

6.1.1.2.3 CUSTOMER COMMUNICATION

• Is there a written procedure to handle customer complaints?	☐	☐	☐	☐	**7.2.3**
• Does the organization take appropriate steps (e.g., surveys) to determine and implement effective arrangements for communicating with customers in order to get customer feedback?	☐	☐	☐	☐	**7.2.3**

6.1.1.3 DESIGN & DEVELOPMENT

6.1.1.3.1 DESIGN & DEVELOPMENT PLANNING

• Does the organization have a written procedure for the planning & control of design & development of product?	☐	☐	☐	☐	**7.3.1**
• If yes, does the procedure clearly specify the responsibilities and authorities involved in the design & development?	☐	☐	☐	☐	**7.3.1**
• Are the design & development stages identified with a clear mention of review, verification, & validation at the appropriate stages?	☐	☐	☐	☐	**7.3.1**
• Is there a system to update the planning output with the progress of design & development?	☐	☐	☐	☐	**7.3.1**

6.1.1.3.2 DESIGN & DEVELOPMENT INPUTS

• Does the review result in the determination of inputs for product requirements? Are the inputs for product requirements reviewed for adequacy?	☐	☐	☐	☐	**7.3.2**
• Are records of such reviews being maintained?	☐	☐	☐	☐	**7.3.2**
• Does the review include functional and performance requirements as well as the statutory and regulatory requirements?	☐	☐	☐	☐	**7.3.2**
• Does the review also consider the information derived from previous similar designs and other requirements essential for design & development?	☐	☐	☐	☐	**7.3.2**
• Is there a procedure to ensure the completeness and unambiguity of the requirements and that they are not in conflict with each other?	☐	☐	☐	☐	**7.3.2**

6.1.1.3.3 DESIGN & DEVELOPMENT OUTPUTS

• Is there a procedure to verify that the outputs are meeting the input requirements for design & development?	☐	☐	☐	☐	**7.3.3**
• Are the design & development outputs approved prior to release? Is the procedure available?	☐	☐	☐	☐	**7.3.3**
• Does the output of design & developments cover the appropriate information for purchasing, production, and service provision?	☐	☐	☐	☐	**7.3.3**
• Does it contain the product acceptance criteria?	☐	☐	☐	☐	**7.3.3**
• Does it specify essential characteristics for the product's safe & proper use?	☐	☐	☐	☐	**7.3.3**

6.1.1.3.4 DESIGN & DEVELOPMENT REVIEW

• Is there a formal procedure for systematic review at suitable stages?	☐	☐	☐	☐	**7.3.4**
• Does this procedure evaluate the results in terms of fulfilling requirements?	☐	☐	☐	☐	**7.3.4**
• Does this procedure propose necessary actions when problems are identified?	☐	☐	☐	☐	**7.3.4**
• Does the procedure identify the personnel responsible for such reviews?	☐	☐	☐	☐	**7.3.4**
• Do the personnel involved in the review include representatives of functions being reviewed at suitable stages?	☐	☐	☐	☐	**7.3.4**
• Are the records of all such reviews and corrective actions established and maintained?	☐	☐	☐	☐	**7.3.4**

6.1.1.3.5 DESIGN & DEVELOPMENT VERIFICATION

• Is there a system to verify and ensure that design & development outputs have satisfied input requirements?	☐	☐	☐	☐	**7.3.5**
• Are records being kept & maintained for verification and any necessary action carried out during the verification process?	☐	☐	☐	☐	**7.3.5**

6.1.1.3.6 DESIGN & DEVELOPMENT VALIDATION

• Does the validation being carried out ensure that resultant product is capable of fulfilling the intended use?	☐	☐	☐	☐	**7.3.6**
• Are these validation studies planned and prearranged (repeatedly meeting the acceptance criteria)?	☐	☐	☐	☐	**7.3.6**
• Are validation studies completed prior to delivery?	☐	☐	☐	☐	**7.3.6**
• If no, are these validation studies completed prior to implementation of the product?	☐	☐	☐	☐	**7.3.6**
• Are records of validation results maintained?	☐	☐	☐	☐	**7.3.6**
• Is the recording of any necessary actions, in response to validation studies, documented?	☐	☐	☐	☐	**7.3.6**

6.1.1.3.7 CONTROL OF DESIGN & DEVELOPMENT CHANGES

• Is there a change control procedure?	☐	☐	☐	☐	**7.3.7**
• Does this procedure cover design & development changes?	☐	☐	☐	☐	**7.3.7**
• Is there a formal procedure for review, verification, and validation of changes identified during design & development?	☐	☐	☐	☐	**7.3.7**
• Are the changes identified documented and records maintained?	☐	☐	☐	☐	**7.3.7**
• Does the system allow for approval of these changes prior to incorporation or implementation?	☐	☐	☐	☐	**7.3.7**
• Is there a formal procedure for the evaluation of the effect of the changes on the constituent parts and delivered product?	☐	☐	☐	☐	**7.3.7**
• Are the records for reviews of changes and any necessary actions maintained?	☐	☐	☐	☐	**7.3.7**

6.1.1.4 PURCHASING

6.1.1.4.1 PURCHASING PROCESS

• Does the organization have written purchasing specifications?	□	□	□	□	**7.4.1**
• Is there a procedure of vendor approval to ensure that purchased products conform to specified requirements?	□	□	□	□	**7.4.1**
• Does the organization have a vendor rating system?	□	□	□	□	**7.4.1**
• Is there a procedure of approval of vendors (vendor's facility & systems audits) prior to inclusion in the list of approved vendors?	□	□	□	□	**7.4.1**
• Does the organization have criteria for the selection and evaluation of vendors?	□	□	□	□	**7.4.1**
• Do these criteria include reevaluation of the vendors periodically?	□	□	□	□	**7.4.1**
• Are records of the results of evaluations and any necessary actions arising from the evaluations maintained?	□	□	□	□	**7.4.1**

6.1.1.4.2 PURCHASING INFORMATION

• Do these specifications include the process procedure, equipment, and product requirements for approval?	□	□	□	□	**7.4.2**
• Where appropriate, does the specification include qualification of personnel?	□	□	□	□	**7.4.2**
• Are the quality management system requirements also mentioned in the purchasing specifications?	□	□	□	□	**7.4.2**
• Does the organization review the adequacy of specified purchase requirements prior to their communication with a supplier?	□	□	□	□	**7.4.2**

6.1.1.4.3 VERIFICATION OF PURCHASED PRODUCT

• Is receiving inspection performed on purchased materials?	□	□	□	□	**7.4.3**
• If verification is performed on the supplier's premises, is the arrangement mentioned on the purchase specification along with method of product release?	□	□	□	□	**7.4.3**

6.1.1.5 PRODUCTION & SERVICE PROVISION

6.1.1.5.1 CONTROL OF PRODUCTION & SERVICE PROVISION

• Are production & service provisions planned?	□	□	□	□	**7.5.1**
• Are descriptions of products made available during production?	□	□	□	□	**7.5.1**
• Are work instructions clean & available on site?	□	□	□	□	**7.5.1**
• Is suitable equipment being used?	□	□	□	□	**7.5.1**
• Does the provision include the availability of monitoring & measuring devices to control the process?	□	□	□	□	**7.5.1**
• Does the provision for production & service include release, delivery, & post-delivery activities?	□	□	□	□	**7.5.1**

6.1.1.5.2 VALIDATION OF PROCESSES FOR PRODUCTION & SERVICE PROVISION

Question					Clause
• Are all the processes validated?	☐	☐	☐	☐	7.5.2
• Does validation demonstrate the ability of these processes to achieve planned results?	☐	☐	☐	☐	7.5.2
• Does the organization possess defined criteria for review and approval of the processes?	☐	☐	☐	☐	7.5.2
• Is there a system for approval of equipment and personnel qualifications?	☐	☐	☐	☐	7.5.2
• Is there an arrangement or system to use a specific method and procedure?	☐	☐	☐	☐	7.5.2
• Does the organization have a revalidation program?	☐	☐	☐	☐	7.5.2
• Is there an established arrangement for the requirement of records?	☐	☐	☐	☐	7.5.2

6.1.1.5.3 IDENTIFICATION & TRACEABILITY

Question					Clause
• Does the organization possess a system to identify the product by suitable means throughout product realization?	☐	☐	☐	☐	7.5.3
• Is the product status, with respect to monitoring & measurement requirements, identified by the organization?	☐	☐	☐	☐	7.5.3
• Are there systems and controls in place to uniquely identify the product for traceability purposes?	☐	☐	☐	☐	7.5.3

6.1.1.5.4 CUSTOMER PROPERTY

Question					Clause
• Are there procedures to exercise care for the customer's property?	☐	☐	☐	☐	7.5.4
• Does the organization have procedures to identify, verify, and protect customer property?	☐	☐	☐	☐	7.5.4
• Are records maintained for any customer property lost, damaged, or otherwise found unsuitable for use?	☐	☐	☐	☐	7.5.4

6.1.1.5.5 PRESERVATION OF PRODUCT

Question					Clause
• Does the organization preserve the conformity of product during internal processing and delivery to the intended destination?	☐	☐	☐	☐	7.5.5
• Does the preservation process include identification, handling, packaging, storage, and protection?	☐	☐	☐	☐	7.5.5
• Does the organization apply the preservation to the constituent parts of a product?	☐	☐	☐	☐	7.5.5

6.1.1.6 CONTROL OF MONITORING & MEASURING DEVICES

Question					Clause
• Are monitoring & measuring devices required to provide evidence of conformity of product identified?	☐	☐	☐	☐	7.6
• Is the process for monitoring & measurement consistent with monitoring & measurement requirements?	☐	☐	☐	☐	7.6
• Is the equipment used for the purpose calibrated?	☐	☐	☐	☐	7.6

• Are the calibration standards traceable to NIST standards?	☐	☐	☐	☐	**7.6**
• Is there a procedure to assess and record the validity of previous measuring results when the equipment is found not to conform with requirements?	☐	☐	☐	☐	**7.6**
• Are appropriate actions carried out on equipment and product affected by the conditions mentioned above?	☐	☐	☐	☐	**7.6**
• Are records of the results of calibration and verification maintained?	☐	☐	☐	☐	**7.6**
• Does the organization utilize computer software for monitoring purposes?	☐	☐	☐	☐	**7.6**
• Is the ability of computer software to satisfy the intended application confirmed?	☐	☐	☐	☐	**7.6**
• Are the appropriate calibrations being carried out prior to initial use and reconfirmed as necessary?	☐	☐	☐	☐	**7.6**

Wrap-up session date: ______________________________

Attendees: ______________________________

____

____

____

Observations/Comments:

____

____

____

____

Audited by:__________________

Key: **EX = Excellent, AD = Adequate, PO = Poor, U.S. = Unsatisfactory**

(From *Quality Management Systems Requirements*, 3rd ed., ISO 9001:2000, BSI, London, 2000. With permission.)

APPENDIX 6.5

ISO 9001:2000 AUDIT CHECKLIST

Facility: ______________________________

Address: ______________________________

Audit date: ______________________________

Auditor (head): ______________________________

Audit Team Members (member 1): ______________________________

(member 2): ______________________________

(member 3): ______________________________

Purpose of audit: ______________________________

General Requirements	Rating (EX) (AD) (PO) (U.S.)	Ref. Clause for Int'l. Standard ISO 9001:2000
7.1.1 MEASUREMENT ANALYSIS AND IMPROVEMENT		
7.1.1.1 GENERAL		
• Is there a program within the organization to plan and implement the monitoring, measurement, analysis, and improvement process?	☐ ☐ ☐ ☐	**8.1**
• Does the monitoring, measurement, analysis, and improvement process demonstrate conformity of the product?	☐ ☐ ☐ ☐	**8.1**
• Does the monitoring, measurement, analysis, and improvement process demonstrate conformity of the quality management system?	☐ ☐ ☐ ☐	**8.1**
• Does the monitoring, measurement, analysis, and improvement process ensure continuous improvement and effectiveness of the quality management system?	☐ ☐ ☐ ☐	**8.1**
• Does the monitoring include determination of applicable methods, including statistical techniques and the extent of their use?	☐ ☐ ☐ ☐	**8.1**
7.1.1.2 MONITORING AND MANAGEMENT		
7.1.1.2.1 CUSTOMER SATISFACTION		
• Does the measurement of performance include monitoring information relating to customer perception concerning whether the organization has fulfilled customer requirements?	☐ ☐ ☐ ☐	**8.2.1**
• Is there a procedure to determine how to obtain the information about customer perceptions?	☐ ☐ ☐ ☐	**8.2.1**
7.1.1.2.2 INTERNAL AUDIT		
• Is there a procedure to conduct internal audits at planned intervals to determine whether the quality management system conforms with the planned arrangements (see 7.1)?	☐ ☐ ☐ ☐	**8.2.2**
• Does the organization conduct internal audits at planned intervals to determine whether the quality management system conforms with the requirements of this international standard?	☐ ☐ ☐ ☐	**8.2.2**
• Does the organization conduct internal audits at planned intervals to determine whether the quality management system conforms with the quality management system requirements established by the organization?	☐ ☐ ☐ ☐	**8.2.2**
• Does the organization conduct internal audits at planned intervals to determine whether the quality management system is effectively implemented and maintained?	☐ ☐ ☐ ☐	**8.2.2**
• Is there a plan for an audit program, taking into consideration the status and importance of the processes and area to be audited as well as the results of previous audits?	☐ ☐ ☐ ☐	**8.2.2**

• Does the audit program define the audit criteria, scope, frequency, and methods?	□	□	□	□	**8.2.2**
• Does the selection of auditors and conduct of audits ensure objectivity and impartiality of the audit process?	□	□	□	□	**8.2.2**
• Does the audit procedure ensure that auditors not audit their own work?	□	□	□	□	**8.2.2**
• Are the responsibilities and requirements for planning and conducting audits and for reporting results and maintaining records (see 4.4.4) defined in a documented procedure?	□	□	□	□	**8.2.2**
• Does the management responsible for the area being audited ensure that actions are taken without undue delay to eliminate detected nonconformities and their causes?	□	□	□	□	**8.2.2**
• Do follow-up activities include verification of the actions taken and the reporting of verification results (see 8.5.2)?	□	□	□	□	**8.2.2**

7.1.1.2.3 MONITORING AND MEASUREMENT OF PROCESSES

• Does the organization apply suitable methods for monitoring and, when applicable, measurement of the quality management system process?	□	□	□	□	**8.2.3**
• Do these methods demonstrate the ability of the processes to achieve planned results?	□	□	□	□	**8.2.3**
• Are the corrections and corrective action taken appropriate to ensure conformity of the product when the planned results are not achieved?	□	□	□	□	**8.2.3**

7.1.1.2.4 MONITORING AND MEASUREMENT OF PRODUCT

• Does the organization monitor and measure the characteristics of the product to verify that product requirements are fulfilled?	□	□	□	□	**8.2.4**
• Is there a procedure to monitor and measure the product characteristics at appropriate stages of the product realization process in accordance with planned arrangements (see 7.1)?	□	□	□	□	**8.2.4**
• Is there evidence of conformity with the acceptance criteria maintained?	□	□	□	□	**8.2.4**
• Do the records indicate the person(s) authorizing release of product (see 4.2.4)?	□	□	□	□	**8.2.4**
• Does product release and service delivery not proceed until all the planned arrangements (see 7.1) have been satisfactorily completed, unless otherwise approved by relevant authority and where applicable by the customer?	□	□	□	□	**8.2.4**

7.1.1.3 CONTROL OF NONCONFORMING PRODUCT

• Does the organization ensure that product which does not conform with product requirements is identified and controlled to prevent its unintended use or delivery?	□	□	□	□	**8.3**

• Is there a documented procedure to describe the controls and related responsibilities and authorities for dealing with nonconforming product?	☐	☐	☐	☐	**8.3**
• Does the organization deal with nonconforming product by taking actions to eliminate the detected nonconformity?	☐	☐	☐	☐	**8.3**
• Does the organization deal with nonconforming product by authorizing its use, release, or acceptance under concession by a relevant authority and when applicable by the customer?	☐	☐	☐	☐	**8.3**
• Does the organization deal with nonconforming product by taking action to preclude its original intended use or application?	☐	☐	☐	☐	**8.3**
• Are records of the nature of nonconformities, and any subsequent actions taken including concessions obtained, maintained?	☐	☐	☐	☐	**8.3**
• Is there a procedure to ensure a corrected nonconforming product is subject to reverification to demonstrate conformity with requirements?	☐	☐	☐	☐	**8.3**
• Does the organization take action appropriate to the effects, or potential effects, of the nonconformities when nonconforming product is detected after delivery or use has started?	☐	☐	☐	☐	**8.3**

7.1.1.4 ANALYSIS OF DATA

• Is there a procedure to determine, collect, and analyze appropriate data to demonstrate the suitability and effectiveness of the quality management system?	☐	☐	☐	☐	**8.4**
• Does the organization determine and evaluate where continual improvement of the quality management system can be made?	☐	☐	☐	☐	**8.4**
• Does the evaluation of continual improvement include data generated as a result of monitoring and measurement and from other relevant sources?	☐	☐	☐	☐	**8.4**
• Does the analysis of data provide information relating to customer satisfaction (see 8.2.1)?	☐	☐	☐	☐	**8.4**
• Does the analysis of data provide information relating to conformance with product requirements (see 7.2.1)?	☐	☐	☐	☐	**8.4**
• Does the analysis of data provide information relating to characteristics and trends of processes and products, including opportunities for preventive actions?	☐	☐	☐	☐	**8.4**
• Does the analysis of data provide information relating to suppliers?	☐	☐	☐	☐	**8.4**

7.1.1.5 IMPROVEMENT

7.1.1.5.1 CONTINUAL IMPROVEMENT

• Is there a system to ensure that the organization continually improves the effectiveness of the quality management system through the use of the quality policy, quality objectives, audit results, analysis of data, corrective and preventive actions, and management review?	☐	☐	☐	☐	**8.5.1**

7.1.1.5.2 CORRECTIVE ACTION

- Does the organization take action to eliminate the cause of nonconformities in order to prevent recurrence? ☐ ☐ ☐ ☐ **8.5.2**
- Are corrective actions taken appropriate to the effects of the nonconformities encountered? ☐ ☐ ☐ ☐ **8.5.2**
- Is there a documented procedure established to define requirements for reviewing nonconformities (including customer complaints)? ☐ ☐ ☐ ☐ **8.5.2**
- Is there a documented procedure established to define requirements for determining the causes of nonconformities? ☐ ☐ ☐ ☐ **8.5.2**
- Is there a documented procedure established to define requirements for evaluating the need for action to ensure that nonconformities are secure? ☐ ☐ ☐ ☐ **8.5.2**
- Is there a documented procedure established to define requirements for determining and implementing action needed? ☐ ☐ ☐ ☐ **8.5.2**
- Is there a documented procedure established to define requirements for records of the results of action taken (see 4.2.4)? ☐ ☐ ☐ ☐ **8.5.2**
- Is there a documented procedure established to define requirements for reviewing corrective action taken? ☐ ☐ ☐ ☐ **8.5.2**

7.1.1.5.3 PREVENTIVE ACTION

- Does an organization determine action to eliminate the causes of nonconformities in order to prevent their occurrence? ☐ ☐ ☐ ☐ **8.5.3**
- Is there a system to ensure that preventive actions are appropriate to the effects of the potential problems? ☐ ☐ ☐ ☐ **8.5.3**
- Is there a documented procedure established to define requirements for determining potential nonconformities and their causes? ☐ ☐ ☐ ☐ **8.5.3**
- Is there a documented procedure established to define requirements for evaluating the need for action to prevent occurrence of nonconformities? ☐ ☐ ☐ ☐ **8.5.3**
- Is there a documented procedure established to define requirements for determining and implementing action needed? ☐ ☐ ☐ ☐ **8.5.3**
- Is there a documented procedure established to define requirements for reviewing preventive action taken? ☐ ☐ ☐ ☐ **8.5.3**

Wrap-up session date: ______________________________

Attendees: ______________________________

____

____

Observations/Comments:

____

____

____

____

Audited by:_______________

Key: **EX = Excellent, AD = Adequate, PO = Poor, U.S. = Unsatisfactory**

(From *Quality Management Systems Requirements*, 3rd ed., ISO 9001:2000, BSI, London, 2000. With permission.)

7 ISO 14001 and Best Industrial Practices*

Syed Imtiaz Haider, Ph.D.

7.1 INTRODUCTION

Concern about the state of the natural environment has deep historical roots, but the nature and scale of these concerns and their economic and political importance have grown and changed considerably over the past few decades. Business is often seen as the enemy of the environment, the polluter. There is, however, a tide of change, and it is now increasingly being seen as the essential partner, as part of the solution and not the problem, the provider of the wealth and resources needed.

International Environment Day, International Earth Day, World Environment Day in different countries around the world over the past years clearly demonstrate increasing environmental awareness at the international level. Environmental legislation and enforcement measures are becoming ever stricter, and increasing emphasis is being placed on the environmental management of businesses.

Fully environmental management systems can have a major impact in ways that have been widely proven to be both cost saving and environmentally responsible. Environmental management industrial practices can be divided into the following:

1. Energy use: Lighting, ventilation, electrical equipment and machinery, the solar option
2. Other environmental impacts: Water use, laundry, boilers, waste, recycling, ozone depleting substances, hazardous substances, stationery and office supplies, office equipment, transport, external influences, miscellaneous
3. Environmental management initiatives: Energy management systems, training, awareness and responsibilities, purchasing, and the total-cost approach
4. Summary

Items 1, 2, and 3 above contain recommendations for action, including no-cost, low-cost, and larger-cost measures. No-cost measures should be implemented immediately because they do not require any financial outlay (other than staff time). Low-cost measures are defined as approximately U.S. $500 and below, and could also be implemented quickly. Larger-cost measures involve a cost-benefit analysis (CBA) to work out respective payback times before implementation.

* Portions of this chapter have appeared in Seddon, S., Environmental Management Report, Gulf, UAE, July 2000. Reprinted with permission.

7.2 ENERGY USE

7.2.1 LIGHTING

Choosing, controlling, and maintaining a lighting system is an important step in achieving good energy management. Lighting is one of the major uses of electricity, and research shows that many businesses could reduce their lighting costs by at least 30%.

7.2.1.1 Recommendations and Guidelines

If lighting is not fulfilling an important aesthetic or decorative purpose, consider the use of energy-saving light bulbs. A low-energy bulb uses about one quarter of the electricity and lasts eight times as long an ordinary bulb, making the energy-saving bulb a staggering *32 times* more energy efficient than the ordinary bulb. Furthermore, bulb replacement costs are reduced.

Compact fluorescent bulbs should be used wherever possible because they use only 20% of the electricity of incandescent bulbs.

Ensure that all fluorescent bulbs are 26-mm diameter slimline compact fluorescent types (fitted with reflector hoods in appropriate areas to increase light output). These bulbs save 30% of the energy used by the older 35-mm fluorescent bulbs.

Ensure that these bulbs all have electronic or high-frequency ballasts (starters for fluorescent tubes) that use 20% less electricity than bulbs without these ballasts (old-style coil starters). They also reduce stroboscopic effects that can damage the eyes.

Bulbs without high-frequency ballasts are not worth switching off (from an energy-saving point of view) unless they will not be switched on again for at least 30 minutes.

Ensure that bulbs that are no longer working are either taken out or replaced *immediately* because they are still using the same amount of electricity even though they are not emitting any light. Large savings can be achieved doing this alone.

Tungsten spotlights should be replaced with tungsten halogen lamps that use less energy. Tungsten filament lamps should be replaced with compact fluorescent lamps that are more energy efficient.

If a lighting system is more than 5 years old, check that light levels are correct, using a light meter.

Ensure that lights are used only in areas and rooms where there is not sufficient natural daylight to light the work area, while maintaining the correct lux levels (for more information on Lux levels, check the Chartered Institute of Building Services [CIBS] lighting guide). Lux levels might be unnecessarily high and could be reduced, which would save on energy costs. Recommended lux levels are as follows:

100 — corridors, changing rooms, bulk stores, auditoriums
200 — dining rooms, foyers, entrances
500 — general offices
750 — meat inspection
1000 — mortuary
1000 to 5000 — hospital operating theatre

If appropriate, consider installing motion-detecting outside security lights to reduce energy waste by coming on only when necessary. Photoelectric light sensors could also be used.

Replace lamps as a group, at planned intervals. As a rule, lamps should be replaced when their output has depreciated by approximately 30% (this can be checked using a light meter). The energy used to power an inefficient bulb exceeds that of a new one.

It is usually most economical to replace starter switches on fluorescent bulbs in groups, every two lamp lives. Replacing lamps at the same time minimizes labor costs.

Passive infrared (PIR) sensors should be considered for restrooms. These are not very expensive to install and prevent lights being left on.

Ensure that in open plan areas, light switches are labeled to facilitate specific use, so that unnecessary lights are not switched on and used.

Ensure that sufficient switches are available to enable staff to control their own lighting.

Follow the latest approach and switch off all unnecessary lights when leaving the building at the end of the day. A main isolation switch could be positioned near exit doors so that the last person leaving can turn everything off.

The most cost-effective control system will depend on the occupancy patterns of the building, the main activities in each area of the building, and the amount of natural light available (research would be needed to gather these data).

Establish a regular cleaning schedule for lamps, shades and fittings, windows, walls, and ceilings

7.2.2.2 Ventilation

Slowly but surely, people are beginning to realize that an enormous amount of energy — not to mention the financial drain caused by electricity bills — is being wasted because of ineffective air-conditioning and refrigeration.

7.2.2.1 Recommendations and Guidelines

Gas-fired air-conditioning systems are more efficient and considerably cheaper in the long run than electricity-powered systems (although the tariff for gas is around six times that of electricity in the United Arab Emirates (UAE) at the moment, this is likely to change over the coming years). Energy consumption can be reduced by up to 85%, and the emission of ozone-depleting chlorofluorocarbons (CFCs) is also reduced. Gas-fired systems are also safe and reliable. Furthermore, because they have no moving parts or compressors, gas-fired systems are largely maintenance free. Other suggestions:

- Reduce the air volume being handled to the minimum required.
- Ensure that the air-conditioning running hours are kept to a minimum; when possible, switch it off in office areas not being used.
- Ensure that fibrous insulation on chilled water pipes and air-conditioning ducting is well maintained.

- Ensure that ducts, evaporators, condensors, and cooling towers are kept clean.
- Ensure that filters are changed regularly.
- Avoid simultaneous heating and cooling. Keep windows and doors closed when cooling is on so that the system is not working harder than is necessary.
- Fit blinds or reflective window film to reduce the inside heat caused by solar gain, without reducing the amount of natural light being allowed in.
- Consider installing an optimizer on air-conditioning.

Is the air temperature sometimes colder than it needs to be? Check that staff, customers and visitors are happy with the room temperature. During extremely hot summer months, there can be a tendency to overcool.

7.2.3 Electrical Equipment and Machinery

Power-factor correction units fitted to the incoming main supply equalize induction and prolong the average motor-drive life expectancy across a range of high-power demand equipment.

7.2.3.1 Recommendations and Guidelines

7.2.3.1.1 Computers and Printers

Ensure that all computers are fitted with a temporary suspend power-down function that automatically switches to an energy-saving mode when not in constant use. This not only saves energy but also suspends the emission of harmful radiation from the computer itself. Remember, however, that the power-down mode still uses more electricity than when the machine is switched off completely.

Ensure that staff follow "no-tech" procedures and turn off their computers at lunchtimes or during long periods when not in use, and certainly at the end of the working day.

Switch the printer on just before using it and switch it off just after finishing with it. Do not turn printers on at the same time as computers and then simply leave them on.

When next purchasing computer equipment, consider buying low-energy, flat screen monitors and low-energy printers that also have an auto-switch-to-idle mode.

It is cheaper and faster to photocopy rather than print multiple copies on a laser printer.

Leaving computers and printers on unnecessarily not only wastes energy directly, but it also places an extra demand on space-cooling systems.

Do not leave electrically operated calculators on when not in use.

7.2.3.1.2 Photocopy Machines

Ensure that the photocopy machine is switched off at the end of the day. *A photocopy machine left on overnight consumes enough electricity to make 1000 photocopies.* Consider fitting a timer on photocopy machines to prevent them from being left on overnight.

After finishing using the photocopier, switch the machine to stand-by/idle if there is no one waiting to use it.

Reduce the amount of photocopying by circulating documents or using notice boards or e-mail. Encourage staff to use double-sided copies whenever possible.

Keep spoiled copies for personal scrap use.

7.2.3.1.3 Stand-Alone Refrigerators and Freezers

Locate refrigerators and freezers away from heat sources, and schedule a defrosting program if needed.

Power-factor correction plugs can be fitted to these appliances to reduce consumption of electricity by 5 to 15%.

A minimum temperature or a target temperature range for the freezer should be specified. Having a freezer too cold is not a problem in terms of food hygiene, but it is using more energy (i.e., money) than is necessary.

Ensure that the door seals on refrigerators and freezers are still good.

7.2.3.1.4 Dishwashers

- Maximize the value of each dishwasher load with correct stacking.
- Clean the filters and service dishwashers regularly.
- Ensure that the dishwasher incorporates adequate insulation.
- Consider using sanitizing liquids and water softeners to reduce boost temperatures.
- When buying a new dishwasher, check its energy consumption and consider buying one with a heat-recovery cycle.
- Consider if the wash water can be recycled.

7.2.3.1.5 Point-of-Use Water Heating

Consider point-of-use water heaters for areas such as toilets, washing sinks, and locations remote from the hot-water storage tanks. Point-of-use water heaters can be wall-mounted both above and below sink units — and out of sight for plush executive designs.

7.2.4 The Solar Option

Has the option of solar power been explored, or even considered? Two weeks' solar radiation corresponds to the Earth's entire store of fossil fuels and uranium. With approximately 3500 hours of sunshine per year in countries having extensive deserts, the potential for solar power is massive. Most large solar panels worldwide are for water heating, but the use of (desert temperature-proof) photovoltaics (PV) that converts light directly into electricity for virtually any electrically operated device as well as for telecommunications also exists.

7.3. OTHER ENVIRONMENTAL IMPACTS

The following environmental impacts are identified as potentially significant in terms of giving rise to large quantities of waste or polluting emissions, or having the potential to breach environmental legal requirements.

7.3.1 Use of Water

A global water crisis could be just 20 years away, according to a recent survey by the *Financial Times*, with the UAE and other Gulf countries at particular risk. The UAE is the second largest consumer of water in the world, despite having one of the lowest annual rainfalls, and it is estimated that $3.5 billion will be needed over the next 20 years to make the country self-sufficient in water. Dubai Electricity & Water Authority (DEWA) recently launched a water conservation plan that aimed at tackling the problem of waste water — from leaks in the distribution system and from waste of individual consumers. Other parts of the world are not as much at risk, but still need to be aware of increasing concerns for availability and cost of potable water.

If the following policies are properly implemented and expenditure over a similar time span is monitored, then the savings to be made can be calculated. Research shows that the average company could save 15% of its water and effluent bill by minimizing water use.

7.3.1.1 Recommendations and Guidelines

Measure and monitor consumption (cheap meters can be bought if need be), concentrating on the main areas of use.

7.3.1.1.1 Inside Buildings

Encourage male staff to use the urinals provided for urinating and not the toilet cubicles, as is often the case. Flushing away one person's urine wastes an enormous amount of water over time. Cleaning time would also be reduced.

The use of the Biomat system in the men's urinal toilets is a very useful and cost-effective system. The Biomat is an antiseptic, scented mat that is placed in the urinal. It kills odors, maintains hygiene levels, and does not require the use of water for flushing. The urinal is cleaned and washed out at appropriate intervals.

Passive infrared (PIR) sensors are also a cost-effective way to save water for urinals. Research shows that companies using the older raised cistern (flush tank) for urinals can reduce water use by up to 70%. The PIR sensors typically use long-life batteries lasting 3 to 4 years (PIR sensors can be extended to control lighting and fans as well as water supplies).

Fit Hippos (or even a plastic water bottle filled with water or sand) in toilet cisterns to reduce flush volumes. If flushing is subsequently ineffective, then the levels of water or sand can be adjusted. Toilets plumbed after 1993 will probably have a 7.5-liter cistern. Through fitting Hippos in toilet cisterns, approximately 30% of the water will be saved with each flush regardless of the size of the cistern. Up to 190,080 gal (864,000 liters) of water a year can be saved from 40 flush toilets (not including urinals).

Test for leaking toilets by adding food coloring to the tank. If any color appears in the bowl after 30 min, then the toilet is leaking. A leaking toilet can waste up to 200 gal of water per day.

To reduce water waste from sink taps (and shower heads?), consider fitting spring-loaded return valves, in-line flow restrictors or spray taps, or timer or sensor-controlled taps that work just as effectively as normal taps and dispense water only when required by operatives. (Low-flow, high-velocity showers use water efficiently. Typical water use for a shower is 35 liter, with power showers using substantially more.) Research shows that companies can reduce water use up to 50% using flow restrictors on sink taps alone.

Water waste in food service areas is notoriously high — do not wash hands or kitchen equipment under running water.

Ensure that dripping or leaking taps are repaired quickly — *a tap that drips twice per second wastes 1200 gallons over a year.* Leaks and burst pipes can be extremely costly, as the following data show:

- A ½-inch water pipe loses 50 gallons per minute (gpm)
- A ¾-inch pipe loses 110 gpm
- A 1-inch pipe loses 210 gpm
- A 2-inch pipe loses 850 gpm
- A 3-inch pipe loses 1900 gpm
- A-4 inch pipe loses 3400 gpm

If possible, check the water meter while no water is being used. If the dials are moving, then there is a leak in the system. Report faults immediately.

When possible, collect the water from the air-conditioners in a rain barrel to use for outside watering (if appropriate).

Ensure that hot water is stored at 60°C (about 145°F). Storing water below this temperature increases the risk of Legionella. Storing water above this temperature is unnecessary and wastes energy — reset the immersion heater thermostat if necessary.

Pipe work should be lagged because this ensures that hot water pipes provide hot water and cold water pipes provide cold water (especially important in summer).

Water Management Initiatives

- Find out how much the organization is paying in water and effluent charges
- Carry out a water-use survey
- Agree on a target for water saving
- Estimate potential savings for reducing water use and effluent generation
- Identify other benefits from saving water
- Decide how much is worth spending on water-saving projects
- Train staff (if necessary) on water management
- Identify and evaluate appropriate water-saving devices and practices
- Identify project costs
- Consider the impact of the water-saving measures on your processes and services
- Implement (and monitor) no-cost, water-saving devices and practices immediately, and cost-effective, low-cost measures quickly
- Communicate successes and savings
- Obtain feedback from employees

Don't pay for water more than once, i.e., purchase costs, pumping costs, maintenance costs, capital costs, treatment costs, disposal costs.

7.3.2 Boilers

7.3.2.1 Recommendations and Guidelines

Consider the installation of gas-fired, condensing boilers — they are the most efficient type of boiler available. They have an efficiency rate of over 90%. If your boiler is 15 years old or unreliable, replace it — you could save 20% on your fuel bill. Replacing it with a condensing boiler could save you over 30%.

Fit a good insulating jacket to your hot water tank(s) if it is not foam covered. Ensure that all pipe work, valves, and pipe joints are insulated and that insulation is replaced after pipe work repairs.

Review the loading and operating patterns of your boiler. Check how often your boiler operates on partial load. In larger premises, condensing boilers can be combined with high-efficiency conventional boilers to provide base load and part load capacity, respectively.

Adjust boiler thermostats to ensure that the minimum number of boilers is on, and consider installing an automatic sequence controller.

7.3.2.1.1 Optimizers

- Check to see if the turn on/turn off system of your boiler is being optimized
- Ensure that internal and external sensors are working
- Consider installing an optimizer on 30-kW boilers and higher (if appropriate) to achieve the most efficient possible setting

The boiler should be serviced at least once a year (especially for gas-fired boilers to ensure efficient combustion) and flue emissions checked. An oil-fired boiler servicing should take place twice a year (or more frequently if indicated by high flue gas temps). Only certified and approved persons should examine and test boilers (and other pressure vessels).

When a portable extension light is used in a confined space (e.g., a boiler room), it should not be operated at more than 24 V.

7.3.3 Waste

The Clean Up the World campaign aims at making individuals, companies, and communities more aware of local environmental issues, particularly the importance of reducing waste, recycling, and waste management.

7.3.3.1 Recommendations and Guidelines

7.3.3.1.1 Permits

When appropriate, ensure that a permit is held for the discharge of all waste to the sewer, the land, or the marine environment; they can be obtained from regulatory agencies. It would be useful to keep these on file for at least 2 years. Check that all waste contractors are licensed to take the waste (if appropriate).

A condition of this permit is the periodic monitoring and reporting of the quantity and quality of this waste. If the quality of the waste exceeds prescribed standards, it may need to be disposed of as hazardous waste (see section on hazardous waste below).

7.3.3.1.2 Waste Reduction Initiatives

- Find out what waste is produced on your site
- Ask if any of this waste can be reduced and by whom
- Make someone responsible for reducing waste
- Establish waste collection points for recyclable items (see next section — Recycling)
- Provide separate collection bins for each category of waste
- Join a local waste minimization club or start one with companies on neighboring sites

For example, 14 companies in Merseyside in northwestern England are saving £4.2 million per year as part of a waste minimization club scheme, with half the projects having paybacks of less than 1 year. Eleven companies in Humberside, (eastern England) saved £1 million in their first year of membership in a waste minimization forum. Within 5 years, their forum expects to help another 140 companies. A group of companies in the Midlands (England), from a range of industries including engineering, chemical, foundries, food and drink, metal finishing, and glass, shared ideas and are saving more than £2 million/year.

The following tips will also help reduce waste:

- Encourage employees to reduce waste, for example, by donating a percentage of any savings raised to a local charity
- Use the minimum packaging needed
- Return product packaging to suppliers, agree on reduced or recycled packaging with customers and suppliers
- Minimize or reuse your own product packaging (if appropriate)
- Reuse cardboard cartons and plastic bags
- Compact nonrecyclable by-products to make disposal more manageable

7.3.3.1.3 Waste Water (See Also, Water Use)

Environmental protection agencies are working to create necessary awareness among all waste generators and to encourage them to adopt waste minimization techniques, including water reuse, and to control all discharges to the land and marine environment.

- The basis of an effective waste minimization plan is a detailed wastewater audit study. Such a plan will dramatically reduce water costs, as well as the costs of treatment and disposal.

- A wastewater minimization plan should consider
  - Means of avoiding excessive water use and wastewater generation
  - Means of reducing the strength of contaminants entering the waste stream
  - Means of water reuse and recycling
  - Unexplained water use outside production hours

7.3.3.2 General

Waste chemicals, oil, etc., should not be poured down the sink — this enters the water system and is environmentally damaging. Cleaning staff should be made aware that if such a procedure is necessary, waste liquids should be poured into toilets, draining into the sewer system.

Producing waste is like throwing money down the drain.

7.3.4 Recycling

Close the recycling loop: Collecting materials for recycling is only the beginning of the chain. *If you are not buying recycled, you are not recycling.*

7.3.4.1 Recommendations

Before considering recycling, there are other issues that need to be addressed. Ask yourself, "Can I reduce my use of this item?" and "Can I reuse this item before throwing it away?" First REDUCE, then REUSE, and only then RECYCLE.

Paper is a natural resource that is reusable and *can be recycled up to five times.* Large amounts of office waste paper could be collected in a separate container and taken away for recycling instead of being thrown away with the general rubbish.

Recycled stationery can be expensive unless a lower grade of whiteness is accepted, though most general stationery items are used in-house and there is little justification to use products made from virgin materials when recycled alternatives exist. For example, Mitsubishi, a company certificated to ISO 14001 with JSA (the leading Japanese certification body), has a policy on the use of recycled paper. For example, its business cards are made from 100% recycled paper and it is able to advertise this at the bottom of each card along with the ISO 14001 certification mark, clearly displaying its environmental commitment.

Case Study

Company:	Dubai Duty Free	Industry sector:	Trading
Location:	Dubai, UAE	No. of staff:	Approx. 700

Dubai Duty Free (DDF) installed recycling bins for glass and plastics at its office and warehouse complex on the occasion of the third National Environment Day. The bins were provided by DDF's environmental management program, which already ensures that more than 25 tons of paper and cardboard are sent for recycling every month.

> Mr. Colm McLoughlin, managing director of DDF, says, "We applaud the efforts and vision of President His Highness Sheikh Zayed bin Sultan Al Nahyan in implementing National Environment Day. As the very first [Duty Free] in the world to be awarded the ISO 14001 certification, DDF takes its responsibility to the environment very seriously" (*Gulf News*, January 2, 2000).

Paper could also be reused, because most paper sent to waste has been used only on one side. Faxes and internal memos, for example, could be sent on paper that is still unused on one side. Paper documents that require circulation could be enclosed in reusable envelopes, etc. For example, in the early 1990s, London Electricity developed a policy on paper that aimed to close the paper cycle by expanding waste paper collection, minimizing waste paper, increasing the use of scrap paper, imposing a surcharge on paper usage, and setting clear performance targets.

Common problems of recycled products include

Cost: Some environmentally preferable office supplies are more expensive. However, cost is generally linked to volume. If your supplier is aware that you are serious about using these products and you guarantee the volume, you should be able to negotiate improved rates. Some recycled products are cheaper than virgin equivalents. Increased market demand will also help to reduce costs.

Labeling: The labeling of environmentally preferable products varies considerably between the different stationery retailers. Some are confusing, and you have to search pretty hard to find them. Names of products can also be misleading.

Quality: Not all green products work! It is important to test them before distributing them throughout the office. For example, though some correction fluids work well, others do not. Some recycled Post-its stick, whereas others do not. The performance of the product varies among the different brands. It is important to be selective about the products you introduce to ensure that they are fit for the purpose. There is no point buying a recycled product that has to be disposed of because it doesn't work.

Perception: The perception of recycled products can be poor. However, the quality has substantially improved, and it is important to question the need to have products that are brilliant white for in-house use. There are recycled alternatives available of good quality. However, there can be problems using ink pens with low-quality note pads — the ink smudges. A more practical alternative is to use a better product; you could compromise by using 50% recycled content. The market is improving: look out for new products!

Laser printer toner cartridges can be refilled or remanufactured (although not all cartridge types are recyclable). Also, check that the printer warranty is not affected and that the supplier will guarantee to cover the cost of repair due to cartridge failure. The cost of recycling and using remanufactured toner cartridges can be very cost-effective, saving around 20% against new cartridges.

Case Study

Company:	Stephenson Harwood	Industry sector:	Solicitors
Location:	City of London, UK	No. of staff:	350

Three hundred and twenty HP11 toner cartridges were used per annum. They were collected for recycling (Stephenson Harwood was receiving £4 per cartridge), but the company was not buying back remanufactured cartridges.

Results:

1. Remanufactured cartridges provided by Tonerflow (reputable supplier) were introduced with no loss of quality.
2. Cost saving achieved: £3200 per annum.
3. Environmental benefit: recycling loop closed by using remanufactured units.

DeskJet cartridges can be refilled, which reduces waste and represents a considerable saving compared with buying a new product — an average of 50%. Ensure that the supplier handles these refills. Consider the ability to recycle the cartridges when purchasing new printers.

Plastic vending cups used for drinking water could be kept and reused a second or third time by staff instead of being thrown away after the first use. Pottery mugs and cups or recyclable paper cups could be better alternatives. Using 250,000 plastic cups over a period of a year equates to 1 ton of waste! Reusing and recycling cups reduce the amount spent on purchasing them and also reduce the amount of waste sent to landfill. Other paper-saving suggestions are

- Recycled toilet paper is available at little or no extra cost and at an acceptable level of quality.
- Trash could be collected in recycled refuse sacks.
- Trash collected for recycling can be taken to a municipal recycling center.

7.3.5 Ozone-Depleting Substances

7.3.5.1 Recommendations and Guidelines

The major controlled substances banned by regulatory agencies (they must cease to be produced or imported according to international legislation of the Montreal Protocol) include

CFC-11 — Trichlorofluoromethane
CFC-12 — Dichlorodifluoromethane
CFC-111 — Trichloroethane (a.k.a. methyl chloroform)
CFC-113 — Trichlorotrifluoromethane
CFC-114 — Dichlorotetrafluoromethane
CFC-115 — Chloropentafluoromethane
R-500 — Dichlorodifluoro/difluoromethane
R-502 — Chlorodifluoro/chloropentafluoromethane
Halon 1211 — Bromochlorodifluoromethane
Halon 1301 — Bromotrifluoromethane

Halon 2402 — Dibromotetrafluoromethane
CCl4 — Carbon tetrachloride
CH3CCl3 — Methyl chloroform
CH3Br — Methyl bromide
HBFC — Hydrobromofluorocarbons

7.3.5.1.1 Refrigeration and Air Conditioning

Ensure that old refrigerators are properly disposed of so that CFCs are removed (refrigerant gases, of course, affect the environment only when released).

New equipment should be specified to contain R134a (HFC) or, ideally, a hydrocarbon gas (Calor). R134a is generally the preferred substitute but, although it does not affect the ozone layer, it does have a severe global warming effect, raising some concerns. Other preferable substitutes include R125, R143a, and R22, which have low ozone depletion potential and are available on the market. R22, however, is a Class II ozone-depleting substance, and its use will eventually be phased out under the Montreal Protocol. Ammonia is sometimes used in primary circuits but can cause safety problems. It is recommended that future options be monitored and that measures be taken to ensure minimal leakage of gases through regular maintenance. Losses (i.e., amounts needed to top off systems during servicing) should be tracked.

New equipment must not contain CFC-11, -12, -113, -114, -115 (see above) or other halogenated CFCs. All existing air-conditioning and refrigeration equipment using the aforementioned controlled substances must be

- Maintained leak-free
- Supplied with gases from existing supplies in Dubai or recycled sources
- Converted to use-approved alternative refrigerants

The venting of controlled refrigerants during equipment maintenance is not permitted. Recovery, recycling, and reuse of refrigerants should be practiced during repair and maintenance.

7.3.5.1.2 Dry Cleaning

All products containing CFC-113 or 111-Trichloroethane (also known as methyl chloroform) that are ozone-depleting substances could not be used in new equipment after January 1, 1996.

Owners of existing equipment had to investigate alternatives and inform the regulatory agencies within 2 years of the selected alternative and the deadline for decommissioning all equipment utilizing controlled substances.

Alternative substances having low ozone-depleting potential include, but are not limited to trichloroethane, perchloroethylene, and methylene chloride. These substances should be used in dry cleaning activities.

Companies should examine whether there is a need to clean items at all and whether water-based caustic systems can be used, before considering vapor- and solvent-degreasing systems.

7.3.5.1.3 Fire Protection Systems

Halons 1211, 1301, and 2402 cannot be used in any new fire protection system from January 1, 1996. Alternative fire suppressant substances already available should be used in newly built fire protection systems.

The venting of halons during repair and maintenance of existing fire protection systems is not allowed. Existing large premises should install equipment to recover, recycle, and reuse halon.

All halon-filled cartridges or cylinders for fire extinguishers should be regularly maintained and periodically serviced only in qualified premises having halon recovery equipment to minimize leakages.

Keep a regular maintenance schedule for fire protection systems.

7.3.6 Hazardous Substances

There is a strong link between health, safety, and environmental issues. All these issues, for example, must be considered when assessing cleaning contracts. Selecting more environmentally benign substances reduces environmental effects and health and safety risks — inside and outside of the building. The major concerns are the use of acids, alkalis, bleaches, and solvents (particularly if mixed). All of these cause air and water pollution, so correct use and disposal are important.

7.3.6.1 Recommendations and Guidelines

Try to avoid the use of hazardous chemicals. For chemicals that you have to use, make sure that material safety data sheets (MSDS) are obtained from the supplier and kept on file for 2 years and that personnel are trained in their correct use.

7.3.6.1.1 Acids

Acids are widely used in cleaning operations. The most commonly found are hydrochloric and phosphoric acid. Sulphuric acid is used in many agents for clearing blocked drains. The hazards vary depending on the concentration and, therefore, adequate training in correct use is essential. Concentrated acids must not be diluted with water — this can cause an explosion.

7.3.6.1.2 Alkalis

Alkalis are often used for dishwashing machines, removing greasy deposits, paint stripping, and for cleaning concrete. Common alkalis are sodium hydroxide (caustic soda), sodium metasilicate, and borax (sodium borate decahydrate). They must not be mixed with acids — this can cause an explosion.

7.3.6.1.3 Bleach

Chlorine bleaches are mainly found in cleaning products in the form of sodium hypochlorite. They are highly toxic, especially in water, and can form cyanogen compounds such as cyanic acid. When mixed with acid, the poisonous gas, chlorine, is given off, which is extremely dangerous.

7.3.6.1.4 Solvents

Solvents are used in many cleaning preparations and pose special problems. Their volatility and reaction on the skin can lead to rapid absorption into the body and skin lesions. Some are highly flammable, and most can be decomposed by heat into highly toxic products. Solvents retain their hazardous properties when mixed with other substances. They are used in pure form for specialized tasks, for example, chewing gum removal. Solvents cause emissions of volatile organic compounds (VOCs), which contribute to photochemical smog and can be powerful ozone depleters.

7.3.6.1.5 Phosphates

Phosphates are an environmental concern, but do not have health or safety implications. They are used in dishwasher powders, multipurpose cleaning agents, and scouring cleaners. An excess of phosphorus (in this case phosphates) combined with excess nitrogen contributes to eutrophication (a chemical process that depletes the oxygen in water), encouraging algal bloom and the formation of foul-smelling substances, such as hydrogen sulphide and ammonia. These substances are highly polluting and contaminate drinking water.

Appropriate storage of hazardous substances will reduce or remove the potential for spillage to areas where damage could be great. When possible, keep hazardous and flammable substances securely away from other materials to prevent spills.

Ensure that all members of staff are aware that they *must not* dispose of any spilled chemicals or soiled water into surface water (rain) drains or sinks.

If appropriate, maintain a chemical/oil spillage response procedure (with appropriate equipment for dealing with each spill).

Full chemical analysis must accompany all applications for disposal of hazardous waste. Hazardous waste should be transported only by approved transporters in accordance with recommendations of regulatory bodies.

Discarded containers that once held chemicals or any hazardous substances are classified as hazardous waste because they contain chemical residues. Companies must follow the above procedures in applying to the regulatory agencies for permission before such containers can be transported for off-site disposal, cleaned, sold, stored, or reused elsewhere.

Waste chemicals, oils, paints, etc. may be disposed of in the general waste provided they do not make up more than 1% of the overall weight. Any larger quantities should be disposed of as hazardous waste. Some exceptions:

- Calcium fluoride, which is considered inert at all concentrations.
- All asbestos cement products are classified as nonhazardous, but must be handled and disposed of in accordance with EPSS Technical Guidelines (No. 48).
- PCBs.

A table of exemptions (Table 7.1) is shown below:

TABLE 7.1
Exemptions

Waste Component	Elutrable Fraction (g/m3)	Concentration (mg/kg)
Heavy Metals		
Arsenic	5	500
Cadmium	0.5	50
Chromium	5	3000
Copper	10	1000
Cobalt	—	500
Lead	5	3000
Mercury	0.1	10
Nickel	2	1000
Tin	—	500
Selenium	1	100
Zinc	50	5000
Total Heavy Metals	—	10,000
Inorganic		
Cyanides	10	500
Fluorides	100	5000
Asbestos-containing wastes (excluding asbestos cement products)		10,000 (1%)
Organic		
Phenols		
Petroleum Hydrocarbons	—	50
(> C9)		
(< C9)	—	20,000 (2%)
Organochlorine	—	2000 (0.2%)
Compounds	—	10
PCBs		5
Nonchlorinated pesticides	—	50

7.3.7 Stationery and Office Supplies

Environmental issues for paper are numerous and include

- Loss of natural habitats to environmentally damaging, intensive tree farming
- Pollution during manufacture (e.g., bleaching agents, effluent, optical brightening agents (OBAs))
- Energy usage
- Waste disposal: landfill and incineration

7.3.7.1 Recommendations and Guidelines

Most stationery retailers stock a range of products labeled "environmentally friendly." However, it is important to establish the criteria for this label, because some suppliers' labeling systems are misleading. When purchasing "environmentally friendly" items, your purchasing agent should obtain accurate, detailed information.

Your purchasing agent should work with stationery suppliers to increase the takeup of environmentally sound products. Bars should be put on the purchase of certain products to ensure that environmentally sound options are the only ones used. This can be done at no extra cost. Most stationery companies will source specific products on your behalf if they do not already stock them. You may also want to use specialist suppliers when possible.

Investigate waste reduction. There is often the potential to reduce the cost of office supplies by improving control and making more efficient use of our resources. For example, box files can be relabeled and reused, and a number of products are not strictly necessary.

An IT-based requisition procedure should be used to save time and paperwork. Links with stationery suppliers are becoming increasingly sophisticated because of technological developments, and there are a number of systems available (electronic data interchange [EDI], for example). Your purchasing department should have a direct computer link with suppliers, who can also give you detailed printouts for improved control.

Ensure that any recycled copier paper chosen is compatible with the office equipment and that recycled paper will not affect the equipment warranty. Select chlorine-free paper with the B, C, D classification and aim for at least 75% recycled content.

Filing products made from recycled products and board include ring binders, dividers, lever arch files, suspension files, box files, record cards, folders, memo pads, Post-its, and shorthand pads.

7.3.8 Office Equipment — Fixtures and Fittings

7.3.8.1 Recommendations and Guidelines

Avoid tropical hardwoods and check the sources of raw materials; furniture and other fittings can contain timber from unmanaged forests. Sustainably managed and grown temperate hard- and softwoods from North America, Europe, and Russia are preferable. Consider wood substitutes. Other suggestions include

- Avoid products in which CFCs are used as a blowing agent
- Avoid solvent-based wood preservatives (often organic volatile chemicals) and lacquers (which contain approximately 15% solvents)
- Avoid adhesives containing formaldehyde (resin glue used in chipboard furniture)
- Consider design for reuse and end-of-life disposal
- Investigate packaging — recyclable content, recyclability, and retrieval

For example, in 1995, Abbey National (U.K.) built an office to house one of its management divisions. The company conducted a full "green audit" of a short list of office furniture producers against criteria that included the source of raw materials, the environmental impact of manufacturing processes, the recycling of waste and the impact of the product once it becomes obsolete, the type and use of packaging materials, the commitment to energy-efficient manufacturing and distribution, and the health and welfare of the employees. Abbey National expected to pay extra for the environmentally friendly office furniture, but it was found that the producers who complied with Abbey National's environmental criteria had been able to achieve major efficiency savings, and were therefore able to maintain competitive prices.

7.3.9 Transport

The Federal Environment Law that came into force in the UAE in February 2000 contains 101 articles. Of these 101 articles, 28 refer to the types of fines to be imposed on offending vehicles. Dubai Municipality hopes to help set an example. "By taking this step [of introducing unleaded petrol in all its light vehicles], the Dubai Municipality would like to be a model for other local and federal departments [and private organizations] to follow suit. We also hope that gradually the government would totally ban the use of leaded fuel all over the country, in order to keep our environment safe and clean." — Qassim Sultan, Director-General of Dubai Municipality, March 22, 2000.

7.3.9.1 Recommendations and Guidelines

Selection and allocation. What needs to be considered when selecting vehicles for the company's fleet?

- Fuel efficiency of the vehicle
- Purchase cost of the vehicle
- Cost and type of fuel
- Maintenance requirements
- Range between tank refills
- Likely distances of trips to be traveled by the vehicle
- Overall performance
- Environmental and public health considerations
- Automatic transmission vs. manual transmission (a vehicle with an automatic transmission uses up to 10% more fuel than a similar model with a manual transmission)

Dubai Municipality is encouraging the use of lead-free gas because it is significantly less harmful to our health and also the health of the environment. Virtually all the vehicles produced in the last 10 years have been designed specifically to use unleaded fuel, without the need of a catalytic converter.

Vehicles using diesel emit more particulate (smoke) emissions and nitrogen oxides than gasoline-run cars with catalytic converters, but ultra-low sulfur diesel that reduces smoke emissions is available.

7.3.9.1.1 Servicing

Making sure company vehicles are regularly tuned and serviced (90% of inefficient vehicles can be retuned in just 15 minutes) can save money and reduce exhaust emissions.

- Daily service: quick visual check of body, tires (a 2 psi drop in tire pressure increases fuel consumption 3%, and tires underinflated by 7 psi waste half a gallon of fuel per tank), lights, windscreens, and mirrors
- Weekly service: check brakes, screen-wash levels, oil, and water under the hood
- Monthly service: physical check of tire pressure (overinflated tires have a shorter life and can be dangerous), full check of all under-hood levels, check for exhaust leaks, check that steering is true (i.e., not pulling to one side), check for service requirement against time or mileage

Keeping company vehicles serviced can account for up to 18% of total fuel saving and will reduce the possibility of unexpected breakdowns.

7.3.9.1.2 Training and Driving Style

Driving training can account for up to 11% of total fuel saving. Train all drivers in economical driving techniques, including

- Driving smoothly, avoiding harsh acceleration and heavy braking.
- Using the gearbox efficiently in manual transmission vehicles to maintain revs in the midrange (1500 to 2500 rpm). Research in The Netherlands by Novem (Netherlands Agency for Energy and the Environment 1996) has found that correct use of gears can reduce nitrogen oxide emissions by over 20% while reducing fuel consumption by an average of 15%.
- Avoid "pumping" the accelerator or revving the engine unnecessarily — this wastes fuel.
- Plan all trips in advance.
- The use of air-conditioning in a vehicle can increase fuel consumption by around 15%. Do not use it unnecessarily.
- Use of roof racks, open windows, sunroofs, etc. increases aerodynamic drag and so contributes to increased fuel use. Remove unnecessary roof racks.

7.3.9.1.3 Vehicle Use

- Provide incentives to drivers to save fuel
- Reduce the need to travel (telesales, telecommunications, etc.)
- Plan and schedule trips appropriately
- Maintain the vehicle in good condition

For example, DHL Worldwide Express is another major company certificated to ISO 14001. This illustrates clearly how even a potentially heavily polluting business can make itself more "green" by controlling and reducing its negative environmental impacts and striving for continual improvement in this area.

7.3.10 External Influences

7.3.10.1 Recommendations and Guidelines

Create a company environmental policy and request suppliers, contractors, and others to comply with the requirements of this policy.

7.3.11 Miscellaneous

7.3.11.1 Recommendations and Guidelines

Fitting of soap dispensers can prove cost-effective and reduces waste packaging.

Environmentally friendly batteries that do not contain heavy metals such as mercury, cadmium, or lead (which cause air pollution when burned in a refuse incinerator) are widely available. For example, British Telecommunications (BT) plc is one of the largest users of nickel-cadmium rechargeable batteries in the UK. In advance of European Community (EC) legislation, BT initiated a scheme to recycle all spent nickel-cadmium batteries used by the company. In 1 year, BT recycled 5.9 tonnes (metric tons) of these batteries that they were able to avoid sending to landfill. Furthermore, 55,000 lead acid batteries recovered from telephone exchanges were recycled.

7.4 ENVIRONMENTAL MANAGEMENT INITIATIVES

7.4.1 Energy Management Systems

To be successful, energy management should involve the following steps:

1. Responsibility
2. Energy audit
3. Action plan
4. Employees
5. Finance
6. Energy monitoring
7. Yardsticks
8. Consumption targets

7.4.1.1 Responsibility

A continuous improvement team could be set up to handle the following tasks:

- Monitoring the company's energy use
- Maintaining detailed energy consumption records
- Determining energy targets, yardsticks, and/or benchmarks
- Analyzing records for management meetings
- Reviewing energy performance against targets
- Being aware of energy technology and initiatives
- Investigating potential energy-saving schemes
- Preparing economic justifications for capital projects
- Communicating energy issues throughout the company

7.4.1.2 Energy Audit

To conduct a simple energy audit:

- Assemble all fuel bills for the last year
- Identify the main issues for energy around your site
- Calculate energy used per operation or unit production
- Check how well energy is being used against industry norms
- Maintain a documented schedule of checks for the efficiency of machinery and equipment
- Identify possible energy-saving opportunities or measures
- Evaluate the cost and payback of each measure
- Set priorities for the opportunities identified

7.4.1.3 Action Plan

- Define a simple energy policy
- Assign responsibilities
- Involve all employees
- Set up an energy monitoring system
- Conduct an energy audit
- Take action on no-cost measures immediately
- Set energy consumption targets or yardsticks
- Appraise low-cost and larger cost projects

7.4.1.4 Involve Employees

Employees need to be made aware of

- Why and how energy is consumed at work
- Why it is important to reduce energy costs
- How their behavior affects energy use
- What effect saving energy will have on them

Make full use of promotional material such as posters, stickers, awareness sessions and videos, and leaflets (e.g., distributed in wage packets) to convey these messages. Also, consider including energy management in the supervisor's tasks and providing specific energy efficiency training for staff who have the greatest influence on energy use, e.g., plant operators, janitors, security staff, cleaners. Make energy efficiency part of the employees' code.

Identifying improvements made encourages staff to participate in minimizing energy use.

7.4.1.5 Finance

- Work out the simple payback time for each measure.
- Include nonenergy savings in your payback calculation.
- Decide if a more detailed financial appraisal is needed.
- On large capital projects, such as a new lighting system, remember that the existing system may need updating for other reasons, e.g., if it is no longer safe, so some capital expenditure might have to be laid out anyway.
- When energy savings are made, try to allocate a portion of the savings for reinvestment in further energy efficiency projects.

7.4.1.6 Energy Monitoring

- Train two or three of your staff to read meters.
- Have them read your meters on a regular basis.
- Cross check the accuracy of the meter readings.
- Train staff to enter meter readings into a computer.
- These readings should take place to coincide with your management statistics period. Meter reading takes only 5 or 10 minutes and could signal major losses or possibly unsuspected ways of saving energy.
- Meter readings should be analyzed as soon as possible and actual energy use should be compared against consumption targets, yardsticks, and industry norms.

7.4.1.7 Yardsticks

Common yardsticks are

- Energy use per square meter of floor space.
- Energy use per unit (AED) of staff cost or turnover.
- Energy use per unit (AED) of production.
- Choose the yardstick that is most appropriate to your company and use it to track energy usage. This will allow differences in energy performance to be quickly identified and investigated.
- Benchmarking compares your performance with those of the best performers or competitors. The aim is to close the gap quickly and reach the benchmark.

7.4.1.8 Consumption Targets

To set consumption targets you need to ensure that

- All data relevant to your energy use is collected.
- You know what period of time this data covers.
- The targets you set are SMART (specific, measurable, achievable, realistic, time-based).
- All are aware that unachievable targets will quickly demotivate staff.

Kilowatt hours can be extrapolated by converting energy units, using the following formulas:

Natural gas (therms) × 29.31
Natural gas (cubic feet) × 0.303
Gas and oil (liters) × 10.58
Light fuel oil (liters) × 11.2
Medium fuel oil (liters) × 11.3
Heavy fuel oil (liters) × 11.4
Coal (tons) × 7800
Anthracite (tons) × 8200
Liquid petroleum gas (LPG) × 6.96

7.4.2 Access to Legislative Information

It is important to have access to information about legal requirements in case some are missed or change over time, or new ones are in force.

7.4.2.1 Recommendations and Guidelines

Establish a register of legislation that lists all relevant environmental regulations pertinent to company activities (based on existing knowledge and supported by a literature review of specialist journals, if appropriate).

7.4.3 Training, Awareness, and Responsibilities

Training and awareness are of the utmost importance. The successful implementation of an organization's environmental objectives depends on the hard work, support, and enthusiasm of its employees along with the strong commitment and support of senior management.

The training of all staff, especially staff with key responsibilities, is vital when introducing new processes, systems, or management techniques into the workplace. Training procedures should be established and maintained for general environmental awareness and competence for all staff, and for specific environmental roles for the staff with key responsibilities.

7.4.3.1 Recommendations and Guidelines

Emphasize the importance to staff of minimizing, where appropriate, energy and water use and negative environmental impacts. Invite representatives from environmental regulatory bodies to come and speak on important environmental, energy, and water issues.

Different members of staff should be made responsible for different environmental issues, e.g., one employee concentrates on minimizing energy use, one on water minimization, one on waste reduction, etc.

Environmental issues should be included in employee orientation training.

Encourage staff to make suggestions about areas of work with which they are familiar. Staff should be made to feel that their contributions to energy/environmental management within the organization are welcome and valued — do not ignore their recommendations. Employee ownership of the working environment is a theme of increasing contemporary importance: *one volunteer is worth two pressed people.*

Emphasize the importance of conformance with an environmental policy (and the requirements of the EMS). Highlight the potential consequences of departure from specified operating procedures.

Establish a register of environmental effects at your company site and rank them in order of significance (updated periodically). Identify *causes* of environmental effects and *actions* to control them. Corrective actions can often be extremely simple, including improved housekeeping.

Be proud of your stance on environmental management and the environmental achievements that you make. Publicize and promote them to all interested parties including customers, suppliers, regulatory bodies, banks, insurers, the media, competitors, shareholders, local community groups, local residents, and employees. Highlight the environmental benefits of improved personal performance. This could be in the form of a monthly newsletter or bulletin. For example, the engineering company ABB, certificated to ISO 14001 with SGS in the Middle East, decided to increase the awareness of ABB's corporate environmental strategies. The president's environmentally focused message in the 1999 report was translated into 23 languages, and 200,000 copies were distributed throughout ABB. Their environmental program, environmental news, and copies of environmental speeches and articles by ABB's management are all published on the Web.

You can also relate your current operations to global environmental concerns to demonstrate the importance of this issue.

Internal auditor training might also be useful for several staff who could carry out an annual audit of each department to ensure compliance, peak performance, etc.

7.4.4 Purchasing: The Total Cost Approach

Dealing with suppliers who are committed to sound environmental performance can make your job easier and help you achieve your purchasing objectives. Equally, poor performance from your suppliers weakens your own environmental position. This is particularly important when you are purchasing products with a high environmental impact or profile. Suppliers and contractors represent significant environmental risks, for example, in cases of noncompliance with regional, national, or even international environmental legislation.

Many companies see "green" purchasing as having an inevitable cost premium attached, but this does not have to be the case. Considering environmental criteria in purchasing decisions provides a change in focus away from purely cost-based decisions. More fundamental questions need to be asked, such as whether you need a particular item in the first place — a highly cost-effective and environmentally sound option is to use less and, therefore, order less.

7.4.4.1 Recommendations and Guidelines

To consider the total cost approach, analyze the life cycle of a product. The technique of life-cycle analysis provides a structured approach and creates a greater awareness of the full environmental impact of a product. It takes into account all stages of a product's life cycle from the sourcing of the raw material to end-of-life disposal. It is unrealistic to expect to be able to conduct detailed life-cycle assessments on all products, but the principles do provide a useful approach to integrating environmental criteria into purchasing decisions. Key issues will be

- Sourcing of raw materials
- Manufacturing processes
- Packaging
- Distribution
- Usage
- Reuse, recycle, disposal

The true cost of waste disposal is often underestimated. Many times it costs more to dispose of a product than it does to buy it in the first place. Substantial reductions in waste disposal costs can be achieved by considering recycling and disposal in purchasing decisions.

Another issue is the cost of distribution. Companies should pursue a policy of using only local suppliers whenever possible.

Effective communication between those involved in purchasing and waste management could reduce disposal costs and promote a better understanding between departments. Increased awareness of the environmental impact of company activities will highlight the link between purchasing and waste management, and reduce negative environmental effects.

Consideration of recycling and ultimate disposal of a product in purchasing decisions can significantly reduce costs. If a product can be reused or recycled at the end of its life, the volume of waste is reduced which, in turn, affects waste disposal costs.

The replacement of a hazardous material with a more benign substance means that it can be treated as general waste, rather than hazardous waste that needs separation, cutting back on the added cost of a hazardous waste contractor.

Using the total cost approach will highlight the cost effectiveness of investing in a more expensive product initially to reduce costs in the long run, for example, by buying energy-efficient light bulbs. For example, BBC Worldwide has a purchasing policy that aims to consider all aspects of a product's life cycle, increase the use of recycled products, minimize the use of natural resources, publicize their environmental policy and purchasing objectives, integrate environmental criteria into purchasing decisions, and require suppliers and manufacturers to demonstrate environmentally sound practices in keeping with BBC's environmental policy.

7.5 SUMMARY

It is important for companies to realize the potentially wide benefits to be gained from the application of a systematic approach to environmental management, whether or not formal recognition of the resulting system to an international model is sought (e.g., ISO 14001 certification with r, y, or z).

Development and implementation of an environmental management system (EMS) is, simply, the application of internationally well-established management concepts, principles, and practices to the organization's handling of environmental matters. To be effective, environmental management systems should have a defined relationship with other environmental management tools, as well as other management disciplines utilizing similar skills, including quality management systems (ISO 9000 QMS) and health and safety, within which an EMS can be integrated.

Initiatives such as EMSs and environmental auditing have provided identifiable improvements in reducing the environmental impact of organizations. However, many experts argue that a more radical realignment of the management and objectives of organizations and economies is required for significant steps toward sustainable development to be taken.

> Business has in many cases responded positively to pressure from governments and consumers. Large firms, especially, are exploring new concepts — such as industrial ecology, design for environment and sustainable communities — which go along with a shift in emphasis from pollution abatement to pollution prevention and avoidance and a broad sustainability perspective. To move these concepts from idea to reality there is a common need for firms to change, not only their internal organization, but also their relationships with those outside the traditional boundaries of Western firms, including other firms in the supply chain, local communities, official agencies and customers. This illustrates that the difficulty of industrial transformation lies not just in the need for more eco-efficient technologies and processes but, even more so, in the need for radical changes in the organisation of firms, institutions and the production system in general. The challenge to sustainable production is a challenge to society as a whole.
>
> D. Wallace
> *Sustainable Industrialization*
> Earthscan and the Royal Institute of International Affairs, 1996

Management of the environment and the concept of sustainable development are inextricably linked (a concept defined as "a development which meets the needs of the present without compromising the ability of future generations to meet their own needs"). The transfer of technology and the exchange of experience and expertise from the developed to the developing countries are an essential part of the international environmental effort. For the people of the Third World to experience a rise in their standards of living, the industrialized world must use its resources ten times more efficiently than at present.

7.6 DISCLAIMER

Although every effort has been made to ensure that this chapter is accurate and that recommendations are appropriate and made in good faith, we accept no responsibility for inaccuracies or actions taken by a company subsequent to these recommendations.

8 Lean Manufacturing

Adi Choudri

The term "lean" has been coined relatively recently to summarize Japanese manufacturing philosophy, especially as exemplified by the Toyota system. Lean practices have appeared in other forms such as "just-in time" manufacturing, and "synchronous" or "quick response" manufacturing in the sense that the underlying concepts are the same. The survival of an organization, whether profit or nonprofit, manufacturing or service oriented, may ultimately depend on its ability to systematically and continuously eliminate waste and add value to its products from its customers' perspective. Interestingly, lean practices in their simplest form are founded on common sense, and most of them are not even proprietary to any company. The business objective of lean is to make high-quality products at a lower cost with speed and agility (Figure 8.1). This can certainly lead to an expanded customer base, greater business and employment stability, and increased shareholder value. Because we are not talking about a magical approach here, this generally means that the relative success of lean manufacturing in a specific setting depends on how well the cultural, behavioral, and strategic aspects of the corporate entity were addressed during the lean journey. This also means that the vigor and sincerity of people, both hands-on and off-the-floor, will drive and guide the success of the lean approach.

Lean practices are designed to eliminate waste and enhance the value of the company's products to its customers. Lean businesses compete by creating temporary cost, quality, and speed advantages in focused business areas, but they cannot remain stagnant and rest on their laurels because, as mentioned before, these practices can and will be used by competitors probably with lessons learned. The only way to counter this is to develop a corporate mindset where everyone is focused on continuous improvement every day in everything they do leading to customer delight.

Lean manufacturing is not a secret technology in either the product or the process. It can be applied to all kinds of industries and all types of companies, including high volume, job shop, or process. We also know now that the culture and value system of the workforce probably have less to do with the success of lean. The key to lean manufacturing success lies in the careful integration of production and management practices into a complete management system that generates a collaborative atmosphere of mutual trust and respect between management and labor. Many manufacturing and management practices can be implemented individually and may result in cost and quality improvements. Such gradual change is consistent with the lean concept of continuous improvement and is frequently practiced by many corporations during their initial lean journey. However, an accelerating rate of improvement results when the different subsystems of the lean manufacturing system are in place and have been so for several years. For example, it is often found

1-57444-300-3/02/$0.00+$1.50

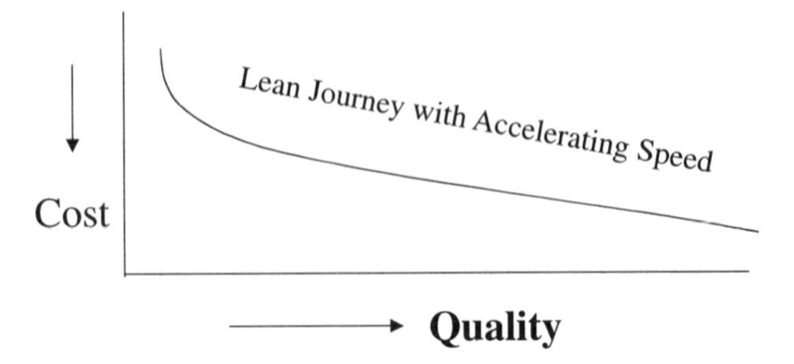

FIGURE 8.1 Quality and cost.

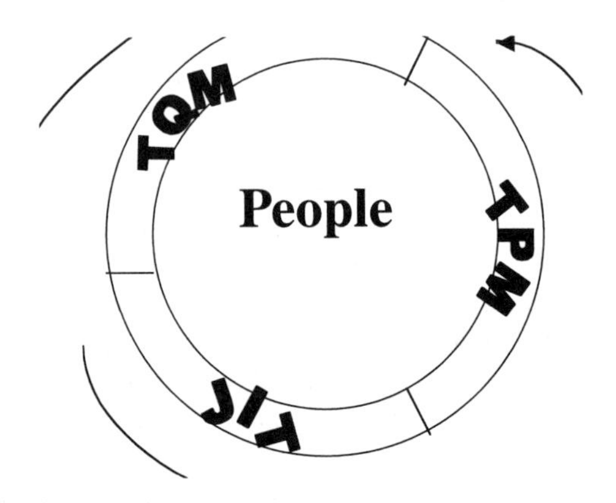

FIGURE 8.2 Lean start wheel.

that sometimes a company will start with a total preventative maintenance (TPM) effort because it was having difficulty with equipment uptime or frequent production disruption due to breakdowns. In some cases (Figure 8.2), the company starts on the lean journey with a total quality management approach to improve yield or process capability and eventually ends up addressing all the subsystems of the lean manufacturing system. Sometimes a company can do a lean self-assessment as shown in Appendix 8.1 to get a feel for where its initial shortcomings are, and develop a lean implementation plan. It is important to note that a manufacturing company eventually needs to address all the different aspects of lean, no matter where it starts its lean journey, and must continue on that path until perfection is reached.

8.1 LEAN MANUFACTURING CONCEPTS AND TOOLS

These concepts and tools can be organized into three levels. The first level encompasses lean manufacturing objectives and basic principles such as value and waste. These are general concepts, which should be taught to all the employees of a lean manufacturing enterprise, and are increasingly being applied to nonmanufacturing support areas such as product development or business processes.

The next level contains lean manufacturing primary management and production strategies used to achieve the objectives and instill basic principles. The strategies are general rules for management behavior, and support one another as well as the basic principles. The third level of lean manufacturing consists of implementation techniques, which are the practices and procedures for implementing and maintaining the strategies. Although these levels are somewhat arbitrary and are not always followed rigorously outside the Toyota production system, it is important to note that each level is built on the solid foundation of the previous level. It helps underscore the point that without the complete system, long-term lean manufacturing success is not sustainable.

Lean manufacturing objectives and principles are adapted from the Toyota production system and over the years have been enhanced by lean practitioners such as Jim Womack, Dr. Schoenberger, and numerous corporations and nonprofit organizations such as Lean Aerospace Initiative at MIT, Lean Enterprise Institute, and others.

8.1.1 Lean Objectives

The basic business objective of a manufacturing corporation is long-term profitability because it is essential to the continued existence of any corporation. To achieve long-term profitability, a company must (1) produce products with quality consistently as high as the best in its class, (2) ensure that production costs are competitive with most manufacturers, and (3) deliver a product–service mix that is competitive with the best in its class as well.

Lean manufacturing helps a company stay competitive by serving its customers better and continuously reducing costs. Lean gives customers the product variety they want, in the quantity they want, and without paying extra for a small-lot size. Lean makes a company flexible enough so that customer demands for change can be accommodated quickly, using lean techniques such as small-lot production.

Why do we need lean manufacturing? Simply, the answer is profit squeeze (Figure 8.3).

In the past, companies simply passed costs on to the customer. The pricing formula was

$$\text{Cost} + \text{Profit} = \text{Price}$$

In today's competitive market, customers insist on a competitive market as well as world-class quality and product features. This means that companies must reduce costs to make a profit:

$$\text{Price} - \text{Cost} = \text{Profit}$$

Lean manufacturing gives a company a key competitive advantage by allowing it to build high-quality products inexpensively because consumers, *not manufacturers*, set prices and determine the acceptability of the products and services they use. Lean manufacturing achieves the above three objectives by adhering to three key basic principles: *definition of value, elimination of waste, and support the worker.*

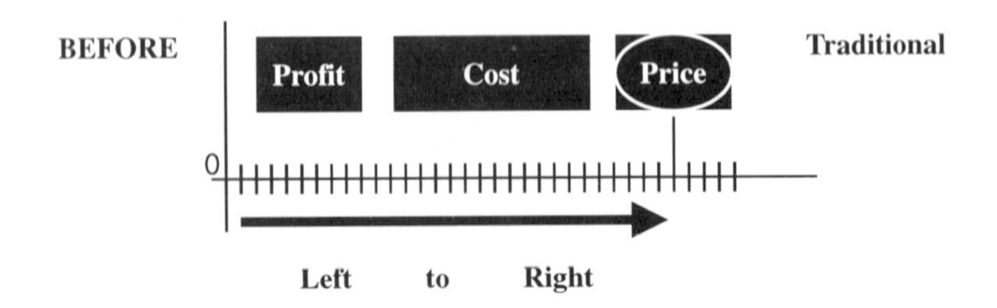

Costs must be targeted

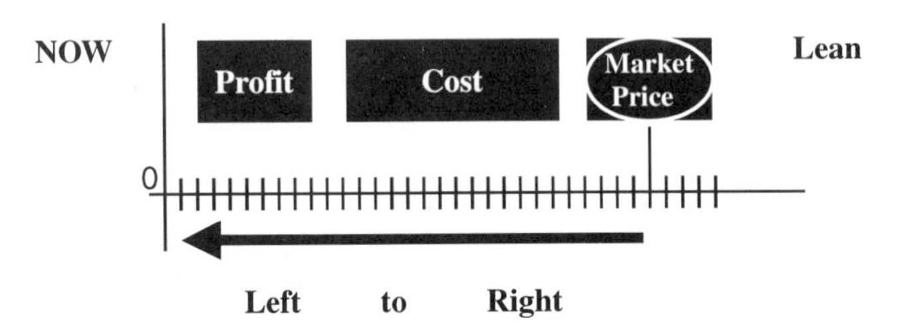

FIGURE 8.3 Price – profit = cost equation.

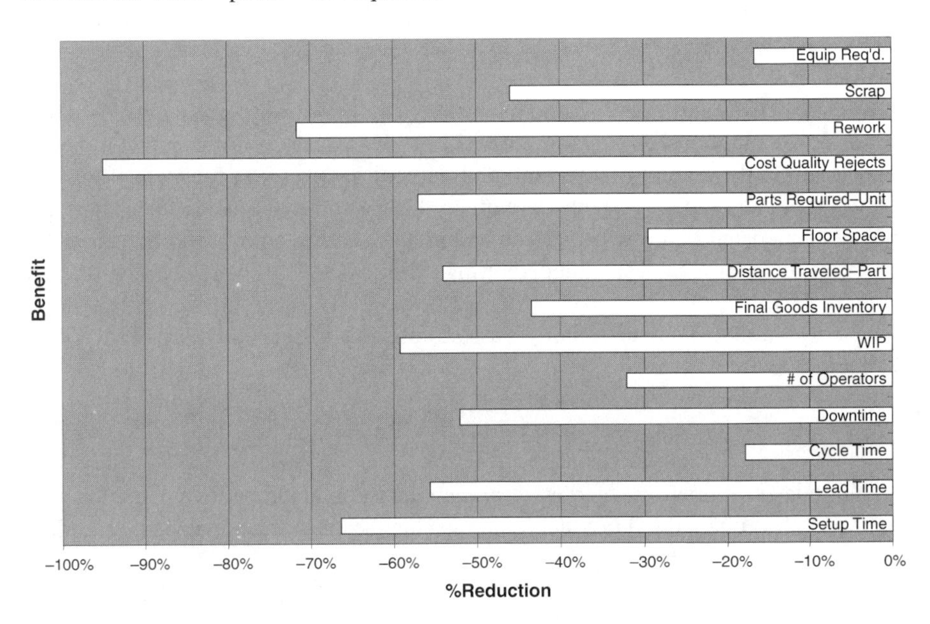

FIGURE 8.4 Typical lean benefits.

These are shown in the basic lean manufacturing model (Figure 8.7). In addition, lean manufacturing can provide significant other benefits as demonstrated in Figure 8.4.

8.1.2 Define Value Principle

Whatever business a company is engaged in, before it starts on the lean journey, it helps to take a hard look at the existing product lines and how they are adding value for its customers. Ultimately, only the customer can define value. Value for a product or service is usually a function of price and the customer's needs or requirements at a given time. Products with a complex customized design and sophisticated processing technologies are of little value if they do not satisfy the customer's needs at a specific price and time.

The employees or the suppliers of the corporation do not decide value, either. A stable workforce and a long-term network of suppliers may be necessary for the lean manufacturing system to work, but they do not define value. With the advent of information technology, especially the Internet, there have been significant advancements in the area of customer relationship management and product customization for individual customers. Several companies have started to define value based on individual customer choices and preferences.

Value must be defined only from the ultimate customer's perspective and should not be skewed by preexisting organizations, technologies, and undepreciated assets or even economy-of-scale considerations. The fundamental question that must be asked about any activity or product feature is whether the customer is willing to pay even a cent more for this processing step or that product feature?

Everyone in the organization will not initially grasp this definition of value; however, this is the first step in the lean implementation process.

8.1.3 Identify Value Stream

Typically, in a manufacturing organization, products and services are provided to an existing base of customers. For any given product line, a value stream can be identified. These are all the specific actions required to bring a specific product or service through the three critical sets of tasks: (1) *information management tasks*, which consist of activities from order taking through detailed scheduling to delivery through its distribution channels to the ultimate customer; (2) *physical transformation tasks*, which convert raw materials to finished product through a series of processing steps; and (3) *problem-solving tasks,* which usually consist of activities such as bid and proposal through product design and prototyping. To keep things simple, a value-stream map for information and transformation tasks should be created for each product or product family. Tools and techniques for value-stream mapping for problem-solving tasks, such as product development, are still emerging and will be touched on briefly later in this chapter. A value-stream map will typically show how various activities are performed to move the final product from supplier to customer. Many of these activities will be value added as well as nonvalue added (waste), which have somehow existed in the organization for a variety of reasons.

8.2 ELIMINATION OF WASTE PRINCIPLE

8.2.1 Definition of Waste

Waste, or *muda,* as it is known in the Toyota production system, is defined as any activity that absorbs resources such as cost or time but adds no value. Waste can be classified in a couple of different ways. Eliminating waste is a basic principle of the lean manufacturing system. To systematically eliminate waste, detailed concepts concerning the nature of the waste and its implication in manufacturing inefficiencies must be taught to every member of the organization. Whether analyzing worker operations, production, or production processes themselves, two fundamental types of waste must be considered: *obvious* (Type I) and *hidden* (Type II).

Obvious waste is something that is easily recognizable and can be eliminated immediately with little or no cost. For example, an operator's time spent cleaning up parts may be absolutely necessary unless arrangements can be made for parts to arrive ready to use.

On the other hand, hidden waste refers to aspects of lean manufacturing that appear to be absolutely necessary under the current methods of operation, technology, or policy constraints but could be eliminated if improved methods were adopted. For example, using X-rays to inspect welds may be needed until welding technology improves.

Either type of waste can further be classified into seven different categories. It is important to recognize and understand these, because equipped with this knowledge, one could simply walk through the shop floor and find many ways to eliminate waste immediately.

8.2.2 Waste of Overproduction

This waste happens when companies produce finished products or work-in-process (WIP) for which they do not have customer orders, or they produce parts faster than required by the downstream process. Companies overproduce for a variety of reasons. Large-lot production, long machine setups, and making up for poor quality are some of them. Part of the root cause of this waste may be the logic of "Just in case somebody needs it," an uneven production schedule, fear of worker idle time, or a misuse of automation, so that parts are produced unnecessarily to justify a large capital investment.

8.2.3 Waste of Inventory

Inventory is an accumulation of finished products, WIP, and raw materials at all stages of the production process. Express inventory is usually a symptom of many other underlying problems such as defects, production imbalances, long setups, equipment downtime, and late or defective deliveries from suppliers. There are major costs associated with excess inventory. First, it hides process problems so people are not motivated to make improvements. Second, when processes make excess inventory, these items must be moved and stored, using up conveyors and forklifts and the time of the people who run them. This transport adds cost but provides no

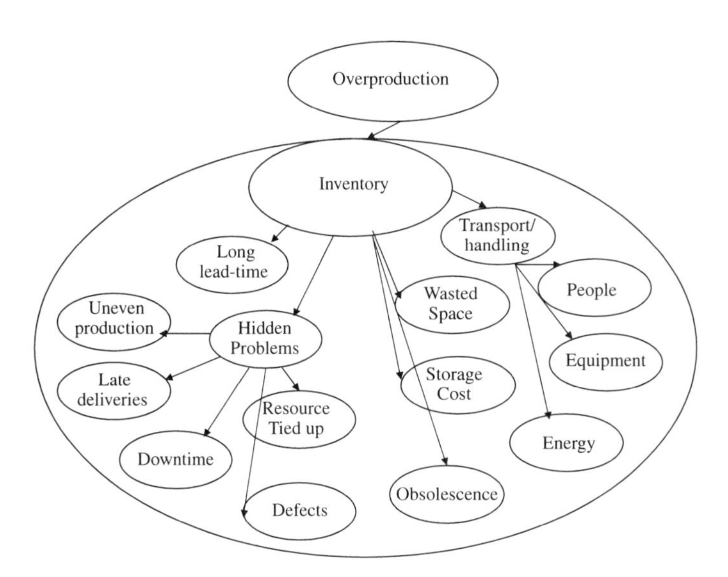

FIGURE 8.5 Impact of inventory.

added value. Third, companies pay to carry this extra inventory in terms of floor space, people to keep track of stores, and other resources such as computer systems and support personnel. Fourth, inventory increases lead time and response time to the customer. Fifth, inventory can lead to handling damage due to excessive transport. Sixth, items can deteriorate over time and become obsolete due to changes in technology or customer demand. Finally, inventory is wasteful in itself because the company uses people, equipment, material and other resources to produce it; as long as that inventory stays in the plant or warehouse, the company is not repaid for its investment in these resources. As a matter of fact, that is why inventory is carried on the books as an asset.

Inventory waste affects every production process that depends on a previous process for parts and materials. The impact of inventory is shown in Figure 8.5. When a plant has many products and processes, each handling items in large lots, the cumulative waste and foregone cost savings can be enormous — it has been estimated at 20 to 40% of a company's revenue. To eliminate this waste, companies use the "pull system" to produce those items in the right amount and at the right time to satisfy customer need. It must be noted that inventory typically exists for a variety of reasons, and those underlying causes must be addressed before an attempt is made to reduce inventory.

8.2.4 Waste of Correction

Correcting or repairing a defect in materials or parts adds unnecessary costs because additional equipment, labor, and material will be needed. Other costs may be a delay in delivering orders to the customer or having to maintain excess inventory to make up for quality problems. Severe quality problems can create lower customer confidence and lead to the loss of future business. Some of the causes of this waste may be weak process control, poor product design, deficient equipment maintenance, inadequate measurement systems, or ineffective worker training. The relationship between this waste and JIT is not always easily understood. Frequently companies undertake major quality or lean initiatives as if they are separate efforts. A lean manufacturing system such as JIT assumes high-quality outputs at all process levels. As a matter of fact, attempting to implement JIT without improving quality could be detrimental.

8.2.5 Waste of Movement

Any material, people, or information movement that does not directly support adding value for the customer is a waste. Poor shop layout, poor workplace organization and housekeeping, wrong work-order information, mislocated material, or excessive inspections can lead to this type of waste. Frequently, "spaghetti maps" or detailed "process maps," as shown in Appendix 8.2, will identify this kind of waste. Both of these techniques follow the material from start to finish and take detailed observation of the movements of both material and people. Appendix 8.3 provides a blank form for collecting distances and cycle time information for a process step.

8.2.6 Waste of Motion

Any motion of people or machines that does not add value to the product or service is a waste. This can lead to operator fatigue or wear and tear on machines and could sometimes lead to injury. Poor process design, an ineffective human-machine interface, bad workplace design, or inadequate planning generally causes this waste.

8.2.7 Waste of Waiting

This is probably one of the most pervasive areas of waste, especially in the factory floor processes, and it happens when people, equipment, or material wait for each other or for information. This can happen as a result of poor quality in upstream operations, a poor or uneven schedule, unreliable suppliers, or poor equipment reliability. Poor communication is also a frequent contributor to this waste.

A related waste is worker frustration or loss of productivity. Lean manufacturing assumes that most people come to work to be productive and add value.

8.2.8 Waste of Overprocessing

Processing efforts or steps that add no value to the product or service from the customer's perspective can lead to this waste. Factors involved can include redundant approvals, poorly defined customer requirements, and redundant steps to make up for lack of process quality. Typing a note on good paper when a quick hand note on scrap paper will do is an example of this. Inspecting a part surface when the surface will later be machined off is another example.

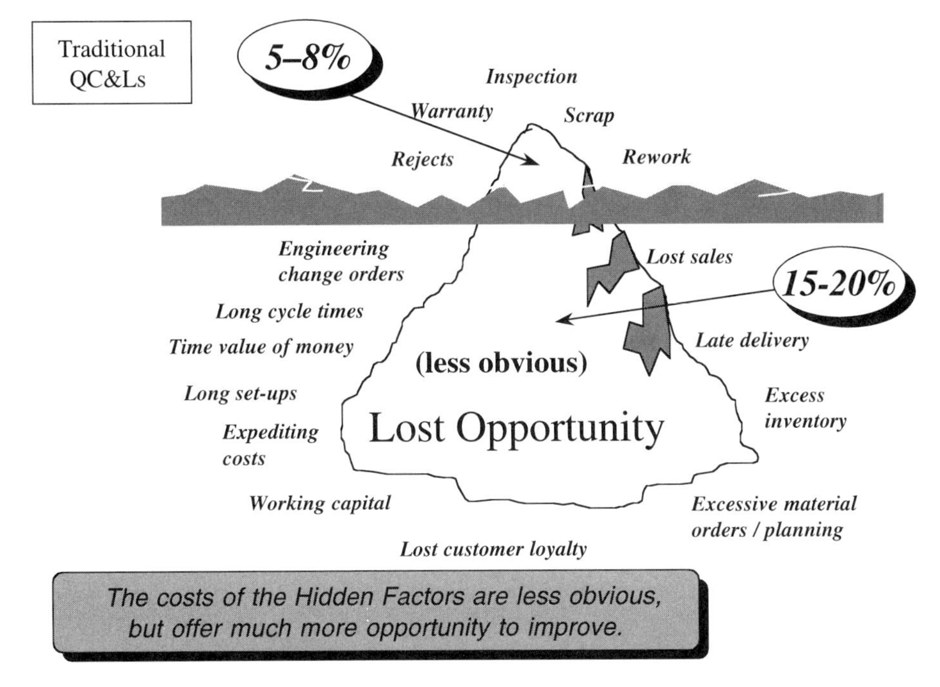

FIGURE 8.6 Traditional QC&Ls.

8.2.9 Impact of Waste

For a variety of reasons most manufacturing corporations do not realize the true impact of all these wastes. It may be due to lack of accounting tools that capture true costs, lack of awareness, or simply an acceptance of the way things have always been done. This is depicted in Figure 8.6.

Closely related to the concepts of waste are two other lean manufacturing concepts: unevenness (*mura*) and overburden (*muri*). A lean manufacturing system is concerned with unevenness in workloads, schedules, material placement, or other aspects of the production process because unevenness contributes to waste and inefficiency.

Similarly, overburdening workers, parts, tools, or machines is also seen as a cause of waste and inefficiency.

8.3 SUPPORT THE WORKERS PRINCIPLE

Supporting the workers involves providing production workers with the tools, training, and management support necessary to do their jobs effectively, combined with a policy that commits to "lay off as the last resort."

Although all employees are part of a lean manufacturing system, production workers' needs take priority. Production workers or service providers are seen as the primary

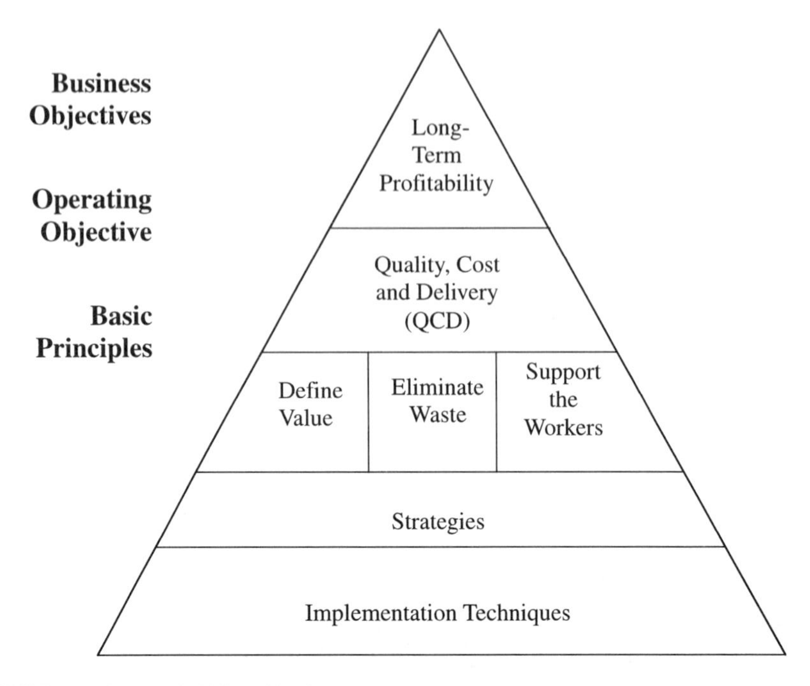

FIGURE 8.7 System building blocks.

value-adding agents because they directly manufacture or assemble parts or provide service. Since other labor does not directly add value to the product, it is justified only if it clearly supports direct production or if it helps tap the creative potential of workers who are directly involved in value-added activities. This principle includes support for work and nonwork needs. The system places high priority on providing good tools, machines that work, parts that fit, and the training required to the job effectively. Beyond work needs, the principle extends to workers' needs for input into decisions which affect them and for recognition and respect. A truly successful lean manufacturing system treats every worker as a valued asset and recognizes the fact that employees at all organizational levels have unique talents and abilities that can make positive and significant contributions to the organization. Providing opportunities for employee involvement and recognition through techniques such as *kaizen* is therefore viewed as an important element in tapping their creative potentials. Thus, lean manufacturing managers and supervisors should be encouraged to build close relationships with workers. Workers are encouraged to know their teammates as individuals and not just coworkers. This encouragement may include off-hours socializing, some of it company paid. This focus on people as the most important asset should be reflected in the way people are hired, trained, and treated.

These three basic principles are implemented by several key strategies and implementation techniques described below in a lean manufacturing system. As illustrated in Figure 8.7, these strategies and techniques form the building blocks of the whole system and will produce only partial and temporary benefits if implemented in isolation. The strategies are general guidelines for management behavior, whereas implementation techniques are specific practices and procedures developed over the years by companies such as Toyota with guidance from pioneers such as Henry Ford, Edwards Deming, and others. For example, the general guideline for the pull-system strategy is to produce only the necessary items, in the necessary quantity, at the necessary time. Techniques used to implement this strategy may include total preventive maintenance, small-lot production, flexible shop layout, level scheduling, kanban, visual controls and standard work. Small-lot production may in turn require quick changeover, and kanban techniques that may require calculating takt time.

8.4 PULL-SYSTEM STRATEGY

In a pull system, the customer process withdraws the items it needs from the supplier process and the supplier process produces to replenish only what has been withdrawn. Pull systems operate with a minimum of buffers and other "safety valves," while ensuring product quality and providing manufacturing flexibility. Such a system cannot function, however, without a management structure that first defines the value system and then supports the workers (value-adding agents) who are expected to operate it. These workers, who are most familiar with the intimate details of each process step, are then trained and encouraged to eliminate waste and find permanent solutions to problems. A well-functioning pull system guides workers on how to identify and eliminate waste, but this strategy must work in tandem with several other lean manufacturing strategies for the overall system to work. For example, production of parts in small quantities is a key technique for a pull system, but it also supports the lean strategy of "build quality into the process." Using small lot sizes for parts means that quality problems are detected quickly before large batches of defective parts are produced. Also, problems must be corrected quickly because in a pull system, minimum buffers are maintained so defects can bring production to a screeching halt. This means that support staff, such as engineers and supervisors, must help the workers without delay.

The goal of the pull-system strategy is to provide the flexibility to rapidly respond to customer demands and eliminate the waste that occurs when upstream processes produce ahead of the needs of the downstream customers. This pull strategy must be extended to all production processes that are linked together within the corporation and eventually to the entire value chain. Since the entire system must still bear the cost of inventory accumulation, this prevents inventory location shifts from production factories to supplier warehouses. More importantly, lean manufacturing does not consider inventory reduction as the primary benefit of the pull-system strategy. Higher quality, increased flexibility, and more efficient space utilization are key benefits.

8.4.1 Kanban Technique to Facilitate a Pull-System Strategy

In a pull system, the coordination of production and the movement of parts and components between processes is critical to avoid either excess or shortages. To achieve this, many companies use a system called *kanban*. This means *cards* or *signal* in Japanese. These visual signals are used to control production in the pull system. Kanban provides the authorization to deliver or produce parts for a process. Pull systems operate by requiring downstream processes (assembly) to withdraw parts from upstream processes (component production or suppliers) only when needed, thus signaling upstream processes to produce only what is necessary (to replace withdrawn parts). In most cases, when parts are used by a downstream process, a kanban card with information on the number and type of parts is detached from the parts container and sent via an in-plant dispatch system to the upstream process. Only upon receipt of the kanban card is the upstream process authorized to produce replacement parts. In some cases, the signal to produce more parts is simply the act of removing needed parts from the staging area, which could be marked by colored tape on the floor, for example. In other cases, such as notifying suppliers, an electronic signal can be sent to authorize the production of another batch of parts. Thus the exact form of the kanban signal may vary, but the upstream process cannot *produce* parts unless it has received the signal to do so.

The main advantage of kanban to the pull system is that changes in production plans due to customer demand changes need to be communicated only to the final downstream (final assembly) process. Changes in final assembly requirements can then ripple through the supply chain by means of kanban signals, which automatically convey the production orders back to preceding processes and throughout the supplier network. This provides the system with the capability to respond flexibly to small changes in demand, fine tuning the frequency of kanban transfers. It also facilitates inventory control because the total number of kanban cards outstanding determines the quantity of work-in-process inventory. Another important efficiency of the kanban system is that hourly workers manually process material requirements and scheduling in the course of performing their regular jobs. Ideally, the kanban technique must be employed systemwide to control production within the factory as well as with the suppliers and customers. However, in reality many companies start just within their own factories and eventually extend it to the supply chain after some experience with the system.

A typical kanban system uses three main types of kanban cards:

- *Move* kanban authorizes a process to get parts from the previous process.
- *Production* kanban authorizes the previous process to produce more parts.
- *Supplier* kanban authorizes an outside supplier to deliver more parts.

Examples of different forms of kanban are shown in Figure 8.8. They all serve the purpose of communicating requirements between upstream and downstream processes.

8.4.2 Level Scheduling (Heijunka) Technique

Leveling of schedules, or *heijunka* as it is known in Japanese, refers to leveling production by both volume and variety. That means if manufacturing is planning to make 8 widget As followed by a batch of 4 widget Bs today, and tomorrow is

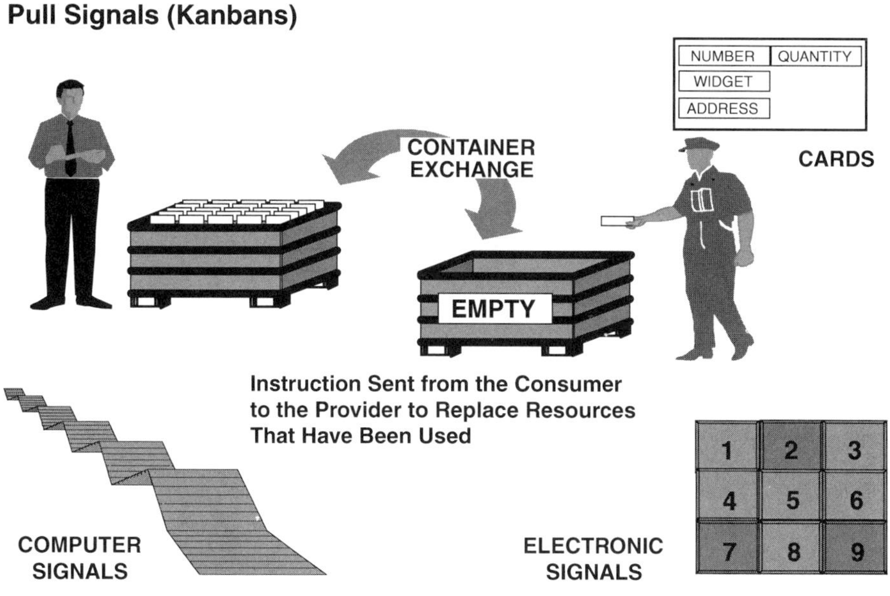

Pull Signals (Kanbans)

NUMBER QUANTITY
WIDGET
ADDRESS
CONTAINER EXCHANGE
CARDS
EMPTY
Instruction Sent from the Consumer to the Provider to Replace Resources That Have Been Used
COMPUTER SIGNALS
ELECTRONIC SIGNALS
1 2 3 4 5 6 7 8 9

FIGURE 8.8 What is kanban?

planning to build batches of 12 As and 6 Bs, then what they really should do is to make 2As followed by a B all day long each day rather than doing 18 As today and 12 Bs tomorrow. This is one of the counterintuitive aspects of lean. This leveling of the schedule accomplishes a steady demand of resources, shortens the lead time of individual product variations, and helps level work requirements throughout the supply chain. Without this technique, pull-system implementation would be extremely difficult, if not impossible. Once the production volume is firmed up, some variation in production mix can be achieved through kanban. A leveled schedule defines the limits of mix and volume flexibility, and it can be used by suppliers to estimate their own resource requirements. This permits the lean manufacturing company and its suppliers to avoid carrying excess materials, machinery, or manpower to meet peaks in demand. However, a lean manufacturing company strives to build a complete mix of each product every day or even every hour if possible. Limiting variations in production mix and volume from week to week is key in a pull system. This permits the company and its suppliers to avoid carrying excess materials, machinery, or manpower to meet peaks in demand. This type of mixed leveling (Heijunka) is carried out with respect to product variations based on models, options, and other features, which can be accommodated at the final assembly level. Without it, the managers of subassembly and upstream parts fabrication processes are required to adopt a just-in-case approach if they are to meet the changing demands of their customers. The combination of level schedule and the kanban system results in tremendous flexibility on a daily or even hourly basis to vary volume, production sequence, lot size, and mix within well-defined bounds.

8.4.3 Takt Time

A key technique to implementing a pull schedule is a calculation called *takt* time. Takt time is the rate at which each product needs to be completed to meet customer demand. It is the beat or pulse at which each item leaves the process. Takt time determines standardized work- and load-balancing requirements and drives many kaizen activities for various upstream operations.

Takt Time = Available Daily Work Time/Daily Customer Requirements

Example: Available Daily Work time = 480 Minutes – 60 Minutes (Breaks) = 420 Minutes
Daily Customer Requirement = 840 Units
Takt Time = 420/1000 = 0.5 Minutes

In other words, a final product must be produced every 30 seconds. This will set the pace of the whole production line. If several products are being produced in the assembly process, then takt time must be calculated for each type and then a repeating smooth pattern of each product type must be scheduled. This process is known as mixed model sequencing. Cycle time is the amount of time required for a single unit to be processed. Cycle time must be equal to or less than the takt time to meet daily customer requirements.

8.4.4 Quick Changeover Technique

The ability to perform quick changeovers from one part or model to another is critical to implementing a pull system in a situation where numerous parts and products are being manufactured. The reason is that rapid changeovers provide the manufacturing capability to produce in small lot sizes as signaled by kanban cards and yet maintain high machine and worker utilization. Quick changeover techniques focus on finding the causes for the equipment to be stopped for a changeover and systematically removing those reasons through teamwork, simplification, standardization, detailed documentation, and continuous improvement of the changeover process. Typically, changeovers are the responsibility of the team operating the equipment; however, other skilled trades and support-engineering personnel must be available when needed. Jigs are fabricated so that those tools can be placed into or removed from machines quickly. Tools and jigs are prearranged in locations beside the machines in which they are to be used. A variety of quick disconnects or locating devices may be needed. A well-trained quick changeover team must be able to perform multiple functions in changeovers without regard to lines of demarcation. This requires substantial training as well as specific labor contract provisions, if applicable, on work rules and job classifications.

One hurdle to quick changeover implementation faced by companies on the lean journey may be that the change necessary to implement quick changeover is not obvious until a pull strategy is in place. Implementation is hard to justify on the basis of direct labor savings alone, although it can free up substantial production capacity. The real benefits of the quick changeover technique tend to appear in areas such as direct labor, reductions in inventory, and improved quality and flexibility. Moreover, the benefits of quick changeovers can often be achieved with little or no capital investment.

8.4.5 Small-Lot Production

A basic concept of the pull system is that the ideal lot size of parts and components is equal to one. The reasoning is that if parts are fabricated and flow together into final assembly and if only one end product at a time is produced, then only one set of parts and subassembly is needed. This results in minimum inventory and one-piece flow and provides maximum flexibility to satisfy customers. However, striving toward this ideal must be balanced with practical considerations of setup and handling costs. Small-lot production also helps the lean manufacturing quality strategy, because problems surface faster and must be dealt with immediately because inventory buffers are not available. Note that if the company stresses equipment efficiencies, then it may prevent small-lot production implementation. One such measurement could be budgets and performance measures based on standard hours rather than actual hours or customer demand. Such measures encourage managers to maximize standard hours by running equipment as long as possible. Not only do such traditional measures discourage frequent setups and small-lot production, they may actually result in overproduction.

8.5 QUALITY ASSURANCE STRATEGY

In lean manufacturing, the basic quality strategy is to build quality into the process itself. Although a variety of techniques, including many Six Sigma quality tools, can be used to implement this strategy, the basic rule for a given process is simple: do not pass on bad output to the next process. The primary focus is on value-adding workers, who are responsible for making sure that only 100% quality parts are passed on to their customers. To do this, inspection procedures must be built into the worker's *standardized work*. In addition, workers must be given authority and responsibility to stop production to avoid passing on bad products. This is facilitated by the *andon* system, which can activate flashing yellow lights or other attention-getting devices to bring support to the worker. When a worker detects a quality problem, it is his or her responsibility to activate the andon device. If the quality problem can be solved within the designated cycle time as required by *takt* time, the andon device is activated again to prevent the production line's coming to a halt. This puts significant pressure on the support team to fix the problem and to prevent repeat occurrences. In addition, workers must be trained in visual inspection techniques, statistical tools, and use of gauges, as well as be supported by a strong preventive maintenance program to assure that equipment works reliably. Visual control and 5S techniques highlight problems and bring quality issues to the forefront.

8.5.1 Poka-Yoke Device (Mistake Proofing)

Another important technique for building in quality is using *poka-yoke* devices. These are simple devices or controls that permit the detection of abnormalities as they occur in the process and shut down the operation if necessary. For example, a limit switch or an electric eye can be positioned so that the machine will not start when the workpiece is loaded incorrectly. This prevents starting an operation that would produce a defect. A variation of this called "action-step poka-yoke device," which helps determine the actions the worker should take. For example, if a worker assembles several different but similar models in a workstation, a simple detecting device can be used to determine the model in the workstation. The system then opens the door to the appropriate parts bin or turns on a light indicating the appropriate part. An important result of poka-yoke devices is that workers are freed from the need to continually supervise equipment and can run multiple machines with a consequent increase in productivity.

The key to an effective poka-yoke device is determining when and where defect-causing conditions arise and then figuring out how to detect or prevent these conditions every time. Workers typically have important knowledge and ideas for developing and implementing poka-yoke devices.

8.5.2 Visual Control and 5S Techniques

A good quality assurance strategy cannot be successfully implemented in a workplace that is cluttered, disorganized, or dirty. Poor workplace conditions give rise to all sorts of waste, including extra motion to avoid obstacles, time spent in searching for needed items, and delays due to quality defects, equipment breakdowns, and accidents. Frequently, a company starts on the lean journey by establishing good basic workplace and housekeeping conditions. Many use the 5S system to improve and standardize the physical conditions of their work areas. The 5S system is a set of five basic principles with names beginning with *S*.

- *Sort:* Teams begin by sorting out and removing items that are not needed in the area. They use a technique called *red tagging* to identify unneeded items and manage their disposition.
- *Set in order:* Next, teams determine appropriate locations for the items they do need. After relocating the items, they apply temporary lines, labels, and signboards to indicate the new positions. The theme here is "*A place for everything and everything in its place.*" This helps identify unnecessary parts, equipment, and other materials. An example of this can be hanging tools required for an area on a color-coded pegboard on a wall near the work area.
- *Shine:* The third *S* involves a top-to-bottom cleanup of the work area, including equipment. Shine also means inspecting equipment during cleanup to spot early signs of trouble that could lead to defects, breakdowns, and accidents.

- *Standardize:* In this phase, teams establish the new, improved conditions as a workplace standard. At this stage, visual controls are adopted to ensure that everyone understands and can easily follow the new standards.
- *Sustain:* The final 5S principle uses training, communications, and measures to maintain and monitor the improved conditions and make it a integral part of everyday workplace behavior. A 5S checklist has been included in Appendix 8.4.

8.5.3 Visual Controls

The technique of visual control or management, also known as management-at-a-glance, is to arrange the workplace so management and workers can tell at a glance if anything is wrong and, if so, what actions need to be taken. Andon lights are an example of visual control. Simple visual graphics at every workplace promote rapid and clear communication and minimize the need for formal reports. Boundary lines painted on the floor show the area of responsibility of each worker or product team. Graphic displays may include information such as work standards, takt time, supplier performance, schedules, work procedures, or attendance records. Visual controls can also be used to manage the flow of parts. For example, where many parts are required, flow racks are used to keep those parts grouped and under control. Clearly marked shelves and color-coded labels are used for each type of part. Visual information also helps prevent mistakes. For example, shaded red and green "pie slices" on a dial gauge give an instant status reading.

8.5.4 Preventive Maintenance Technique

Another key element of the quality assurance strategy is adherence to a strict preventive maintenance system. This may include avoidance of highly integrated and automated systems managed by complex sophisticated controllers when the same results could be achieved through the use of simple, independently controlled machines. This approach is based on the fact that simpler equipment is easier to maintain and modify and that complex equipment is more likely to have more downtime from failures simply due to the laws of reliability. In addition, large expenditures in complex machines can provide a strong incentive to overproduce. Preventive maintenance is an essential part of a lean manufacturing system because there are few inventory buffers to cushion the effects of equipment failures.

Total preventive maintenance (TPM) is a comprehensive, companywide, team-based approach to reducing equipment-related losses due to downtime, speed reduction, and defects by stabilizing and improving equipment conditions. Overall equipment effectiveness (OEE) is a key measure in TPM, and Appendix 8.5 describes how it is calculated. Value-adding workers have a key role in the TPM activity called *autonomous maintenance*. In this activity, workers learn how to perform routine and basic equipment maintenance tasks such as cleaning and lubrication, as well as learning how to watch out for trouble signs or unusual conditions. The knowledge and skill of production workers should be used to help keep the equipment from

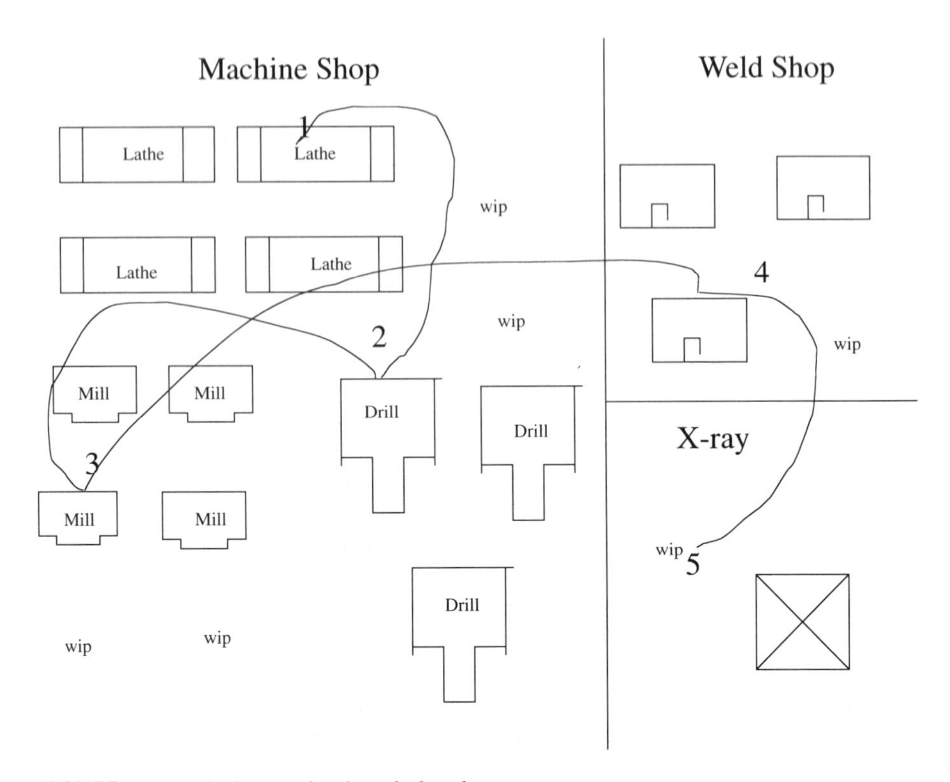

FIGURE 8.9 Typical operation-based plant layout.

breaking down. This, of course, must be done in cooperation with the maintenance staff to create a win–win situation where, ultimately, workers learn more about their equipment, and maintenance personnel are not constantly firefighting.

8.6 PLANT LAYOUT AND WORK ASSIGNMENT STRATEGY

Most traditional production processes use operation-based layouts and contain so much waste in terms of unnecessary material and worker movements that very often changes in the layout of equipment on the plant floor are required to transition to lean. *Operation-based* layouts group the production equipment according to the type of operation performed. For example, all the drill presses may be located near each other. Parts are often processed in large lots and sent to another processing area based on the part routing.

These cause waste due to unnecessary conveyance and excessive inventory and floor space needs. Positioning equipment close together in the order of the processing steps or routers reduces much of the above waste and improves communication among workers as well.

Figure 8.9 shows a typical operation-based plant layout and how routers define the path of travel of a typical part. Actually, drawing a spaghetti map of the routings of various components can sometimes be real eye-opener. A *process-based* layout

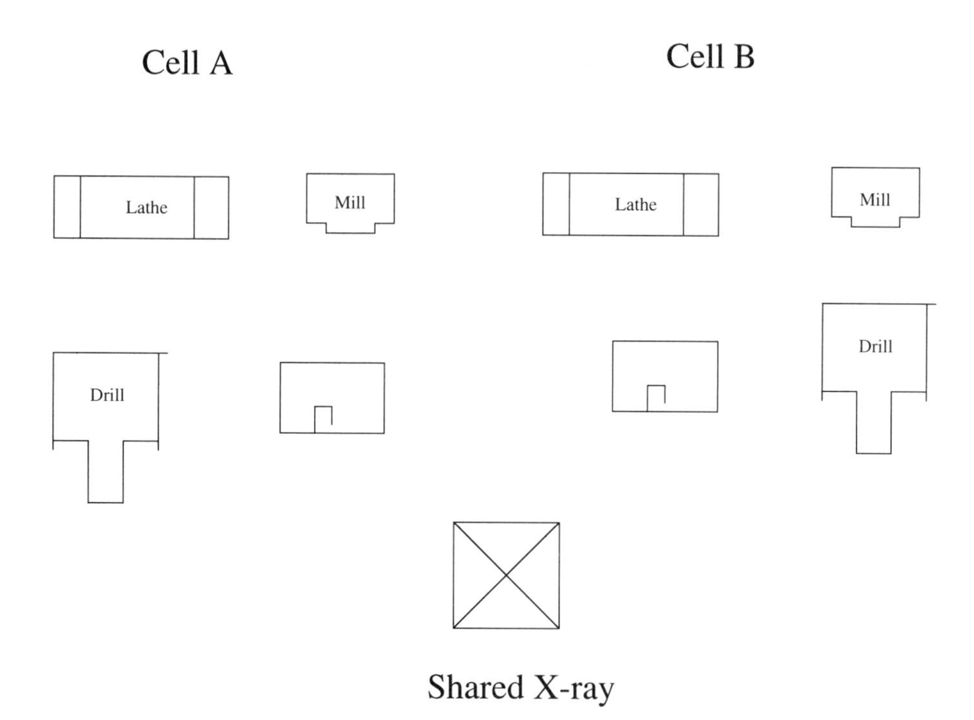

FIGURE 8.10 Process-based layout.

allows material and parts to flow through the process steps in small batches or even a one-piece flow. An example of a process-based layout is shown in Figure 8.10.

This is known as "cellular" or "flow" manufacturing in lean manufacturing terminology. However, implementation of flow manufacturing requires some planning, such as the grouping of parts or products based on their routers. A spreadsheet can be used to get a first-cut grouping. Frequently, a computer simulation of the proposed cell will reveal issues that need to be addressed before the investment in rearranging the equipment is made. Appendix 8.6 shows situations where such simulations may be warranted. This also requires cross training workers who operate in the cell (as opposed to a machine). Workers perform different functions within the cell and the team takes full responsibility for the production unit.

It is possible to implement flow manufacturing with equipment for each operation arranged in a straight line. However, when the worker finishes the last step of the process, he or she must go back to the first step again. To eliminate this waste, flow or cellular manufacturing often uses a U-shaped configuration. This also promotes improved communication among the cell workers.

Machines ideal for flow manufacturing should be small, flexible, and movable so that the new cell can be reconfigured if the customer demand pattern changes. Within the cell itself, multifunction, multiprocess work assignments are designed to eliminate waiting time that occurs when machines go through automatic processing cycles. Used with techniques such as *standardized work* and *job rotation,* they result

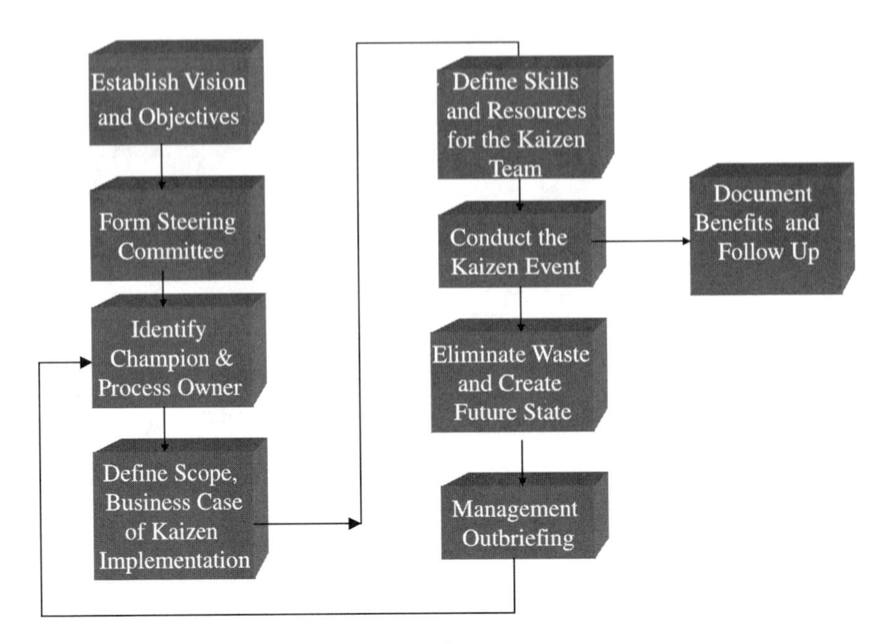

FIGURE 8.11 Generalized kaizen team activity.

in highly trained cell-team members who can then take ownership and responsibility for establishing and improving work routines.

Noted that any existing work rules and job classification issues must be addressed before this technique can be fully implemented.

8.7 CONTINUOUS IMPROVEMENT (KAIZEN) STRATEGY

A key strategy to implement the support-the-worker principle of lean manufacturing is called *kaizen.* Kaizen is the constant elimination of waste through bettering product quality, improving worker safety, and reducing costs, implemented through the collective efforts of employees at every level of the company. This usually involves a team of workers focusing on a specific small or medium step toward increased efficiency. Tools such as brainstorming are frequently used by trained facilitators (see Appendix 8.7). They are generally completed in about a week's time and result from employee ideas that do not require major capital expenditure. Kaizen can be described in baseball terms as a strategy that relies on hitting singles rather than home runs to win the game. Kaizen involves not only the identification of better ways to do things, but also the rewriting and redefining of previously set standardized work. Without standardized work, improvements could be lost and lessons could not be transferred. Kaizen strategy has been very popular in many U.S. companies and has produced outstanding results in many cases. A typical kaizen "event," as it is called, can be planned using a nine-step methodology over the course of a week as shown in Figure 8.11.

Systematic methods are used for generating ideas and evaluating which ideas to implement in the workplace. This methodology should be taught to group leaders

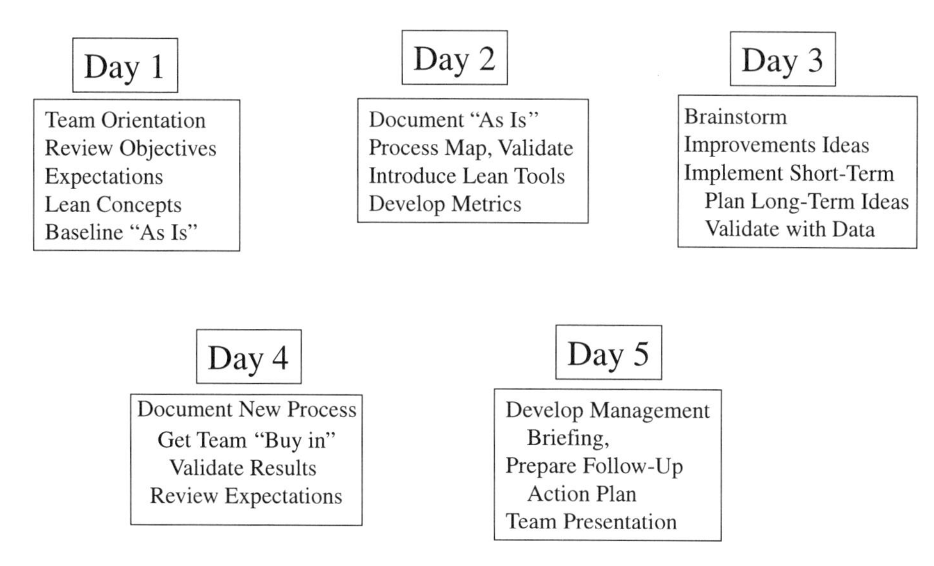

FIGURE 8.12 Typical kaizen event.

and managers, who should lead the kaizen efforts and get workers to actively participate in the process.

Typical kaizen event activities over the course of the week are shown in Figure 8.12.

8.7.1 Standardized Work Technique to Support Kaizen

As processes are improved using the kaizen strategy, it is important to standardize the way work is done. In a pull system, each process must deliver a certain quantity of parts at a certain time, given a certain lead time. If the previous process is unpredictable, the pull system will break down. To establish predictability in processing cycles, *standardized work* must be established for each process. This usually has three parts.

8.7.2 Standard Cycle Time

This is the actual cycle time required to process one part, and should be established by timing the operation from start to finish, including machine cycle time as well as loading and unloading, walking, waiting, and inspection. Process cycle time determines if the process can produce the quantity required according to the takt time. Cycle time must be less than or equal to the takt time. If an unbalanced situation exists, then either kaizen must be planned to bring it under the takt time or the workload redistributed.

8.7.3 Standard Work Sequence

It is not possible to have a consistent cycle time without a consistent work sequence and methods. These usually need to be spelled out and should be readily visible at the workstation. It must be noted here that the kaizen and work standards go hand in hand. Kaizen improves the standard and then the new standard needs to be documented and used by the work teams, and the continuous improvement cycle goes on.

8.7.4 Standard WIP

Sometimes a process step may require a minimum quantity of parts or material to complete a processing cycle in the same manner and same sequence each time. For example, a workpiece may need a cooling period to ensure quality. In this case, several standard WIPs may have to be stored before the next process can start on the workpiece. Any change in the standard WIP quantity indicates a problem in the standard cycle time or work sequence.

8.8 DECISION-MAKING STRATEGY

In lean manufacturing, key strategies to the *support-the-worker* principle are clear performance standards and decentralized consensus decision-making. Workers must know what is expected of them and must be given clear and specific goals and objectives within the bounds set by management. For example, a manufacturing-cell team needs to know how many units to produce and what equipment to use. The team can then determine the exact cell layout, create detailed standardized work routines, and calculate the takt-time target. The assumption behind this is that to make decisions correctly, employees need to know the goals and objectives of the organization. Without that, they can only speculate on the priorities and the trade-offs involved. Decisions also need to be made at the lowest possible level for several reasons: it generally speeds up the decision-making process, it allows decisions to be made by people who are generally more informed about the specifics of a given problem, and the decisions are made by those who have to live day to day with the results of the decisions. Consensus decision-making puts horizontal control on the decentralized decision-making process and makes sure that all the elements work together to make the lean system effective. Consensus decision-making does not mean everyone has to agree but it does mean that the input of everyone who is affected by the change has been taken into consideration. However, once the decision is made, it requires total support from all parties. For this kind of decision-making to be effective, education and training of employees must have top priority. All of these are based on the fundamental lean belief that people are a company's most important assets and need to be developed to their fullest potential, and every employee must be treated with respect.

8.9 SUPPLIER PARTNERING STRATEGY IN LEAN MANUFACTURING

The last but not least important strategy that supports a lean manufacturing system is how a company treats and works with its suppliers. A company may become very lean and efficient within its factory but must still rely on its suppliers for the lean system to function smoothly. Suppliers should be regarded as an extension of the company itself and are expected to follow the strategies outlined above. For example, suppliers must deliver the necessary quantities of defect-free parts at the necessary time to support the pull production strategy. If the suppliers themselves do not implement lean, this may not be possible. That is why lean manufacturing companies frequently assist suppliers in implementing lean strategies and help develop them into reliable partners through careful selection and training. Suppliers are expected to improve performance continuously through their own kaizen efforts. Suppliers are critical to lean because the work content of a product increasingly comes from suppliers; also, they frequently partner in new product-development efforts. Since at least 50% of a typical manufacturing company's value-added assets comes from the suppliers, the full benefit of lean cannot be realized without suppliers initiating lean strategies themselves. There are several techniques to implement an effective supplier partnering strategy.

8.9.1 Small Supplier Network

A small number of first-tier strategic suppliers (out of the existing universe) must be selected. This makes close communication and monitoring easier. It also permits the company to understand its supplier's production processes and promote continuous improvement.

8.9.2 Short-Term Contract/Long-Term Commitment

Informal long-term purchase commitments must be awarded to the suppliers who chose to cooperate and agree to work on a lean strategy. However, short-term, 1-year contracts can also be awarded. The suppliers must be included in the company's long-term strategy and future product development plans. Frequently, this is necessary to reduce the time to market on new products. Cost, quality, and delivery performance targets must be tailored for each supplier, rather than across-the-board percentage reductions or bids by other suppliers. However, to do this effectively, the company must know its suppliers' production processes and materials.

8.9.3 Supplier Assistance

A lean manufacturing company should be willing to provide assistance to its suppliers in the form of lean training and should encourage suppliers to communicate with each other and provide opportunity for additional business. This training could be done informally through daily contacts or formally by subject matter experts in various lean techniques.

8.9.4 Structure for Effective Communication

Once a supplier has been selected, every effort must be made to build trust and to maintain open lines of communication. A supplier development team consisting of procurement, production control, quality control, and financial control representatives must remain in regular contact with the supplier. The purchasing department plays a lead role in supplier relations and communications, and buyers have the long-term responsibility for managing the overall supplier partnership. Some of the lean companies let the workers talk directly with the suppliers regarding quality issues.

If a supplier can meet the company's cost, quality, and delivery specifications, then the supplier is assured of a long-term partnering relationship. Obviously, if suppliers adopt a similar lean strategy and extend that to their suppliers, then the full benefits of lean from the supply chain can be realized.

8.9.5 Supplier Selection and Evaluation

Because suppliers provide a strong link in the lean manufacturing system, it is imperative that their selection and ongoing evaluation process be well planned and executed. Suppliers should be selected based on their ability to meet quality and cost targets. Factors such as technological capabilities, expertise, responsiveness, and past performance must be considered. A supplier's attitude and commitment toward lean, coupled with a desire to be a partner for the long term, can also influence the selection process.

Suppliers should be evaluated at least yearly based on the number of quality problems as well as responsiveness, delivery, and cost performance. They should also be evaluated on their progress with lean and continuous improvement efforts.

8.9.6 Supplier Kanban and Electronic Data Interchange

Kanban techniques to implement the pull system can be extended to suppliers as well, although that requires 100% quality parts, as well as a reliable transportation, pick-up, and delivery system. A communication system such as barcode or EDI must work effectively. Suppliers should also have sufficient manufacturing flexibility to respond to changes in demand within the bounds of a level schedule.

APPENDIX 8.1

Lean Maturity →

Lean Elements ↓

Pillars	Building Blocks	Level 1	Level 2	Level 3	Level 4	Level 5
Process	**Multi-Functional Workers**	Unquestioned support for single skill, single process operations	Single skill, single process operations with some cooperation with operators at adjacent processes	Flow-based cooperative operations; workers capable of helping next worker upstream and downstream	Flexible job assignments with some variation between workers in quality and productivity	Flexible job assignments with little variation in quality and productivity between workers
	Visual Manage-ment	Abnormalities and defects often occur and only create confusion	Abnormalities and defects often occur and are usually resolved in some way	Visual controls highlight abnormalities and defects as they occur	Visual controls actively signal management as abnormalities and defects occur	Poke-yoke eliminates occurrence of abnormalities and defects
	Process Reliability	Factory ships defective products and later deals with customer complaints	Defective products are reworked without process to prevent repeat problems	Factory still produces defective products but analyzes them to reduce repeat problems	Processes do not send defects downstream (self-checking and successive checking)	Factory builds quality in at each process (source inspection and Poke-yoke)
	Level Pro-duction	Processes have no rhythm or synchronization	Each process has its own rhythm; but processes are not synchronized	Overall line is roughly synchronized	Daily production runs; in-line production runs with coordinated cycle times	Completely level production; plant-wide synchroni-zation; no delays anywhere
	Pull Pro-duction	Push production with inventory all over the place	Push production with organized storage sites for WIP	Pull production begins in pilot areas	Pull production with kanban	Pull production with refined kanban

Lean Maturity →

Lean Elements ↓

Pillars	Building Blocks	Level 1	Level 2	Level 3	Level 4	Level 5
Product	Design Process	Engineers and specialists work through design process steps sequentially and toss design "over-the-wall" to manufacturing	The design department still has major responsibility for new product development, but solicits input from other departments	A reengineered development process is formalized; reviews are held on a regular basis but are not necessarily tied to a project plan	The design process continues to be refined to improve quality, cost, and lead time; regular reviews are tied to a plan	Technical and program reviews are fully integrated into the program schedule
	Product Design	Fluid specifications result in many design changes during development; voice of the customer is not heard	Specifications are not frozen early, resulting in many expensive design changes; the customer can be heard, but faintly	The product definition process is focused on the customer; but specifications still can be changed too easily by management	Product definition is timely, with significant customer participation; specifications are difficult to change	The company is capable of anticipating customers' latent requirements in new products
	Risk Management	Risk management is used when there is a technical problem or problem in design	Risk management is treated as a business process and is piloted in at least one major project	Risk scorecards are prevalent for technical issues and are sometimes prepared to track schedule and cost concerns	A disciplined process is used to assess and classify technical, schedule and cost risk according to standard criteria for all projects	
	DF(x)	The company does not utilize value engineering to achieve target cost, and does not practice design for manufacturability, reliability, or serviceability	In a pilot project, Value Engineering is used to achieve target cost; FMEA, DFMA, DOE, Taguchi DOE, DR are also used to achieve other selected goals	The firm extends cost, reliability, serviceability, and technology deployment to major new designs	The firm extends cost, reliability, serviceability, and technology engineering to all new designs	Demonstrates sustained industry leadership; is first to market innovative products

Lean Maturity →

Lean Elements ↓

Pillars	Building Blocks	Level 1	Level 2	Level 3	Level 4	Level 5
Process	S5	Hard for anyone to tell what goes where, when, and how much	Hard for visitors to tell where, when, and how much but workers know	Factory uses outlining and location indicators for visual control	Good location indicators and a clean neatly organized factory	Clean neatly organized factory with mess prevention measures in force
	Standard Operations	Operations procedures are generally left up to each operator	Operation procedures are standardized but not sequenced	Standard, sequenced operations are implemented in most major areas	Standards and work sequence are fully implemented and visually clear	Standards continuously updated based on worker innovations and input from managers and engineers
	Flow Production	Job shop layout	Job shop layout geared for small production	In-line and cellular layouts are geared for single-process, small-lot flow	In-line and cellular layouts are geared for 1-piece flow within and between processes	Full multi-process operations with 1-piece flow
	Information Systems	Multiple systems for given application; systems not integrated	Single system per application; systems not integrated	Single system per application; systems integrated	Common database supports organizational learning and visual control	30 sec information retrieval, real-time updating, workstations at point of use
	Predictable Demand	Frequent changes in production rate	Production baseline changes more than 10% annually	Production baseline changes less than 10% annually	Level demand enables continuous flow	Marketing and capacity planning are integrated
	Make-Ready	Typical make-ready requires 1/2 day or more each time	Make-ready improvements made in some areas	Make-ready improvements made in most areas	All make-ready takes less than 10 minutes	All make-ready is within cycle times of less than 3 min

Lean Maturity →

Lean Elements ↓

Pillars	Building Blocks	Level 1	Level 2	Level 3	Level 4	Level 5
Leader-ship	Direction	Tactical improvement plans are neither clear nor linked to business strategy	Improvement plans are clear and linked to business strategy	Improvement planning is linked informally to the budgeting process	Improvement planning and budgeting are formally linked	Improvement, budgeting, and new product planning are formally linked
	Communication of Objectives	Little or no coordination of improvement plans and activities between different levels of organization	Most departments intend to carry out management's objectives but lack process to turn objectives into concrete plans	Clear instructions from the top with objectives broken down and quantified at every level	Team-based organization with good vector alignment (i.e., clarity plus buy-in)	Total organization can change direction easily in response to changed business conditions
	Metrics	Results oriented Metrics proliferate	Concrete, consistent results- and process-oriented metrics are proscribed from the top down	Open communication produces aligned detailed metrics from the bottom up	Visual management assures adherence to management objectives in daily work	Visual management permits at-a-glance audits and assessments
	Analysis	Management only reviews results	Differences between actual results are reviewed but enabling processes are not carefully analyzed	Structured analysis of results and enabling processes	Feedback and analysis of results, processes, and new ideas are standardized at every level	Feedback and analytical procedures are streamlined through advanced visual management
People	Employee Involvement	Lip service given to employee value	Management of human assets to increase financial performance	Management of human assets to build organizational capability	Knowledge and creativity of each employee are cultivated	Employees are dynamic partners in the enterprise
	Teams	Management plans to develop team-based organization	Pilot activities spark broad interest in team activities	Emerging team culture supports empowerment and cross-functional management	Teams in all major areas mature; support area teams get involved — no going back	Self-managed work teams create HPWO
	Learning Environment	Training plan not integrated with long-term improvement goals	Employees trained to recognize waste and use cause/effect logic	Lean methods are integrated into total training plan and transferred to majority of workforce	Intensive training is a major, permanent initiative funded at 10% of wage build or greater	Employees confidently carry out R&D of new methods
Partnering	Supplier Development	Numerous, unqualified suppliers are managed through armslength dealings	Supplier performance monitored; quality and delivery performance of top suppliers improved	Partnering and long-term agreements are used to create mutually beneficial environment	Lean supplier development program creates extended factory	Supply partners are an integral part of product development

Lean Maturity →

Lean Elements ↓

Pillars	Building Blocks	Level 1	Level 2	Level 3	Level 4	Level 5
Product	Validation	A large number of physical prototypes are used to discover mistakes	A moderate number of physical prototypes are used to discover mistakes	Some analytical prototyping is performed to validate designs as well as discover mistakes	Extensive computer-based analytical prototyping is done resulting in a minimum of physical prototypes used only to validate design and production parameters	
	Product Evolution	The evolution of a product is not planned or managed, leading to the reinvention of many "wheels"	Product evolution plans are piloted, but they represent limited multifunctional thinking	Product evolution plans are prepared by cross-functional teams and communicated to design process stakeholders	Product evolution plans are strongly linked to the company's profit plan and reflect aggressive time-to-market goals and capability	
	Mass Custom-ization	All running design changes as well as full design changes are made in the design department	Model design production line or parallel design process cut design lead times and the number of required drawings in half	A standard design schedule is created and design kanbans are published, but many running changes are still handled by the design department	New designs do not require new drawings Running change requests are handled in the factory, not the design department	Zero engineering The supply chain is fully integrated into modularization

Courtesy of the Lean Aerospace Initiative. For more information visit http://Lean.mit.Edu/Products/products_LESAT.html

APPENDIX 8.2 Process map. Castor oil analysis (before improvement).

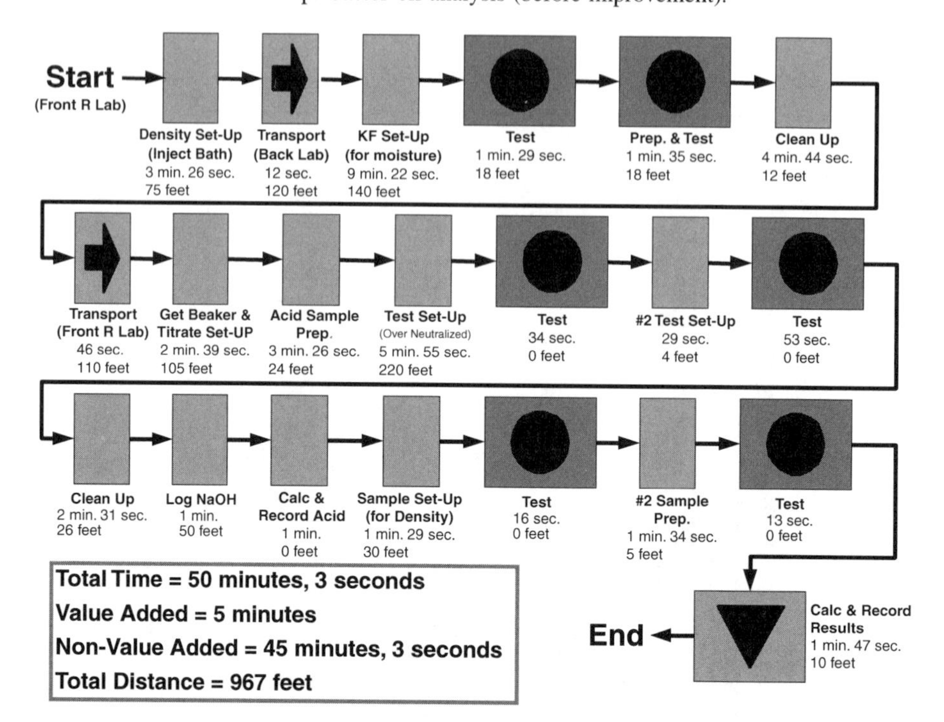

APPENDIX 8.3 Process Map Data Sheet. Spreadsheet format (tabular form).

Spreadsheet Format (Tabular Form)!

Ref#	As Is Process Steps	Inspect	Process	Handle	Store/Wait	Time	Distance	Quantity	V/A	NVA	Eliminate	Combine	Change Place	Change Person	Improve	Comments
1																
2																
3																
4																
5																
6																
7																
8																
9																
10																
11																
12																
13																
14																
15																
16																

APPENDIX 8.4

5S Evaluation Check Sheet

Area ______________________

Number of Problems	Rating Level
5 or more	Level 0
3-4	Level 1
2	Level 2
1	Level 3
None	Level 4

Workplace Scan Diagnostic Checklist

Category	Item	Date Rated			
	Distinguish between what is needed and not needed				
Sort	Unneeded equipment, tools, furniture, etc. are present				
	Unneeded items are on walls, bulletin boards, etc.				
(Organization)	Items are present in aisleways, stairways, corners, etc.				
	Unneeded inventory, supplies, parts, or materials are present				
	Safety hazards (water, oil, chemical, machines) exist				
	A place for everything and everything in its place				
Set in Order	Correct places for items are not obvious				
	Items are not in their correct places				
(Orderliness)	Aisleways, workstations, equipment locations are not indicated				
	Items are not put away immediately after use				
	Height and quantity limits are not obvious				
	Cleaning, and looking for ways to keep it clean and organized				
Shine	Floors, walls, stairs, and surfaces are not free of dirt, oil, and grease				
	Equipment is not kept clean and free of dirt, oil, and grease				
(Cleanliness)	Cleaning materials are not easily accessible				
	Lines, labels, signs, etc. are not clean and unbroken				
	Other cleaning problems (of any kind) are present				
	Maintain and monitor the first three categories				
Standardize	Necessary information is not visible				
	All standards are not known and visible				
(Adherence)	Checklists don't exist for all cleaning and maintenance jobs				
	All quantities and limits are not easily recognizable				
	How many items can't be located in 30 seconds?				
	Stick to the rules				
Sustain	How many workers have not had 5S training?				
	How many times, last week, was daily 5S not performed?				
(Self-discipline)	Number of times that personal belongings are not neatly stored				
	Number of times job aids are not available or up to date				
	Number of times, last week, daily 5S inspections not performed				
	Total				

APPENDIX 8.5 Overall equipment effectiveness (OEE). This is the major metric of TPM. It takes into account not only downtime due to breakdowns, but also downtime due to setup, reduced operating speed, idling, and lost time due to defects and rework. For machine tools, an OEE of 40% is typical, while 85% is considered world class.

OEE = Availability x Performance (speed) x Quality Rate

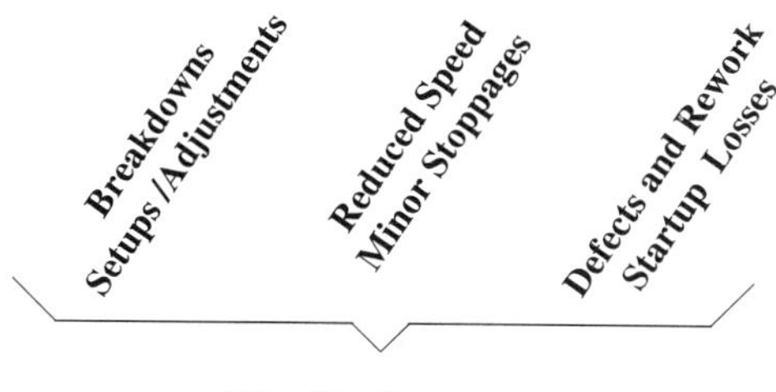

Big Six Losses

Example:

Scheduled Run Time (24 hr - offshifts, lunches, breaks meetings, housecleaning)		**6.5 hr**
minus: - Breakdown Time	**- 0.6 hr**	
- Setup Time	**- 1.2 hr**	
- Minor Stoppages Time	**- 0.4 hr**	
- Rework/Defects Time	**- 0.2 hr**	
= Effective Time (time cutting, welding, etc.)	**=**	**4.1 hr**

OEE = Effective Time / Scheduled Run Time = 4.1 / 6.5 = 63%

APPENDIX 8.6

When Is Simulation the Right Tool for the Job?

- Simulation Should Be Used When Considering:
  - Production Rate Increases
  - New Process Lines
  - Batch or Lot Size Changes
  - Kanban or Pull Systems
  - New Product Lines into the Process Center
  - Shift Schemes
  - Future and Present Bottlenecks
  - "What If " Scenarios for Major Changes to the Product Line
- Simulation Is Not a Good Tool for:
  - Evaluating Shop Floor Standards
  - Effects of Overtime Usage
  - Effects of System Constraints Such as MRP or Bar Coding

APPENDIX 8.7

Brainstorming

The Process Used to Generate Ideas

- Round Robin: Take 5 minutes for everyone to list their ideas on a piece of paper. Everyone participates as their turn comes up. If no idea, person passes. Continues until no more ideas. Ideas posted on flip chart.

- Crawford Slip Method: Everyone writes their ideas down on post-its. Post-its are collected and placed on wall or flip chart by categories. Good method if anonymity desired.

9 Measurement System Analysis

Edward A. Peterson

9.1 WHY PERFORM A MEASUREMENT SYSTEM ANALYSIS?

9.1.1 The Value of Measurement System Analysis

Measurement system capability is becoming increasingly important in the management of developing and producing hardware. With competitors working to dramatically improve quality, measurement system capability is being pushed to its literal limits. Managers today need to examine the gauge capability *before* embarking on a process or design improvement. Efforts to improve or change your manufacturing processes will not improve gauge capability. The data derived by the gauge may be misdirecting your efforts and driving changes that will have no real effect on product quality. Money being spent to measure product and time and effort spent to improve processes may be misdirected by data coming from poor gauges.

The broad-based push in the 1980s for Six Sigma quality led to significant strides in identifying and controlling key process parameters while testing the capabilities of the measurement system. As these process improvements became more prevalent, customers began to expect the new levels of quality from producers, further fueling the improvement cycle. Extensive analysis and controls were put in place to improve manufacturing performance. As this drive to improve quality moved into the 1990s, the measurement systems began to come under fire. Suddenly, one of the primary sources of special-cause variation in the data was the measurement gauge — unbeknownst to the manufacturers.

As background to the analyses that are discussed in other chapters of this book, it will be helpful to briefly examine their use of measurement data.

"Six Sigma," "Design for Manufacture Assembly/Design for Six Sigma," "Process Analysis," and "Robust Design" all use process capability as a primary element in their analyses. Process capability expresses the variation in the measured data relative to the allowable tolerance. Processes are potentially selected during design that will have a high capability of producing the product. During manufacturing, processes with poor capability are selected for process improvements using Six Sigma methods.

"Statistical Process Control" tracks process variation over time. Calculating expected variation with control limits, actions are initiated to correct perceived special-cause variation.

1-57444-300-3/02/$0.00+$1.50

"Design of Experiments" is a structured analysis wherein inputs are changed and differences or variations in output are measured to determine the magnitude of the effect of each of the inputs or combination of inputs.

Each of these analyses depends on subtle changes in the data from the measurement system. Further, if the system does not have sufficient discrimination to detect variation in a process, then it is unacceptable to use for any of the analyses. Your measurement device has now become at best a go/no-go gauge. If the system cannot detect variation in processes due to special causes, then it is unacceptable as a process control tool or to determine process capability. If you are examining process capability, tracking process quality, analyzing attributes before and after changes, and especially if you are designing experiments, you must know if your gauges have the ability to measure the characteristics of interest and their associated variations. The variations found in each of these analyses may be principally due to variations from the gauge itself!

With the expansion of these types of analyses in producing products, the importance of the difference between gauge calibration and measurement system capability became apparent. The first did little to improve variations inherent in the gauge and the measurement process, only to ensure the gauge was on its nominal operating target. Yet while manufacturers had extensive calibration methodologies integrated into functional activities, minimal systematic processes or procedures existed to validate measurement system capability. Many manufacturers now made a critical mistake, assuming that buying better gauges with higher resolution, at considerable expense, would resolve this problem. Many case histories on coordinate measurement machines dispute this type of solution. These gauges would be calibrated and have their resolution verified, only to still have the same problems of variation that existed before. It was only after extensive analysis with gauge studies that it became apparent that the total measurement system needed to be considered and that the gauges were just a part of the measurement system. How you apply a gauge and the processes with which you use it are just as important in determining gauge capability. The manufacturers of tomorrow will have measurement system analysis and capability as primary tools in their push to compete with better products.

These new perceptions are reflected by the changes in ISO auditing. Covered elsewhere in this book are discussions on the current version of 1SO 9000. These audits move from the singular view of calibration processes and procedures into the arena of measurement system analyses. Please review these sections to ensure your understanding of the changes to auditing in this important area.

This chapter examines the fundamentals of the entire measurement process and its impact on the data generated by it. It covers the properties of the measurement system, including bias, linearity, and stability, along with reproducibility and repeatability (R&R) studies. It covers the minimal requirements for discrimination and resolution of gauges. Reviews of gauge analyses break down variance of the data into the separate components of gauge and part variation, along with further separations into the gauge components of repeatability and reproducibility and what these mean to you. Attribute and variable data systems are reviewed and the differences in analysis techniques will be identified. This chapter does not cover calibration or gauge selection; these subjects have generally been well researched in the past. The topics are covered in terms nonstatisticians can understand and make use of and give real examples of case histories to validate the conclusions.

9.2 THE BASICS OF MEASUREMENT SYSTEM ANALYSIS

9.2.1 DATA AND YOUR MEASUREMENT SYSTEM ... WHAT'S IT ALL ABOUT?

Drs. W. E. Deming and Mikel Harry both professed that data should drive the decisions of an organization in improving quality and competitiveness. Such data-driven approaches are dependent on the quality of the data and, therefore, the quality of the measurement system that supplies the data. The quality of the measurement system is based on characteristics of discrimination, bias, and variation. But to discuss each appropriately, we must first introduce the two basic types of data — attribute and variable data.

Attribute data are qualitative elements of a product or process that cannot be measured or quantified. They are usually classified as go/no-go, good/bad, or pass/fail, and are most commonly gathered via visual inspections by trained personnel. Inspections of solder joints, defects in propellant cast via X-ray, and the number of errors in a data entry form are examples of attribute data. The perceived defect is either there or it isn't. Statistical properties for the variation in the identification and classification of defects in this type of data are discussed later in this chapter.

An attribute gauge study has multiple benefits. It will help you determine if different inspectors used the same criteria in identifying good product from bad. It will help you correlate your inspection standards to those of your customers' requirements. It will aid in discovering where training is needed, if standards are vague or ill defined, and how frequently you are misclassifying your product quality.

Variable data are quantitative elements of a product or process that can be represented numerically. Height, weight, length, diameter, temperature, pressure, etc. are all elements that can be gathered and represented numerically. Most of these types of data are gathered using a measuring device: a gauge. For these types of data, the ability of the instrument to detect changes (discrimination) and the statistical properties of bias and variation are discussed.

A variable gauge study also has multiple benefits. It will provide data on the percentage of the total perceived variation that is coming from the gauge. It can compare multiple devices to each other and multiple operators to each other. It will separate those elements of variation in the gauge into repeatability and reproducibility, generally associated with gauge variation and operation variation, respectively. The gauge study should be used to evaluate suspect gauges and before the introduction of new gauges to ensure that these measurement devices will provide the information needed.

In the gathering of data, it is not the sole responsibility of the gauge to produce high-quality data. The gauge, whether it is human as in visual inspections, or a measuring device, is only a part of the measurement system. The parameters of the measurement process and the environment in which the measurement tools are used are just as critical for the ability of the gauge to produce high-quality, useful data.

I have been involved in many examples of organizations using coordinate measurement machines (CMMs) with their heavy granite platforms and ten-digit levels of significance. One particular example concerned formed and machined metals that had an in-house manufacturing aid used in process feedback, but final buyoff was on the CMM. Not only did the CMM fail to sufficiently assess part variability but would misclassify parts, and the process of inspection literally took as long as the time needed to produce the valued-added elements to the part! This misclassification on a gauge with high-resolution capabilities was also noted in a 1963 book by Nathan Cook and Ernest Rabinowicz.*

9.2.2 Properties of a Measurement System

Variable data have the need for a measurement system that can measure or detect the changes in the product or process. The most fundamental of these properties is the ability to detect small changes in a product characteristic, known as discrimination or resolution. Without this, part variability and gauge variability cannot be discerned. This detection capability is relative to the amount of change occurring in the product, not the overall dimension of the product. Large physical products that have very small variations from part to part need to have a gauge that is capable of detecting these changes. When a process tries to produce identical parts, the variability that occurs needs to be detectable by the gauge. For the range of data about a particular characteristic, or similarly that of the specification limits, it is desirable for the gauge to produce ten (10) different increments or categories. The different types of gauge discrimination abilities are covered below.

- If, for numerous parts, all the measures are exactly the same, then the gauge is not meeting this basic requirement. It is incapable of detecting variation in the parts. Not only is the gauge not useful for any of the discussed analyses, but its usefulness in measuring the product has to be seriously questioned. If the product has such a small variability relative to the specification, then why measure the parts at all? Note that one such reason is a requirement to measure by a customer.
- If, for numerous parts, the measures can be grouped into two to four categories, then simplified sorting can be done with the gauge. The gauge has now improved to the level of an expensive go/no-go gauge with an inability to produce control charts or perform statistical analyses and testing.
- If, for numerous parts, the measures can be grouped into at least five categories, then the gauge is meeting a minimum requirement for adequate discrimination. The gauge has the ability to be used for control charts and other analyses. A note of caution: This assumes that the gauge contributes minimally to the variability being perceived. It may be that the variability being measuring is that of the gauge. This is the overall context of this chapter.

* N. H. Cook and E. Rabinowicz, *Physical Measurement and Analysis,* Addison-Wesley, Reading, MA, 1963.

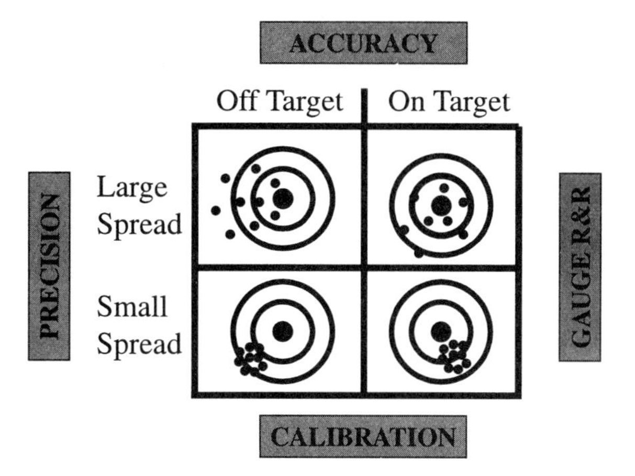

FIGURE 9.1 The two basic attributes of a gauge: accuracy and precision as related to calibration and the gauge R&R study.

If your gauge can at least meet the minimum requirement for resolution, you can now examine the statistical properties of bias and variation. Figure 9.1 shows that these elements are represented by the aspect of being *on target* and the *spread* of the gauge data. Using a single part measured repeatedly, the gauge is on target if it accurately reflects the measurement of the part with the average of the multiple measures. The spread of the data reflects the precision in the gauge to minimize the variation being introduced with the multiple measurements.

Calibration is located at the bottom of the chart and represents the method to ensure that the accuracy or bias of the data is being represented. Gauge R&R, or measurement system analysis, is listed to the right of the chart and is used to evaluate the precision of the gauge to minimally identify variability in the data.

9.2.3 Variable Data — Bias/Accuracy

The bias or accuracy of the data produced by the gauge is reflected in its ability to represent the value of a known measurement. Mathematically, the total mean of the measurements is the actual mean of multiple products being measured *plus* the mean value of bias in the gauge. This is represented in Equation 9.1 below, where the mean is shown as the Greek symbol μ.

$$\mu \text{ total} = \mu \text{ product} + \mu \text{ gauge} \tag{9.1}$$

In basic terms, this says that the value of the product characteristic is at best the actual value of the characteristic. Calibration attempts to make the bias, or mean value, of the measurement system equal to zero so that we get the true value of what we are measuring — ignoring for the moment variations in individual measures.

Calibration methods and procedures are extensively used in manufacturing environments today. Whole organizations and systems have been developed to deal with errors in the accuracy of a gauge, starting with the national organization chartered with producing standards for this use in the United States: The National Institute of

Standards and Technology (NIST). NIST is part of the Commerce Department and serves as the repository for the nation's physical and chemical measurement standards. These known values are used to verify and calibrate gauges. Metrology departments within a company's organization should be charged with the responsibility of ensuring that company standards trace back to the NIST standards. This department must have established procedures for the control and use of gauges, including time frames for each gauge to be brought in for calibration. This time frame is called a calibration interval and is established through historical information that relates time to failure for a gauge. Even without known data failures, gauges should be brought in periodically to be checked, as well as when damage is suspected. As noted, this aspect of gauges is typically well understood and is not addressed further in this text.

Stability is a measure of the bias of a gauge over time. Calibration will correct bias changes due to stability, which is the most frequent type of calibration issue. As indicated in the discussion on bias, this attribute is typically well understood.

Linearity is a measure of the bias of a gauge when measuring over the operating range of the gauge. Such ranges include measuring various concentration levels of chemical components, output levels in electronic signals, or lengths of materials. If by examining bias over the operating range, a change in the bias of the measurement device is discerned, then an evaluation of the acceptability of this gauge needs to be done.

9.2.4 Variable Data — Precision

The precision or variation of the data produced by the gauge is reflected in the spread of the data. As shown in Figure 9.1, when taking multiple measures of the same characteristic, different results will occur even though the gauge has been calibrated. The range or spread of these data gives an intuitive idea of the measurement system's precision. Mathematically, the variation of the measurements is the actual variation of multiple products being measured plus the variation introduced by the gauge. This is represented in Equation 9.2 below, where variation is shown as the Greek symbol σ.

$$\sigma^2 \text{ total} = \sigma^2 \text{ product} + \sigma^2 \text{ gauge} \tag{9.2}$$

Here again, the best we can hope for is that the perceived variation is solely due to the variation in the products being measured. A gauge study will determine the amount of the variation that is attributable to the product and the amount attributable to the gauge.

The dashed line represents the actual spread of the product values and the solid line represents the measured product values that have the additional variation introduced by the gauge. When the additional variation due to the gauge is large relative to the actual product values, improving the process will have limited results, as shown in Figure 9.2.

Though measurement system contributions of this relative magnitude may seem unusual, it is not as uncommon as you might imagine, as shown in actual results

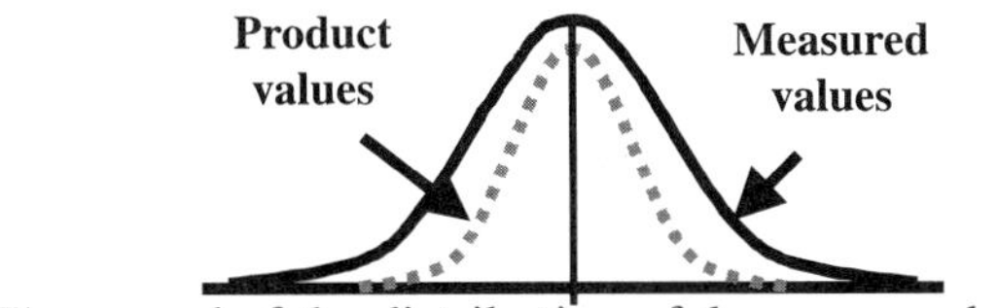

The spread of the distribution of the measured product

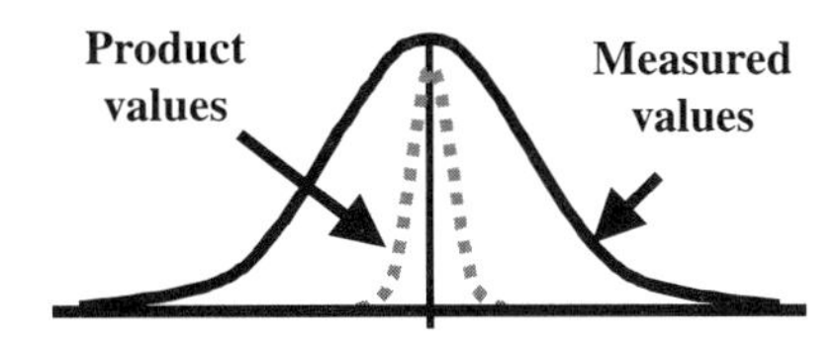

Initial product and measurement system variability

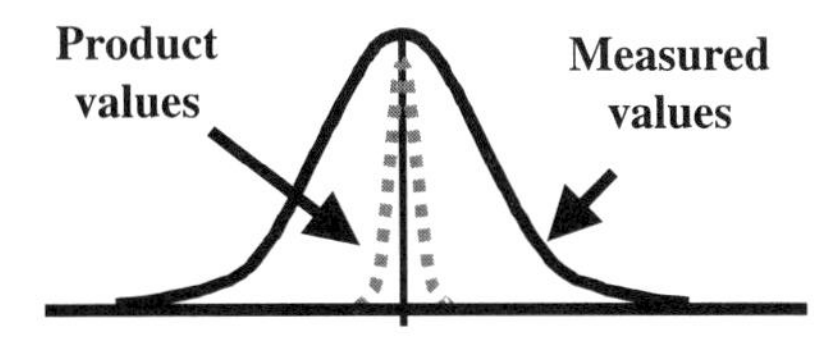

New product and measurement system variability
after reducing product variability in half

FIGURE 9.2 Product variation. (a) The spread of the measured parts includes the actual spread of the product plus the additional spread introduced by the gauge. (b) If the initial product value spread was small relative to the perceived spread in the measured values, then (c) reducing product variability in half would have only a minor effect on what we see in the measured values.

later in the chapter. Note that the variation of the gauge will not be affected as you act on the process, whether in a Six Sigma process-improvement activity, as a result of responding to perceived special causes with statistical process control, or with changes to the input parameters in a designed experiment! These are the driving issues for measurement system analysis, and variability introduced by the gauge must be understood.

9.2.5 Why There Is Variability

It is not intuitively obvious why the measurement process can add so much additional variation to the product data. Given that most gauges are straightforward to use, their resolution is seemingly obvious, and clear procedures have been developed. In this way, the measurement process is much like the manufacturing process: there are many subtle variables at play. Measurement system analysis is much like an experiment in manufacturing to discover the sources of variation in a product. Using

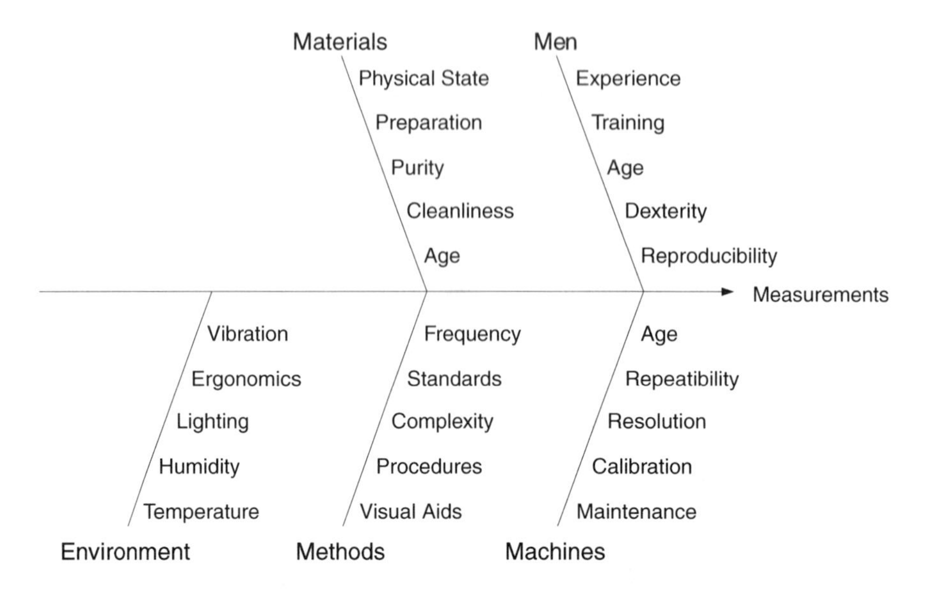

FIGURE 9.3 A fishbone diagram showing many potential causes of variability in the measurement system.

a simple quality tool such as the fishbone diagram in Figure 9.3, we can gain some insight into some of these possible contributors.

This list may differ for the inspection or test processes for electronics, chemicals, or machining, but the idea of many subtle variables acting on the measurement process is the same. And just as in manufacturing processes, the source and magnitude of this variation will be a mystery unless we undertake a measurement system analysis.

In order to organize this analysis, we will classify the variation into its various sources. Figure 9.4 breaks the variation in measured values into two basic components: actual variation in the product values and the additional variation introduced by the measurement system. The measurement system variation is then further broken down into its two basic components: variation of reproducibility and variation of repeatability.

9.2.6 Variable Data — Types of Variation for Measurement Systems

Analysis techniques will separate the components of variation into two types for the measurement system: repeatability and reproducibility.

Repeatability is an evaluation of the variation in trying to repeat a measure. Using the same gauge, the same inspector, examining the same characteristic on the same product, do we get the same value? Interestingly, the answer is "often not." Figures 9.5 and 9.6 depict this attribute. The real question we have to ask is, "How much of the total variation is due to this particular aspect and how significant is it?"

Reproducibility is an evaluation of the variation with multiple inspectors measuring the same parts. Using the same gauge, if multiple inspectors measure the

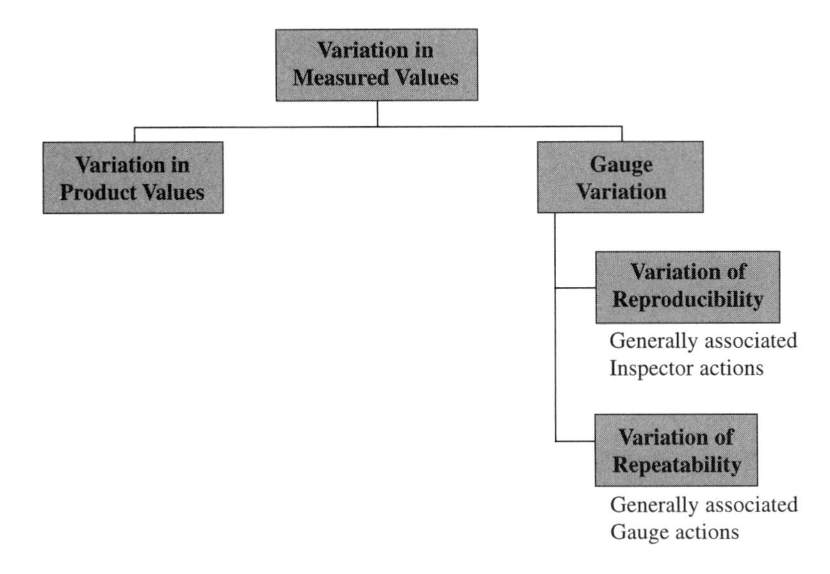

FIGURE 9.4 The elements of variation in measurements. Measured variation in parts is composed of both variation in the parts and variation introduced by the gauge.

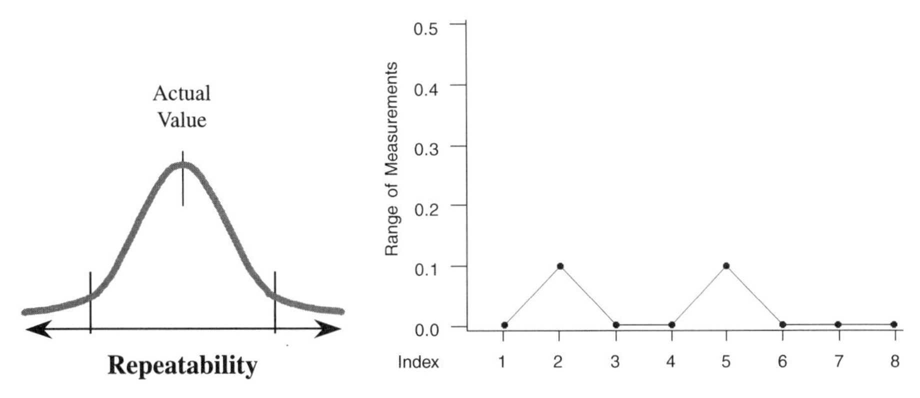

FIGURE 9.5 Repeatability for variable data. Same inspector with same gauge gets a range of different measures on the same part with repeat measures.

FIGURE 9.6 Repeatability data. Each plot shows the range of multiple measures of eight different parts. On six parts the inspector always got the same value, so there was no difference. On two parts, the inspector got at most a difference of 0.1 in the values for these parts.

same characteristic on the same products, do we get the same value? Figures 9.7 and 9.8 depict this attribute. Again, the real question is, "How much of the total variation is due to this particular aspect and how significant is it?"

9.2.7 Attribute Data — Types of Variation for Measured Systems

As with variable data, attribute data have the two components of measurement variation or differences: repeatability and reproducibility.

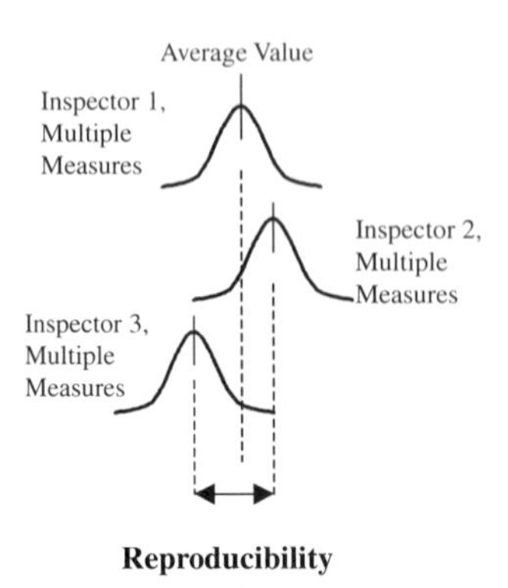

FIGURE 9.7 Reproducibility for variable data. Different inspectors whose averages of multiple measures using the same gauge on the same part get different values.

Average Values for Inspectors
45
44
43
42
41
40
Index 1 2 3 4 5 6 7 8
Inspector 3
Inspector 1
Inspector 2

FIGURE 9.8 Reproducibility data. Each plot shows the average of multiple measures on eight different parts by three different inspectors. The range of high to low on each part is about the same for the eight parts. Inspector 3 always seems to have the highest value.

TABLE 9.1
Repeatability for Attribute Data

Product Knowledge		Inspector Appraising		
Part #	**Attribute**	**1st**	**2nd**	**3rd**
1	Good	Good	Good	Good
2	Good	Good	Good	Good
3	Bad	Bad	Bad	Bad
4	Bad	Good	Good	Good
5	Bad	Bad	Good	Bad
6	Good	Good	Bad	Bad

Repeatability is an evaluation of the attribute inspection system to get the same result each time it inspects the same characteristic on a product. Note that this does not mean the inspection result is correct, only that it consistently gets the same result each time. A matrix depicting this attribute is shown in Table 9.1.

For the first three parts, the inspector repeats the inspection result and identifies the quality of the attribute correctly. For the fourth part, the inspector has good repeatability, but incorrectly identifies the quality characteristic. For the last two parts, the inspector has poor repeatability, varying between the classifications for this characteristic.

Reproducibility is an evaluation of whether an attribute system can have multiple inspectors arrive at the same results for the same characteristic on a product. A matrix depicting this attribute is shown in Table 9.2.

TABLE 9.2
Reproducibility for Attribute Data

Product Knowledge		Inspector #1 Appraising		Inspector #2 Appraising		Inspector #3 Appraising	
Part #	Attribute	1st	2nd	1st	2nd	1st	2nd
1	Good	Good	Good	Good	Good	Good	Good
2	Good	Good	Good	Good	Good	Good	Good
3	Bad	Bad	Bad	Bad	Bad	Bad	Bad
4	Bad	Good	Good	Good	Good	Good	Good
5	Bad	Bad	Good	Bad	Bad	Good	Bad
6	Good	Good	Bad	Good	Good	Good	Bad

For the first three parts in this example, the inspectors each match the others and correctly identify the quality of the attribute. For the fourth part, each of the inspectors matches the others, demonstrating good reproducibility, but incorrectly identifying the quality of the characteristic. For the last two parts, inspectors fail to consistently match each other's appraisal as well as their own, demonstrating poor reproducibility.

9.3 PERFORMING A MEASUREMENT SYSTEM ANALYSIS

9.3.1 Plan the Analysis

The analysis of a measurement system is based on statistical methods that require a structured approach to gather and analyze the data. Analysis of variance is the principle method used to separate sources of variation into the various components. Similar to design of experiments (DOE), it requires strict control and observation of the test environment. Improper selection of the data, methods used, or recording and analyzing the data will result in an ineffective analysis, not to mention the loss of time and resources used to conduct the test. The proper preparation for each of the types of data is discussed, but executing the analysis effectively requires a concerted effort on the part of management and personnel who perform the analysis. Inattention to detail can result in the loss of a valuable lesson on the capability of the gauge — and ways to improve its effectiveness.

9.3.2 Which Inspection Processes to Analyze

Measurement system analysis probably cannot be performed on all the inspection points in a manufacturing enterprise. It is both costly and time consuming to perform. In addition, analysis of one inspection system may provide insight that other similar inspection systems may perform as well or as poorly. Initially, critical inspection points, inspection points with high variability and poor quality, measurement systems that are perceived as having suspect capability, inspections costly to perform, and inspection equipment that is expensive to purchase should top the list of your efforts.

You must ensure that these systems are providing the right data and feedback to manage and analyze your production environment. Conversely, if an inspection point is not worth analyzing, is it truly worth the time you are spending to perform the inspections? Ultimately, experience with performing the analysis and your own knowledge of the manufacturing enterprise will probably be the best guide in determining the priorities for this insightful analysis.

9.3.3 Variable Measurement System Analysis — Preparation

The following list provides the basic guidelines for preparing for the analysis. Issues are noted for each of the steps to give you insight about the foundation of the step.

1. Select parts whose characteristics cover the range of the perceived variability of the process.
   - The analysis is a study of the relative amount of variability attributed to parts and gauges. If there is little or no variability in the parts, the relative variability of the gauge will be artificially high.
2. Make sure your gauge meets the basic requirements of resolution and bias.
   - If the gauge has insufficient resolution to measure variability, then the analysis won't work. If your gauge is out of calibration, it may or may not have a significant effect on the analysis; don't take that chance.
3. Collect ten parts to perform the analysis.
   - Analysis of variance, like all statistical techniques, requires a minimum of data to effectively differentiate the component differences. In some cases you may not have that many parts available for an analysis. It may be possible to evaluate different areas on the same part to represent different parts — providing they still reflect the perceived variation in the process. If I did not have ten parts to analyze, I would perform the analysis with what product I did have, review the results, and decide where to go from there, rather that not do the analysis. The alternative is to wait to perform the analysis, and continue to lack insight into your gauge capability.
4. Set up a data collection sheet ahead of time.
   - Analysis takes time and resources away from making product. Do not wait until the start of the test to decide how to collect data. This should be a natural byproduct of deciding which parts, which inspectors, and how many trials you are going to perform. If you do not have a data collection sheet, you probably have not done the necessary background work.
5. Identify the inspectors to perform the measurements.
   - These inspectors should be individuals familiar with the gauge, the parts, and the procedures. The best choice would be people who use this gauge daily. The analysis separates sources of variation. You do not want a major source of variation to be associated with reproducibility only because one of the operators didn't know how to properly operate the gauge.

6. The analyst should run the data collection process and note any observations.
   - Statistical principles appropriately call for blind sampling and randomization of the order of part inspection. But, if your analyst doesn't observe the process, he or she may be missing out on valuable information that will not show in the data. Subtle differences in the methods of the different inspectors may provide insight into improving the gauge performance.
7. Have the first inspector measure all the parts in random order.
   - This mitigates human bias in the sample. The operator should not know the identity of the parts, only the analyst should.
8. Have the next inspector measure all the parts in random order. Continue this process with the remaining inspectors until they have each measured all the parts once.
9. Go back two steps and repeat the inspection of the parts for the planned number of repeat inspections.
   - As a minimum, two complete inspections of all the parts must be performed by each of the inspectors. This provides the basis for the repeatability analysis. The more repeat inspections, the better this analysis will be. Again, time and money are constraints.
10. Analyze the data, review with the team, and take any required corrective action.
    - If the measurement system performs well, the team should know this. If the measurement system performs poorly, the analysis may point you in the direction of the cause, and the team may be able to pinpoint these causes and provide possible corrective action.

9.3.4 Variable Measurement System Analysis — Analysis

Analysis of the data is a rigorous statistical technique and should be performed by a competent analyst with the appropriate tools. It is not the intent of this chapter to train the analyst in these techniques. Various types of outputs are looked at and their potential implications examined, real examples are given and discussed, and in the next section sources for the analysis and software are presented.

In analysis, we want to identify the components of measurement variation and their percents of contribution to the total variation. Our desire is to have the amount of variation introduced by the measurement process be minimal. In Figure 9.9, the "% Study Var" gives us that insight. As a general rule of thumb, we would initially consider the guidelines for these values as shown in Table 9.3.

The tabular output in Figure 9.9 gives a representation of what an analysis might look like. First examining Total Gauge R&R, at the bottom of the table, we see that this value is 71.54%. Remember, the analysis is a comparative one. If there is minimal variability in the parts, then the gauge will have a large relative percent of the variation. But if this low variability in the parts is a result of high process capability relative to the specification, then it may not be necessary to measure the process at all. The suggestion is to not automatically conclude that the gauge must be replaced if its contribution to variability is over 30%. Alternatively, the components of gauge variability, repeatability and reproducibility, may give direct insights into how the measurement

ANOVA Table

Source	DF	SS	MS	F	P
Parts	9	5.14E-06	5.71E-07	40.492	0
Operator	2	3.36E-06	1.68E-06	118.984	0
Repeatability	48	6.77E-07	1.41E-08		
Total	59	9.17E-06			

Gauge R&R

Source	VarComp	StdDev	5.15*Sigma
Total Gauge R&R	9.73E-08	3.12E-04	1.61E-03
Repeatability	1.41E-08	1.19E-04	6.12E-04
Reproducibility	8.32E-08	2.88E-04	1.49E-03
Operator	8.32E-08	2.88E-04	1.49E-03
Part-To-Part	9.28E-08	3.05E-04	1.57E-03
Total Variation	1.90E-07	4.36E-04	2.25E-03

Source	%Contribution	%Study Var
Total Gauge R&R	**51.18**	**71.54**
Repeatability	**7.42**	**27.24**
Reproducibility	**43.76**	**66.15**
Operator	**43.76**	**66.15**
Part-To-Part	**48.82**	**69.87**
Total Variation	**100.00**	**100.00**

FIGURE 9.9 An analysis of variation (ANOVA) breaks down the components of the variation.

TABLE 9.3
Guidelines for Evaluating Gauge R&R

% Study Variation	Rule of Thumb	Other Considerations
<10%	Gauge is acceptable for analysis and use.	
10–30%	Gauge is marginal.	Consider costs for improving the gauge's performance or upgrading gauge. Review importance of application. Review process capability.
>30%	Generally not acceptable.	Review process capability. Are measurements necessary?

might best be improved. In this example, the repeatability value is 27.24%, while the reproducibility value is 66.15%. If reproducibility values dominate the gauge variability, components improvement in inspector performance through training might significantly improve the overall measurement system results.

Graphical results may be easier to understand and may provide greater insight into the causes of measurement system variability. Figure 9.10 shows these same tabular values displayed in a simple bar chart. The total Gauge R&R is displayed at the left, showing the high %Study Var. The next two sets of bars show the

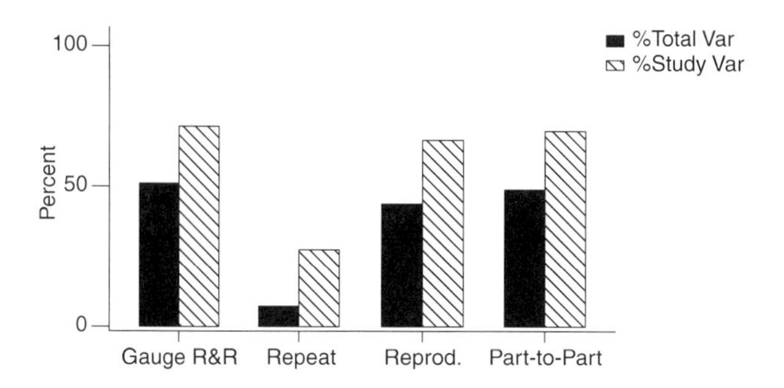

FIGURE 9.10 Components of variation. A graph of the results from the ANOVA table in Figure 9.9.

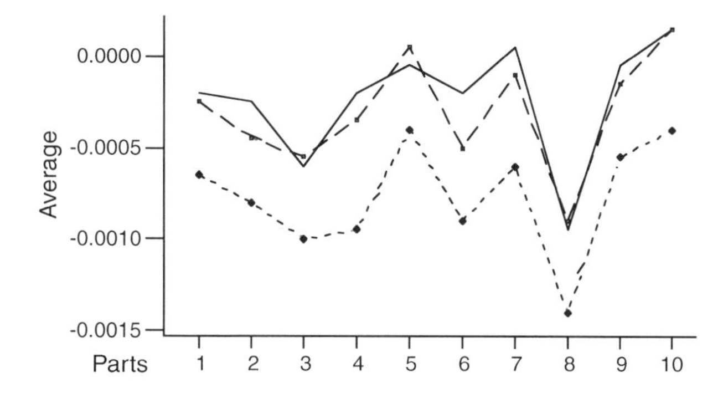

FIGURE 9.11 Gauge R&R: reproducibility. Operator 1, solid line; operator 2, dashed line; operator 3, dotted line.

components of the gauge variability, repeatability and reproducibility. Figure 9.11 shows a more detailed representation of the data of reproducibility. The points on the chart show the average value for the multiple measures of each of ten parts by the various operators performing the inspection. The range for reproducibility on each part remains fairly consistent across all ten parts, with operator 3, the lowest line, consistently getting lower average values for all parts. There appears to be a systemic cause of the poor reproducibility values. Initially one might conclude that operator 3 needs training to better "reproduce" results similar to the other operators. In fact, when observing the measurements by each operator, each of the first two placed the part into the gauge while pushing the part against a hard stop to gain greater consistency in measures. Operator 3 let the part float in the gauge. When questioned about this method, the operator replied that the measurement device should be robust to the process; they shouldn't have to try to overcome ineffectiveness in the gauge. It now *seemed* clear that changes to operator 3's methods would improve our overall gauge performance.

Now let's look at the other component of gauge variability, repeatability, shown in Figure 9.12. The chart shows each operator performing the inspection sequentially, with operator 1 shown in the first ten measures, then operator 2 and operator 3. The R chart

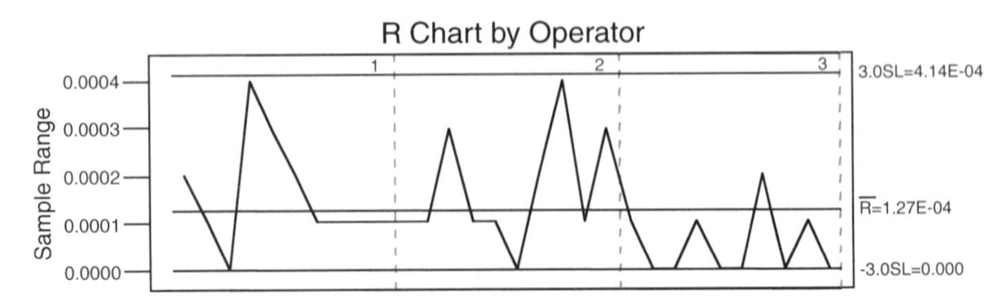

FIGURE 9.12 Gauge R&R: repeatability. Three operators each measuring 10 parts. The chart shows the range of measures for each part.

is a range chart showing the difference between high and low values for repeated measures on the same part. It is clear from this chart that operator 3 demonstrated better repeatability, since on 6 of 10 parts the part measures were identical. Maybe we'd better wait on retraining operator 3! Further discussion with each of the operators and quality personnel led to the conclusion that letting the part float in the inspection device allowed the parts to consistently find the same location in the gauge and thus have greater repeatability. Instead of retraining operator 3, the other operators' methods were changed, resulting in significant improvements in both gauge reproducibility and repeatability. Given the magnitude of the percentage contribution to variation by the gauge, both these improvements were necessary to achieve a viable gauge.

Some generalized rules for the interpretation of these sets of charts are as follows:

- If repeatability is the significant cause of poor gauge capability, and it is consistent across all operators, then the gauge may be the source of the problem.
- If reproducibility is the significant cause of poor gauge capability, with consistent separation of one or more of the operator data sets, then training may be the solution to improving gauge performance.

As seen, you need to take great care in examining all aspects of the analysis, planning and observing the actual measurements, and reviewing the results with all members of the team. It is interesting to note that this solution *would not have been found* if I had instructed operator 3 to perform the measurements in the manner that the other operators perceived was the correct procedure. Using operators or inspectors who actually perform these measures and allowing them to use their current methods and procedures are integral to an effective analysis.

9.3.5 Variable Measurement System Analysis — A Correction Technique

There is a short-term solution to measurement systems that show excessive variation. The measurement system must meet the other requirements of resolution and bias, then, through a signal averaging technique, the amount of variation contributed by the gauge to the resulting measurements can be greatly reduced. This approach is good for pressing immediate needs of delivering hardware and quick studies to improve a process, but is costly to use on a recurring basis for applications such as statistical process control. The technique requires you to take multiple measures of the same characteristic and average them. The contribution of random variation by the gauge will be reduced by the reciprocal of the square root of the number of times you repeat inspection of the part.

$$\text{Reduction in Variation} = \frac{1}{\sqrt{n}} \tag{9.3}$$

where n is the number of inspections performed on the characteristic. If you had four (4) repeated inspections, the gauge error would be reduced by half. Obviously, this would be an expensive approach to quality control and verification.

9.3.6 Attribute Measurement System Analysis — Preparation

The following list provides the basic guidelines for preparing for the analysis. Issues are noted for each of the steps to give you insight about the foundation of the step.

1. Select parts whose characteristics cover the range of the perceived variability of the process. Make sure that almost half are defective.
   - The analysis is a study of the inspectors' ability to detect good from bad and the consistency among inspectors. If all parts are obviously good or bad, the analysis does not represent real conditions and your results will be skewed.
2. Determine the attribute of the characteristic before the start of the trials.
   - One of the comparisons is the determination by the operators against the "known" condition of the characteristic. It may be that employing more effective visual systems, such as higher power microscopes, will be used to make this determination. Another way is to use the most experienced inspector, or group of inspectors, to determine the true condition of the characteristic.
3. Make sure you are using the right inspection equipment.
   - If inspection aids are used, such as microscopes or magnification glasses, ensure that these are at the right power of magnification. Each inspector should perform the appraisals under the same conditions.
4. Collect at least 30 parts to perform the analysis.
   - Attribute analysis, wherein the resulting data are either good or bad, does not have the discriminating potential of measured data with a numeric range of results. Therefore, the analytical techniques are not as powerful and require more data.
5. Set up a data collection sheet ahead of time.
   - This analysis takes time and resources away from making product. Do not wait until the start of the test to decide how to collect data. This should be a natural byproduct of deciding which parts, which inspectors, and how many trials you are going to perform. If you do not have a data collection sheet, you probably have not done the necessary background work.

6. Identify the inspectors to perform the measurements.
   - These inspectors should be individuals who are qualified to perform the inspection, are familiar with the parts and the procedures, and preferably routinely perform this inspection.
7. The analyst should run the data collection process and note any observations.
   - Statistical principles appropriately call for blind sampling and randomization of the order of part inspection. But if your analyst doesn't observe the process, he or she may be missing out on valuable information that does not show in the data. Subtle differences in the methods of the different inspectors may provide insight into improving the inspection process.
8. Have the first inspector measure all the parts in random order.
   - This mitigates human bias in the sample. The operator should not know the identity of the parts; only the analyst should.
9. Have the next inspector assess all the parts in random order. Continue this process with the remaining inspectors until they have assessed all the parts once.
10. Go back two steps and repeat the inspection of the parts for the planned number of repeat inspections.
    - At a minimum, two complete inspections of all the parts must be performed by each of the inspectors. This provides the basis for the repeatability analysis. The more repeat inspections, the better this analysis will be. Again, time and money are constraints.
11. Analyze the data, review with the team, and take any required corrective action.
    - If the measurement system performs well, the team should know this. If the measurement system performs poorly, the analysis may point you in the direction of the cause(s), and the team may be able to pinpoint these causes and provide possible corrective action.

9.3.7 Attribute Measurement System Analysis — Analysis

Analysis of an attribute inspection system is at least initially more straightforward and less technically demanding than the variable gauge. Competent analysts can extend the initial set of values and percentages to include confidence intervals, along with tests for statistical differences. But first, the assessment should look for basic information to understand the status of the system and potentially take preliminary actions.

Figure 9.13 gives a representation of what a tabular output of the analysis might look like. Remember that in the analysis we want to identify the components of measurement variation. Our desire is to have the amount of variation introduced by the measurement process be minimal. We want 100% agreement within inspectors, between inspectors, and with the known attribute condition.

The condition of repeatability, wherein inspectors agree with themselves, is shown in the table as values of 90.9, 100, and 63.6%. Inspector 3 is inconsistent in his or her assignment of values and either training or other improvements in inspection aids may be necessary.

Product Knowledge		Inspector #1 Appraising		Inspector #2 Appraising		Inspector #3 Appraising		Reproducibility Inspectors Agree	Inspectors and Attribute Agree
Part #	Attribute	1st	2nd	1st	2nd	1st	2nd		
1	Good	Good	Good	Good	Good	Good	Good	Y	Y
2	Good	Good	Good	Good	Good	Good	Good	Y	Y
3	Bad	Bad	Bad	Bad	Bad	Bad	Bad	Y	Y
4	Bad	Good	Good	Good	Good	Good	Good	Y	Y
5	Bad	Bad	Good	Bad	Bad	Good	Bad	N	N
6	Good	Bad	Bad	Good	Good	Good	Bad	N	N
7	Good	Good	Good	Good	Good	Good	Good	Y	Y
8	Bad	Bad	Bad	Bad	Bad	Bad	Good	N	N
9	Bad	Bad	Bad	Bad	Bad	Bad	Bad	Y	Y
10	Bad	Bad	Bad	Bad	Bad	Good	Good	N	N
11	Good	Bad	Bad	Good	Good	Bad	Good	N	N
								54.5%	45.5%

Repeatability	Inspector #1	Inspector #2	Inspector #3
Inspector Agrees with Self % =>	90.9%	100%	63.6%
Inspector to Attribute % =>	63.6%	90.9%	45.5%
False Positives =>	1	1	2
False Negatives =>	2	0	0
Mixed Results =>	1	0	4

FIGURE 9.13 Attribute gauge data. Ten parts are reviewed by different inspectors for consistency to the actual attributes, to themselves (repeatability), and to each other (reproducibility).

The condition of reproducibility, wherein inspectors agree with themselves and each other, shows a score of 54.5% in the table. A score with this much disagreement tends to suggest that both improvements in inspection aids and team training may be necessary to gain consistency in the appraisals.

An inspector coming up with numerous false positives wherein bad parts are accepted or, conversely, numerous false negatives wherein good parts are rejected, indicates an individual training need. If multiple inspectors falsely classify the same part, it may be an indication that the defect is not well defined in the training or with the inspection aids. The team should review these details. As with the variable measurement system, sharing the results with the team and the inspectors will probably lead to the best conclusions and recommendations for improving your measurement system.

As is true of the variable measurement system, inspection results can be improved through multiple inspections of the same characteristic. It is important that the inspectors score high in the percentage agreement with the attribute to effectively screen out the good parts from the bad. An inspector having 90% screen effectiveness would require at best two sets of inspections to achieve 99% effectiveness. However, an inspector with 70% screen effectiveness may require four sets of inspections to achieve 99% screen effectiveness. This assumes no systemic failure modes in each of these inspectors' abilities. Again, time and cost become important considerations in multiple inspections of the same attributes. Improving your inspection effectiveness through training, process improvements, and better aids is the more desirable approach.

9.3.8 A Case History

The following actual case history demonstrates a measurement system analysis. The analysis was performed in preparation for a DOE study to improve the process capability of putting composite material on a cylinder. The standard method of process measurement under current use was a Pi Tape® to measure composite buildup on the base mandrel. The real quality characteristic was the percentage of resin by weight in the composite. Previous studies had shown a correlation of diameter growth with resin content. Pi Tapes go around the circumference of the material and translate this directly into a calculation of the growing diameter of the part. This easy and quick measurement tool was employed as a manufacturing aid. Discussions were underway on whether to use the tool in a more formal statistical process control application. The variable measurement system analysis charts are shown in Figure 9.14.

The graphical output uses Minitab software to produce this Gauge R&R Sixpack analysis.

The top left chart in Figure 9.14 shows sufficient resolution. The bottom left chart shows the contribution of the gauge to the total variation as about 80%, the largest contributor being reproducibility! The middle left chart shows the repeatability data, with the possibility of improvement due to the one large spread value for operator 3. But because the larger component of gauge variability was due to reproducibility, the team needed to focus on dramatic improvement in this aspect. After looking at the varying nature of the data and having discussions with the team

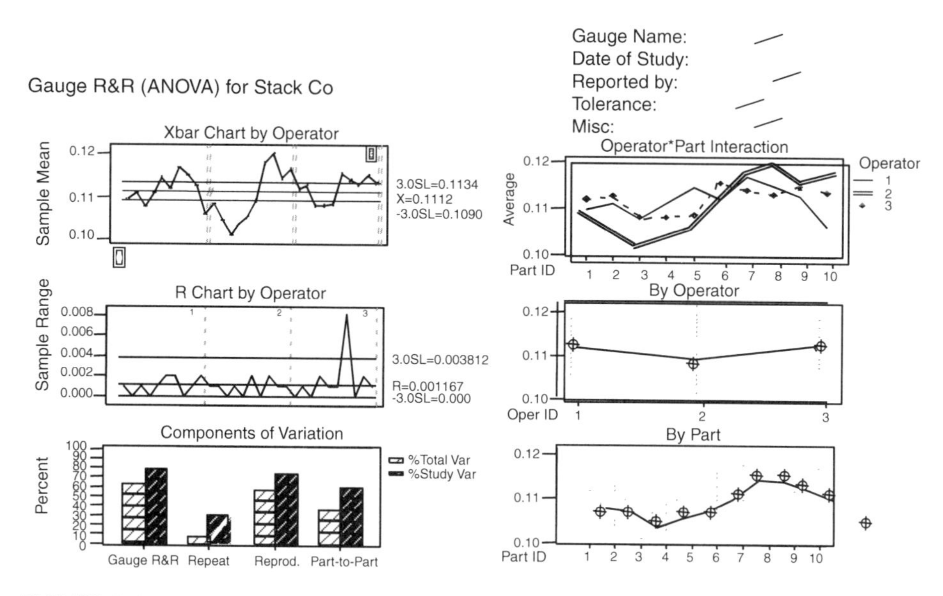

FIGURE 9.14 Measurement system analysis on the Pi Tape for composite build up.

we still did not have a clear solution for improving the performance of the gauge. A more costly and time-consuming approach was to take samples from the process and send them to our quality laboratory for chemical analysis. Days to weeks would be involved in processing the material, along with the expense of laboratory labor and materials. The team finally came up with the approach of weighing a measured section of composite material removed from the mandrel. The weight of the fiber material could be determined ahead of time, as could changes in weight due to added resin. Before proceeding with this approach, we had to show a strong correlation with a more precise chemical analysis in the laboratory. Once this was done, we had a quick, though not as easy, method for measuring changes in percentage resin content. We could now proceed with the DOE study.

There were multiple benefits from this gauge study. First, we avoided performing a DOE using a gauge that was incapable of measuring variation in the product. The results would have been spurious at best, and may literally have misled our efforts to improve the process. The team discussions led to a more effective way of measuring the critical quality characteristic, and this method allowed for successful improvement in the manufacturing process. Additionally, the analysis made clear that statistical process control using the Pi Tape was not feasible. I can only guess where we would be in our efforts if we had not performed this measurement system analysis.

9.4 THE SKILLS AND RESOURCES TO DO THE ANALYSIS

9.4.1 Technical Skills

The technical skills required on the part of the analyst vary greatly with the methods used and the complexity of the study. For applications of variable data systems, at a minimum the analyst should be familiar with the basic statistical elements of mean and variation. If the analyst has software that is simple and straightforward to use in performing the analyses, then he or she should be able to competently get underway with minimal formal training in the foundations of measurement system analysis. Experience is the best teacher, because conditions and solutions vary dramatically. Inclusion of the team in all aspects of the study and potential solutions helps greatly.

For those analysts who perform their own calculations or rely on reference manuals such as those published by the auto manufacturers, a much more extensive statistical analysis background is required, one that includes substantial formal training. Most applicants for variable data systems would benefit from additional training in calculations of statistical process control charts and the various tables that support these calculations. Sorting your way through the various analytical options can be daunting, so give your analyst the time and training necessary to gain confidence in the analytical tools. It is a good idea to have expert technical support available to answer some of the unusual technical issues the analyst will undoubtedly encounter.

9.4.2 Measurement System Analysis Software

Numerous software tools are available to perform analysis on both types of data. The first recommendation would be to examine current in-house production control and quality-control software systems for existing or add-on measurement system analysis capabilities. Current familiarity with these systems will help to speed the process of getting started. The second suggestion would be to seek out combined measurement system analysis training and software programs. In this way, the full features of the software and applications are integrated into the training program on methods of analysis.

Next, I would suggest talking with other organizations that produce similar products about the type of software they employ for their analyses. Involvement with professional and technical organizations helps to facilitate these types of dialogues. Researching trade journals and Internet searches will provide some information and methods of contact for further information on various companies. My own explorations have found the American Society for Quality (ASQ) *Quality Progress* publication to be a good source for identifying various software companies. In particular, pages 104–114 of the June 2000 publication have an extensive list of software providers, along with identifying which quality tools their systems provide. The telephone number and Web site for ASQ are listed below, along with various statistical software systems I have used and found to be effective. I strongly suggest you do your own research, because the types of applications, your skills, and even the frequency of these types of applications enter into determining the software of choice.

ASQ	414-272-8575	<http://www.asq.org>
Minitab, Inc.	814-238-3280	<http://www.minitab.com/>
SAS Institute, Inc.	919-677-8000	<http://www.JMPdiscovery.com>
Intercim, Inc.	512-458-1112	<http://www.intercim.com>

REFERENCE

Measurement System Analysis, Reference Manual, 1995 Chrysler Corporation, Ford Motor Company, General Motors Corporation.

JOURNAL

Quality Progress, American Society of Quality Control.

GLOSSARY OF TERMS

Attribute data: Data that represents the quality of an attribute as good/bad, pass/fail.

Bias: The difference between the average value for a set of measurements on a specific characteristic and the actual or master value for that characteristic.

Common causes: Sources of variation that are inherent in a process, sometimes called noise.

Discrimination: The ability of the measurement system to detect changes, or variation, in the specific characteristic being measured. Also called resolution.

Gauge R&R: Gauge repeatability and reproducibility are the two components of variation associated with the gauge measurement system.

Linearity: Bias in the measurement system that reflects differences over the operating range of the gauge or length of the part.

Measurement system: The entire process used in collecting data on a characteristic of a part. These include the procedures, gauges, software, personnel, and documentation used in the process.

Part screening: Inspecting all the parts from a process, typically with an attribute measurement system.

Part-to-part variation: Variation occurring in the same characteristic on different parts.

Properties of measurement systems: The two statistical properties of measurement systems are bias and variance, sometimes referred to as accuracy and precision.

Repeatability — attribute data: Variation or differences in inspection results when the same inspector appraises the same characteristic on a product more than once.

Repeatability — variable data: Variation in measurements that occur when repeatedly measuring a specific characteristic of a part with a gauge used by one operator.

Reproducibility — attribute data: Variation or differences in inspection results when multiple inspectors appraise the same characteristic on a product.

Reproducibility — variable data: Variation occurring in the averages of the measurements by multiple operators measuring the same characteristic of a part with the same gauge.

Screen effectiveness: The ability of the screen to correctly identify parts as good or bad. Usually noted as a percentage of being able to catch defects correctly.

Special causes: Sources of variation that arise due to special circumstances that can otherwise be controlled and are not inherent to the process. These are also referred to as assignable causes.

Stability: Variation in the measurement system over time measuring a specific characteristic on a master part or the same group of parts.

Variable data: Data that results from measurement and is quantified numerically.

Variance: The statistical term for the spread in the data for a set of measurements on a specific characteristic. Gauge R&R studies focus on identifying the different components causing the spread or variation in the measurement process. The two principal components are part-to-part variation and variation introduced by the measurement system.

10 Process Analysis

Jack B. ReVelle, Ph.D.

10.1 DEFINITIONS

- Activity. A measurable happening that occurs over time.
- Annotation. The process of assigning specific codes or symbols on a process flow chart or process map so as to identify the specific location where defects or errors are created, where excessive cycle time is consumed, where cycle time is most unpredictable, or where unacceptable costs are generated.
- "As-Is" condition. The way a process or system actually functions or operates without regard to whether it is efficient, effective, or competitive.
- Event. A nonmeasurable happening that occurs at a specific time, e.g., the start or finish of an activity.
- Parallel Events. Two or more events that take place simultaneously, i.e., concurrently.
- Parking Lot. A place or location where ideas, concepts, and suggestions for process improvement are recorded when they are conceived for easy reference at a later time, e.g., a white board or easel paper.
- Predecessor Event. An event that must take place prior to the start of a specific event.
- Process. A series of sequentially oriented, repeatable events having both a beginning and an end and which results in either a product (tangible) or a service (intangible).
- Process Analysis. Examination of a process using tools or methods such as process flow charts, process maps, and annotation. The purposes of a process analysis are to expand the process stakeholders' understanding of the entire process from suppliers to customers, including the critical linkages between the quality requirements and performance metrics of both inputs and outputs, and of the ways in which the voice of the customer drives the process.
- Process Analysis and Improvement Network (PAIN). An integrated collection of process flow charts designed to facilitate understanding and enhancement of existing processes, both production and transactional.
- Process Flow Chart. A one-dimensional collection of geometric figures connected by arrows to graphically describe the sequential occurrence and interrelationships of events in a process.
- Process Improvement. Enhancement of an existing process by slightly improving various phases or by redesigning all or most phases.

1-57444-300-3/02/$0.00+$1.50

- Process Map. A two-dimensional version of a process flow chart that also portrays handoffs and receipts of products or services from one person, organization, or location to another.
- Series Events. Two or more events that take place sequentially, i.e., one following or preceding another.
- "Should-Be" condition. The way a process or system should function to be most efficient, effective, or competitive.
- Successor Event. An event that must take place following the finish of a specific event.
- System. A collection of processes, arranged in series or parallel, that has a common beginning and a common end, and which together constitute a program, a project, or an entire organization.

10.2 PROCESS ANALYSIS

10.2.1 PROCESS

What is a process? A process is a series of sequentially oriented, repeatable events having both a beginning and an end and which results in either a product or a service. A product, of course, is something tangible, something you *can* see, taste, or touch. A service is something intangible, something that you *can't* see, taste, or touch, but which you know you've received. For example, delivery of training is a service.

10.2.2 SYSTEM

Well, if that is a process, then what is a system? A system is a collection of processes arranged in series or parallel, and which together constitute a program, a project, or an entire organization. A company, large, medium, or small, is an example of an entire organization. An initiative might be a project such as the initial use of some new software. A program could be an ongoing activity that is done periodically. In any case, whether it is a program, a project, or an entire enterprise, it's a collection of processes.

10.2.3 PROCESS FLOW CHART

Having defined for baseline purposes what a process and a system are, now let's review what we can do to better understand these processes, these basic elements or components of an organization. There are a number of different ways we can analyze a process. The most common and one of the most useful forms is a graphic tool known as a *process flow chart.* This chart is a series of geometric figures — rectangles, diamonds, and circles or various other shapes — arranged typically from left to right, and from top to bottom, connected by lines with arrowheads to show the flow of activity from the beginning to the end of the process.

When a process is being created or an existing process is being analyzed, it is useful to create a process flow chart so that everyone involved, that is, all the stakeholders in the process, can see exactly what is supposed to happen from beginning to end without having to try to imagine it. Each of us may have a picture in our own mind, a graphical portrayal of what the process flow looks like, but the reality may be different. The only way we can be sure we understand that we have a common perspective or outlook on the process is by graphing it as a process flow chart, a linear or one-dimensional process flow chart. I say one dimensional to distinguish it from the two-dimensional graphic that we are going to talk about shortly, known as a process map.

Let's talk about the creation of the process flow chart. Traditionally, people have created process flow charts from the first step to the last. I don't, and the reason is that, when people put this flow chart together, they are looking at processes in the same way they look at them every day, so there is a high potential for missing something. What I suggest people do as we bring them together in a room to create a process flow chart is to start with the end in mind, a concept understood by everyone familiar with Stephen Covey's *The 7 Habits of Highly Effective People.*

We begin by defining the last step or the output of the process and then start asking the question sequentially, "What has to happen just before that?" If we know we have a specific output or step, we ask what must be the predecessor event or events that must take place to satisfy all the needs so that the step we are looking at can take place. So we work backward from the last step to the first step and keep going until someone says, "That's where this whole thing begins." Now we have defined, from the end to the beginning, the process, graphed as a process flow chart.

Some people might question why you want to do it that way. The analogy I use that is very effective is this: suppose I were to ask you to recite the alphabet. You would say A, B, C, D, E, F, G ... without thinking, because you have done it hundreds, perhaps thousands, of times.

But if I were to ask you to recite the alphabet backward, you would probably say Z and have to stop and think what happens before that, what letter precedes Z. What most people do, I have discovered, is first to do it forward to find out what the letter is and then come back and say that the letter before Z is this, and the letter before that is this, and so on. Working the alphabet backward makes people look at it in a way they have never looked at it before, noticing the interrelationships between the predecessor and the successor events.

The same psychology of working backward applies in dealing with our processes, whether we are dealing with a process of building a home, working with accounts payable, developing a flow chart, understanding a process as it relates to training, or whatever the case may be. Establishing the process flow chart from the last step to the first step is a very strong and powerful way to help people understand what their processes really look like.

10.2.4 Process Map

Once the process flow chart has been created and everyone is satisfied that it truly reflects the order in which the events take place with regard to predecessor and successor events, the next step is to create a process map. Earlier I said a process map is created in two dimensions. We are going to use exactly the same steps we used in the process flow chart, except now, instead of just having the flow go from left to right, we take the people, positions, departments, trades, or the functions that are involved in the process. and list them vertically down the left-hand side from top to bottom.

For example, it might be department A, B, or C; person X, Y, or Z; or trades such as concrete, plumbing, or framing. Then, we take the rectangles that we created in our process flow chart and associate them with the various functional areas, departments, persons, or trades listed on the left-hand side. What you see is a series of rectangles being built from left to right and also moving up and down the vertical axis we have created on the left-hand side of our process map. In so doing, we see what might look very much like a sawtooth effect with blocks going up, down, and across. Thus we end up with a view of the handoffs from one person to another, one function to another, or one trade to another, so we can see where queues are being built and where the potential for excess work in process is being created among the various areas of responsibility (listed down the left-hand side).

This gives us a very clear, visual picture of some of the things we might want to consider doing in terms of reordering the various steps to minimize the total number of handoffs that are a part of this process, recognizing that every time there is a handoff, there is a strong potential for an error, an oversight, something left out, a buildup of a queue, the creation of a bottleneck, or the like.

In creating our process map we gain tremendous insights into what we can do to continuously improve our processes. Remember, the order of the steps may have been absolutely vital at one time, but with changes in technology, people, and responsibilities, what we did then may no longer be valid, and we need to periodically assess or review our processes. The use of a process map is an excellent way to do that.

Now, in addition to looking at the process flow chart and process map in terms of the sequence and the handoffs, we can also use the process flow chart and the process map to assess cycle time and value-added vs. nonvalue-added events or steps in the process. The technique I use is to ask everyone in the room to assess the cycle time of the process that was just evaluated using a process map or process flow chart. Does it take 3 hours, 5 days, 10 weeks — whatever? When we get an agreement of 6 to 8 hours or 6 to 8 weeks — whatever the final range may be — we go through and evaluate each individual step, asking how long each step takes. When we have gone all the way through that, we arrive at the grand total of all the individual step estimates and compare that to the estimate that the group has already made of the overall process.

What we frequently find is that the sum of the individual steps is only 20 to 30% of the overall total. That quickly presents an image of a lot of lost and wasted time, costly time that could be used for other, important purposes. If, for example, a process is estimated to take 6 weeks, but the sum of the individual components takes a week and a half, it's obvious that we have some time we can save the company. Now, what needs to be done? Where are the barriers, the bottlenecks in the process that we can study, where can our trades (for example) share responsibility? Instead of having a particular trade come back three, four, or more times to do some little job that takes a half hour, an hour, another trade already on-site could be doing it for them. That is a very effective way of reducing cycle time. Steps can be eliminated and days upon days can be banked for use in more important projects.

10.3 PROCESS IMPROVEMENT

10.3.1 "As Is" vs. "Should Be"

Now let's look at the "as-is" vs. the "should-be" conditions. When we create the first process flow chart or process map of an existing process, we refer to that as the as-is process, i.e., the status of a process as it is currently operating. It gives us a baseline to create the new, revised process that we call the should-be process. Working together, the process improvement team is now able to view the as-is process in juxtaposition with the should-be process that they have created.

Subsequent to the creation of the should-be process map, the team begins to build a bridge from the as-is to the should-be process. The bridge is supported by a series of steps that we must go through to change the process from the as-is way to the way it should be.

A good example of that is the creation of some superhighways where conventional surface roads exist. During the building effort, traffic still has to flow, so as we move from the as-is surface streets to the should-be superhighway, we have to go through a series of steps, closing down and opening various components of the roads to support as much as possible the flow of traffic that never stops. Picture the Los Angeles freeway traffic any time of the day or night. This approach graphically illustrates what we need to do to move from the as-is process map to the should-be process map. These are things we might have otherwise overlooked.

10.3.2 Annotation

Using either a process flow chart or a process map, a process improvement team can easily identify specific locations within a process where events should be monitored to determine the extent of defects, errors, oversights, omissions, etc. Monitoring is usually accomplished using statistical control charts, e.g., X-bar and R, C, P, Np, U, and other charts. Chapter 15 on statistical process control (SPC) presents information on this topic.

Annotation is the development of a listing of defects and variances associated with the process being analyzed. Each known defect or variance is assigned a number by the team. Then the team annotates (assigns) each defect or variance to one or more events on the process flow chart or map. At this point the team evaluates the combined impact of the defects or variances at each event. Based on this evaluation, the team determines where SPC control charts should be physically located on the manufacturing floor, design center, or office. In addition, the team identifies which defects or variances should be counted (attribute/discrete data) or measured (continuous/variable data).

The combined effect referred to above is determined by the quantity of defect or variance identification numbers annotated at each event. Those events with the greatest incidence of identification numbers have a greater need for monitoring using SPC control charts than the events with few or no identification numbers. This is a simple application of the Pareto principle, also known as the 80-20 rule. In this case, 80% of the SPC control charts will be needed to monitor 20% of the process events.

The annotation methodology is also valuable in identifying where, within an as-is process, changes are needed in the creation of a should-be process.

10.4 PROCESS ANALYSIS AND IMPROVEMENT NETWORK (PAIN)

10.4.1 Reasons for PAIN

There are several reasons for using the Process Analysis and Improvement Network (PAIN). Whenever a process exhibits undesirable attributes, it is incumbent upon the process owner, process stakeholders, members of a process improvement team (PIT), or any other interested parties to take timely and appropriate corrective actions to eliminate or at least to reduce the presence or influence of the negative attributes. The most common of these negative attributes are

- Process too long (excessive cycle time)
- Process too inconsistent (excessive variation)
- Process too complex (excessive number of steps)
- Process too costly (excessive cost per cycle)
- Too many errors (poor quality — transactional process)
- Too many defects (poor quality — manufacturing process)
- Insufficient process documentation (for training or benchmarking)

10.4.2 PAIN — Main Model (Figure 10.1)

- Senior management identifies a process critical to success of the organization.
- Senior management establishes a team composed of the process owner, process stakeholders, and process subject-matter experts (SMEs).
- Convene the team with a facilitator experienced in process analysis and improvement.
- Have the facilitator provide a tutorial on the development of an as-is process flow chart.
- Start the development of the as-is process flow chart with identification of the final step in the process and then a backward pass through the process, finishing with its first step.
- Complete the development of the as-is process flow chart with at least two forward passes.
- With the assistance of its facilitator, the team should now convert the as-is process flow chart into its corresponding should-be process map.
- At this point, the process improvement team has a variety of options from which to select, depending upon its objectives. As noted above, there are a number of reasons for PAIN. The following models and discussions are offered to clarify the team's choices.
- When the team completes one or more of the following models, there are three steps remaining to complete the PAIN. These steps are spelled out in the final blocks of the PAIN — main model (Figure 10.1).

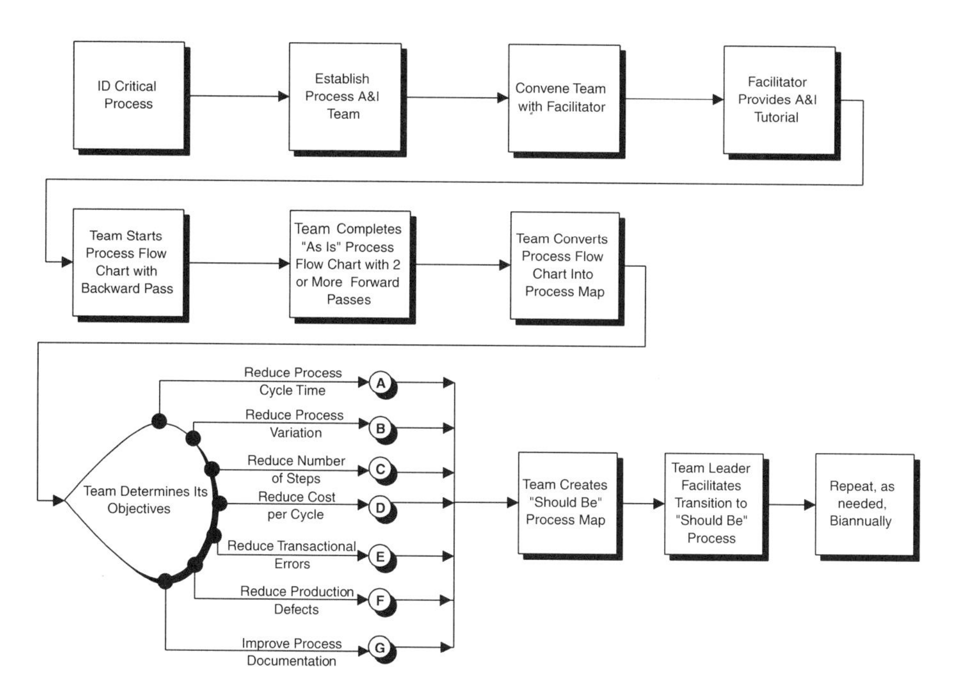

FIGURE 10.1 PAIN — main model. Process analysis and improvement navigator.

10.4.3 PAIN — Models A Through G

PAIN — Model A (Figure 10.2). The objective of this sequence of events is to reduce process cycle time. The process improvement tools, cause-and-effect analysis (also known as the fishbone diagram or the Ishikawa diagram), and force field analysis are explained in numerous books on continuous improvement.

PAIN — Model B (Figure 10.3). The objective of this sequence of events is to reduce process variation. The process improvement tools, cause-and-effect analysis (also known as the fishbone diagram or the Ishikawa diagram), and force field analysis are explained in numerous books on continuous improvement.

PAIN — Model C (Figure 10.4). The objective of this sequence of events is to reduce the number of process steps. This is accomplished primarily by identifying the value-added (VA) and non-value-added (NVA) steps that exist within the as-is process.

PAIN — Model D (Figure 10.5). The objective of this sequence of events is to reduce the cost per cycle of using a process. After determining whether the costs in question are direct or indirect and the pertinent cost categories, the objective is accomplished through the sequential use of several process improvement tools. The improvement tools, Pareto analysis, cause-and-effect analysis (also known as the fishbone diagram or the Ishikawa diagram), and force field analysis are explained in numerous books on continuous improvement.

PAIN — Models E and F (Figures 10.6 and 10.7). The objective of these models is to provide guidance in the reduction of transactional errors and defects (Model E,

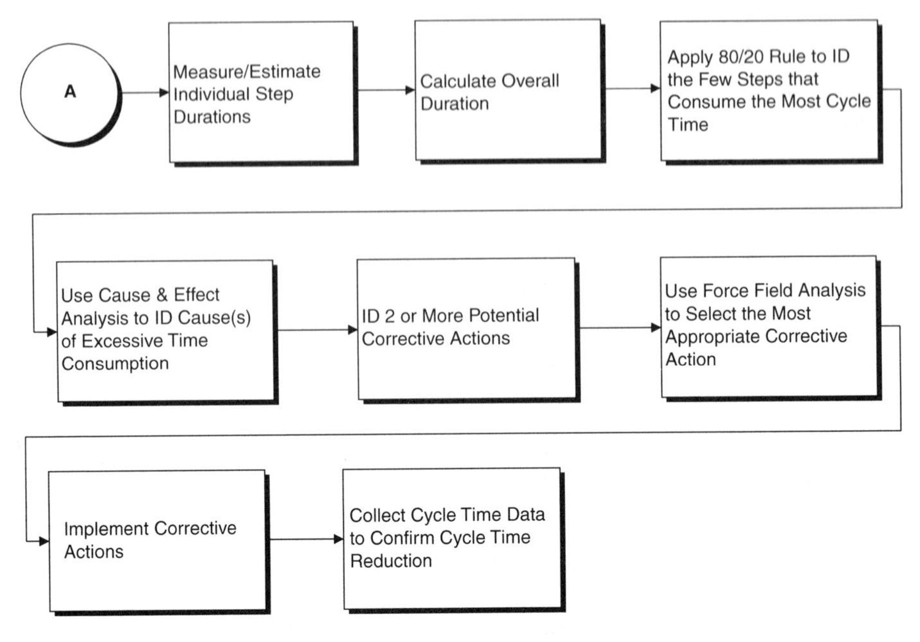

FIGURE 10.2 Objective: reduce process cycle time.

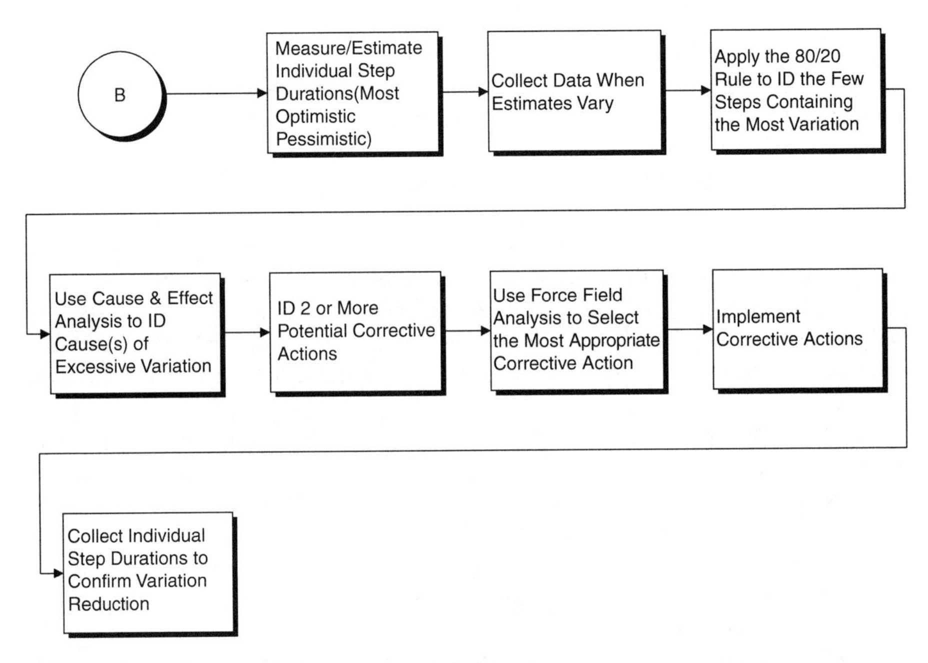

FIGURE 10.3 Objective: reduce process variation.

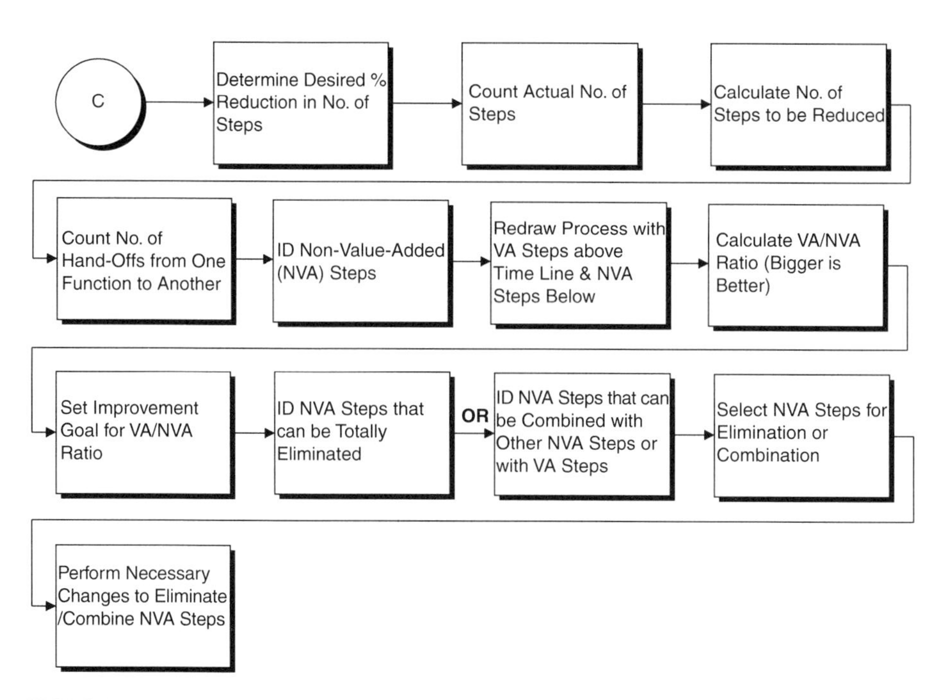

FIGURE 10.4 Objective: reduce number of steps.

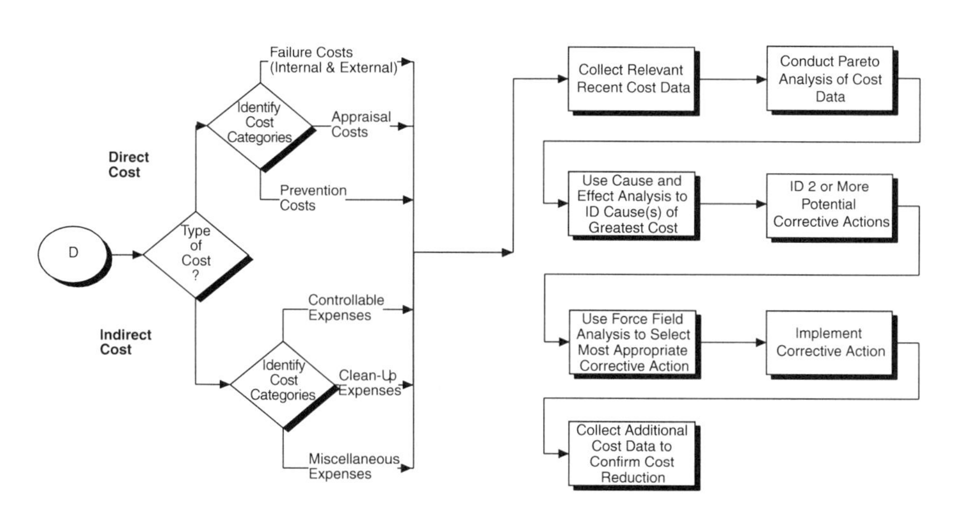

FIGURE 10.5 Objective: reduce cost per cycle.

Figure 10.6) as well as production errors and defects (Model F, Figure 10.7). The model is based on the Deming-Shewhart Plan-Do-Check-Act cycle. The earliest version of the model was created in 1985 as a part of a continuous improvement seminar. When the model is first introduced to a process improvement team, it is important to gain consensus from the team members regarding the rationale of event selection and arrangement.

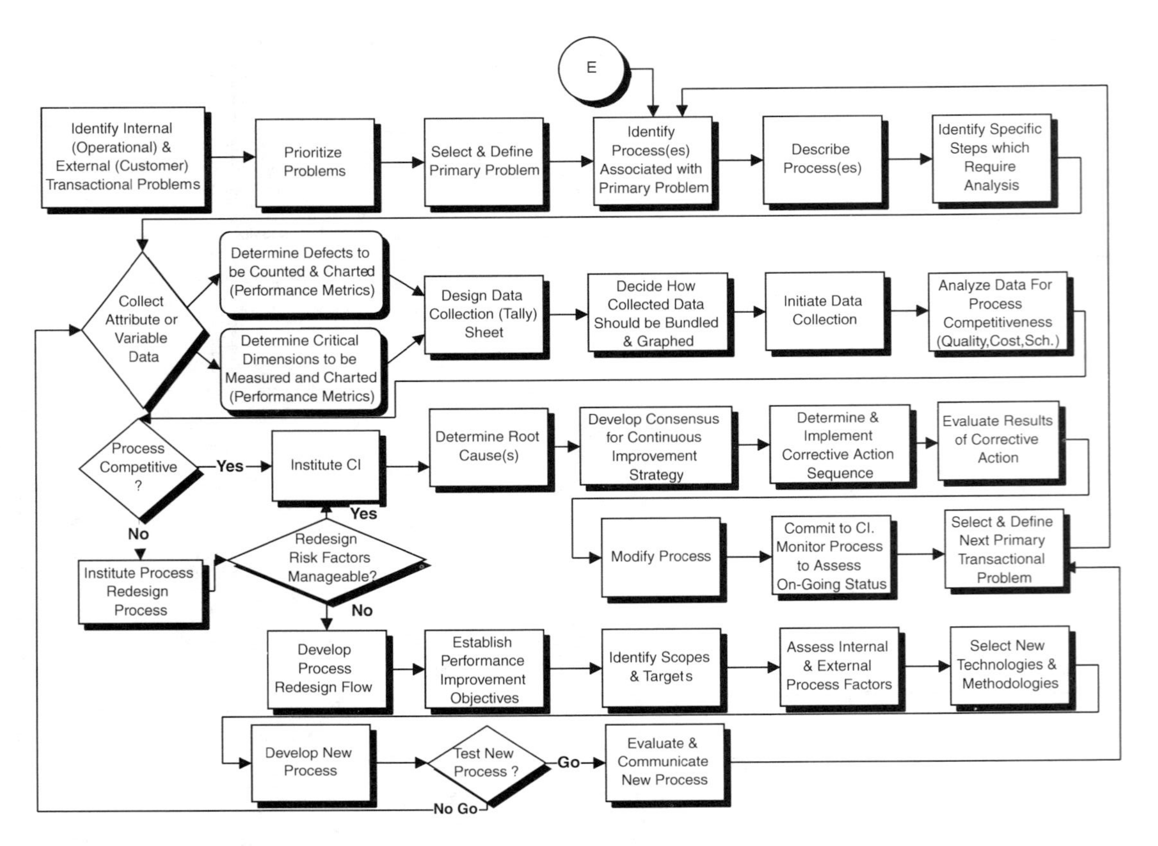

FIGURE 10.6 Objective: reduce transactional errors/defects.

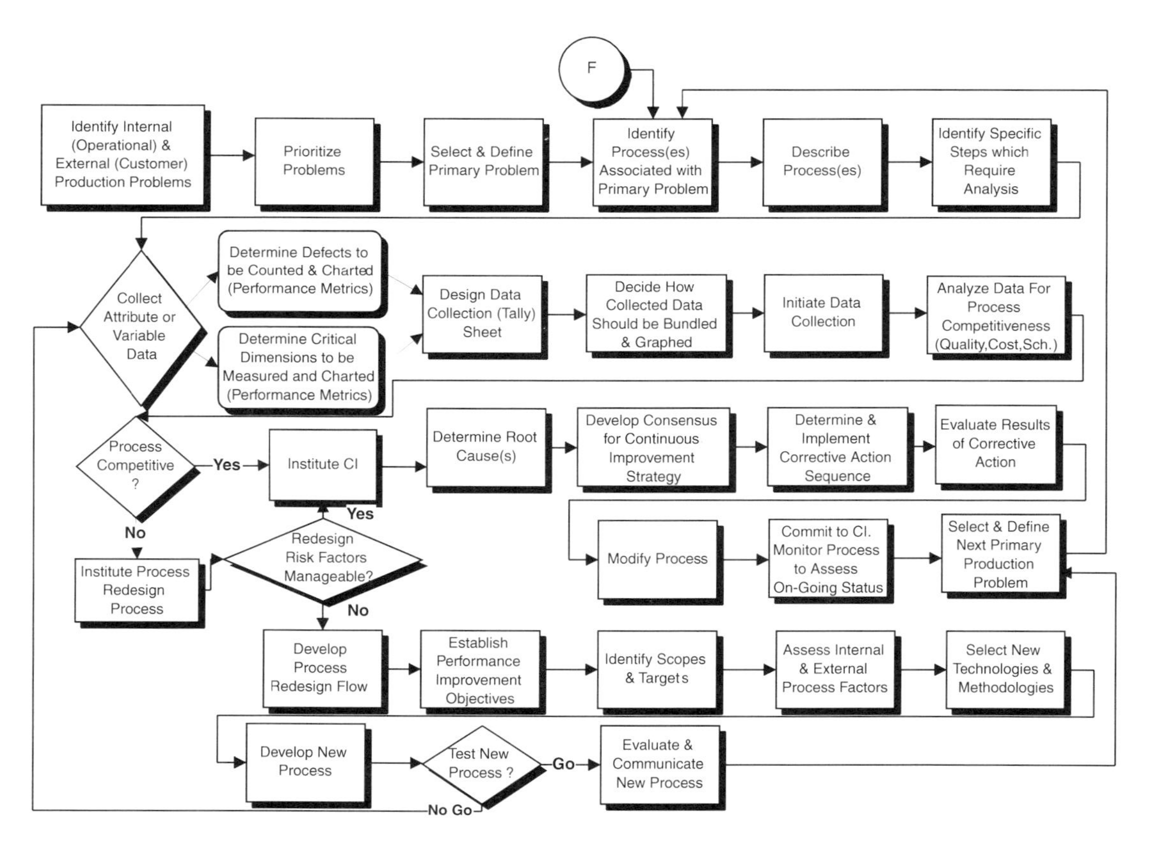

FIGURE 10.7 Objective: reduce production errors/defects.

10.4.4 Phase 1 — Model F

This model (Figure 10.7) is best understood by beginning its examination at the top left and moving to the right or left by following the arrowheads. Phase 1 of the model starts with the identification of both internal (operational) and external (customer) problems. This can be as simple as developing a comprehensive listing of problems drawn from a specific department, multiple departments (also known as cross-functional), a single division, multiple divisions, or from the entire company.

Once the list has been developed, it should be prioritized by rank-ordering the problems. The problem at the top of the list is identified as the primary problem. The next step is to identify the process or processes (if more than one process is involved in the creation of the problem) associated with the primary problem. Then it is necessary to clearly describe the selected process(es). Now magnification is increased so model users can identify the specific steps within the process(es) requiring analysis.

At this point in the problem-solving sequence, it is necessary to make a decision whether to collect attribute data or variable data. Whatever decision is reached, the next step is to decide which performance metrics will be used throughout the remainder of the problem-solving model. If the decision is to collect attribute data, then it is necessary to determine which defects should be counted and charted. If the decision is to collect variable data, then it is necessary to determine which critical dimensions to measure and chart.

At this point the data collection sheet, sometimes referred to as a tally sheet, is designed (with the end in mind being a user-friendly form that is easy to complete and just as easy to summarize). Then, following appropriate discussions, it is necessary to decide how the collected and summarized data should be bundled and graphed. Bundling describes the numerator and denominator of the ratio to be used as the performance metric. This completes phase 1.

10.4.5 Phase 2

As might be expected, phase 2 starts with the collection of sufficient data to be representative of the entire problem. When these data have been collected, it is time to initiate data analysis to determine just how competitive the process really is with respect to quality, cost, and schedule. This completes phase 2.

10.4.6 Phase 3

Phase 3 begins with a decision, i.e., is the process competitive? If the decision is in the affirmative, then we track phase 3-A in which continuous improvement (Kaizen) of the process is appropriate and should be instituted. Continuous improvement (CI) begins with the determination of the root cause(s) of the original problem. Developing a consensus strategy for CI follows the identification of the root cause(s). Next, a corrective action sequence is determined and implemented. Evaluation of the data generated and collected subsequent to the introduction of the corrective actions should reveal the wisdom of the corrective action sequence.

When the results justify doing so, the next step is to modify the process in whatever way the newly collected data indicate is appropriate. At this point a commitment must be made to CI and to monitor the process to assess its ongoing status. Without this commitment, the likelihood of the process's reverting to its original status is virtually 100%. The final step of phase 3-A is to select and define the next problem to be addressed, thus returning our attention to phase 1.

Turning our attention back to the beginning of phase 3, if a decision is made that the process is not competitive with respect to quality, cost, and schedule, then we follow phase 3-B that begins with instituting the process re-design sequence. This brings us to still another decision point, where it must be decided whether the redesign risk factors are manageable.

If it is determined that they are not, then we return to phase 3-A. If, on the other hand, the redesign risk factors are assessed to be manageable, then the next step is to establish specific performance improvement objectives followed by quantification of the target values.

Phase 3-B continues with the assessment of germane or pertinent internal and external process factors. These are the factors that have a high potential of contributing to the success or failure of the process redesign effort.

At this point the team should turn its attention to the selection of new technologies or methods that may replace those used in the existing process. It is at this time that the team should identify the old technologies and methods that will be retained, as well as their new counterparts so as to develop the new process.

The new process is tested using all the steps of phase 2 to decide whether it is as good as or better than the original process. If the decision is favorable, then it is communicated to all the process stakeholders and we return to the final step of phase 3-A. If the decision is a "no go," then we must return to the first step of phase 2.

10.4.7 PAIN — Model G

The objective of this model (Figure 10.8) is to improve process documentation. There are three paths to follow depending on the specific reason for wanting to accomplish this objective. The benchmarking path is provided to assist in making comparisons with other similar processes, either internal or external to the team's organization. The ISO 9000 certification path is provided in response to the directive's expressed interest in maintaining a current file of process flow charts to increase the likelihood of product consistency. The training path is presented to remind a team of the need for current documentation for training new or recently transferred employees.

The facilitator should encourage the team to identify two or more points within the process where specific knowledge of cycle time (elapsed time) is needed. As a rule of thumb, the team should focus on points of handoff between process stakeholders.

With the cycle time points selected, the facilitator should assist the team to create data collection forms, one for each of the selected process points.

Individual process stakeholders, using the newly created data collection forms, will collect 100 to 200 data values. These values should be recorded on the forms.

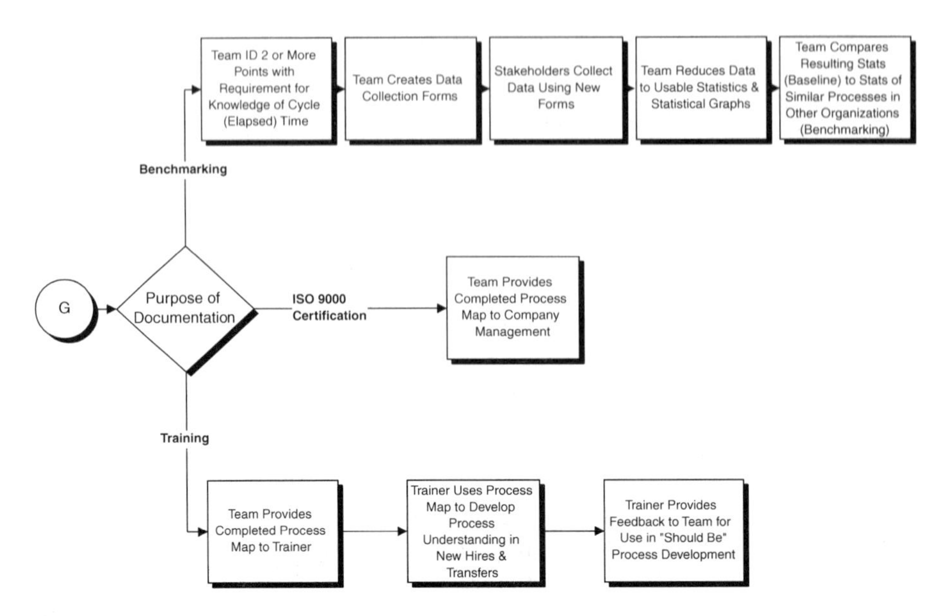

FIGURE 10.8 Objective: improve process documentation.

With guidance provided by the facilitator, the data should be reduced from a mass of values to usable statistics and then converted into statistical graphics by process stakeholders. The resulting graphics will assist the team to better understand their process.

The resulting baseline data are now ready to compare with other data collected from similar processes. The purpose of the comparisons is to determine which of two or more processes generates the desired results, i.e., the shortest and most consistent cycle times at the least cost and resulting in the greatest customer satisfaction.

Appendix A, which follows, provides a user-friendly template for a preliminary, step-by-step process analysis.

APPENDIX A — PROCESS ANALYSIS: STEP-BY-STEP

The following text, as numbered, is intended for use with Figure A.1.

A.1 FULLY DEFINE THE WORK ACTIVITY

- What product or service is created?
- What value-added characteristics are provided?
- What non-value-added characteristics are introduced?
- Which of the five *m*s and an *e* (men and women, material, machine, method, measurement, and environment) are required to conduct the work activity?

A.2 DESCRIBE ALL THE OUTPUTS OF THE WORK ACTIVITY

- What are the tangible products and the intangible services?
- How are the products or services related to specific customer demands, wants, and wishes?
- What are the production rates for each category of output?

A.3 IDENTIFY THE CUSTOMERS OF THE WORK ACTIVITY, i.e., THOSE WHO RECEIVE THE OUTPUT

- Are the customers external, internal, or both?
- Where are the customers located relative to the work activity?
- What are the customers' demands, wants, and wishes?

A.4 DESCRIBE THE QUALITY REQUIREMENTS ASSOCIATED WITH THE OUTPUTS OF THE WORK ACTIVITY

- What are the sources of the quality requirements?
- Can the quality requirements be expressed in terms a customer can understand?
- Are the requirements subject to change according to the demands, wants, and wishes of different customers?

A.5 LIST THE PERFORMANCE METRICS USED TO EVALUATE THE QUALITY REQUIREMENTS OF THE OUTPUTS

- Are the metrics expressed as ratios, e.g., defects per unit, defects per million defect opportunities, process capability index, process performance index, or a Six Sigma quality level index?
- How often are the output performance metrics evaluated for trend information?
- What feedback is provided by customers regarding the quality of the process outputs? How often?

A.6 DESCRIBE ALL THE INPUTS TO THE WORK ACTIVITY

- What inputs are sourced from outside or inside the organization?
- Which inputs are products and which are services?
- Do any of the inputs have shelf lives that must be observed?
- List the suppliers to the work activity, i.e., those who provide the inputs to the process
- Are the suppliers external, internal, or both?
- Where are the suppliers located relative to the work activity?
- Are the external suppliers certified?
- Are the suppliers expected to provide statistical control charts, if appropriate?

A.7 DESCRIBE THE QUALITY REQUIREMENTS ASSOCIATED WITH THE INPUTS TO THE WORK ACTIVITY

- What are the sources of the quality requirements?
- Are the quality requirements subject to periodic modification?
- Are the quality requirements stated in user-friendly terms?
- Are the quality requirements sufficiently demanding to ensure virtual perfection of the inputs to the work activity?

A.8 LIST THE PERFORMANCE METRICS USED TO EVALUATE THE QUALITY REQUIREMENTS OF THE INPUTS

- Are the metrics expressed as ratios, e.g., defects per unit, defects per million defect opportunities, process capability index, process performance index, or a Six Sigma quality level index?
- How often are the input performance metrics evaluated for trend information?
- What feedback is provided to suppliers regarding the quality of their process inputs? How often?

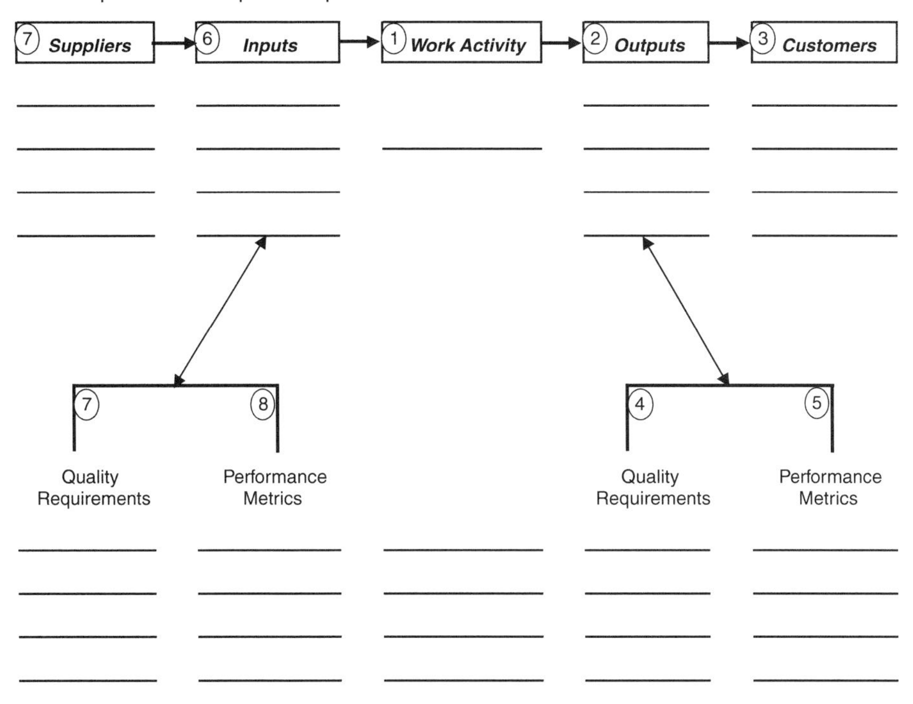

FIGURE A.1 Process examples.

11 Quality Function Deployment (QFD)

Charles A. Cox

11.1 INTRODUCTION

QFD is a way to capture, organize, and deploy the voice of the customer — both the external and internal customers of the organization. QFD has often been associated with product development activities, but has manufacturing applications as well. The QFD concepts and tools are useful to people involved in manufacturing in its long-run and short-run applications.

In a long-run situation, when a new product is designed, QFD requires that the organization's customers including an important internal customer, manufacturing, have input into the design process. The customers' choices and priorities are then converted to technical statements and quantified, which aids the design process. Once the product has been designed, the QFD process is extended to help design the manufacturing process as well. More recently, through integrated process and product design (IPPD), both the product and the process that will be used for producing it are developed in tandem. This results in a much shorter "concept-to-cash" cycle that uses fewer resources for the design and launch. This approach allows greater flexibility and responsiveness to the market.

In the short run, the use of QFD helps the manufacturing team do a superior job of characterizing the process, especially in understanding the linkages between different segments of the process. An important QFD tool, the matrix, when applied as a simple cause-and-effect matrix (see Figure 11.1), shows the process's input–output relationships with the varying strengths between the different inputs and outputs. This structure takes a process map and makes it come alive for ongoing control efforts and further improvement efforts. The figure shows the relationships between ten different inputs in five steps of a plastics molding process to the three key outputs of dimensional stability, uniform density, and smooth finish.

Equipped with a process map and the information in a cause-and-effect matrix, people involved in manufacturing operations can create a process control plan (Figure 11.2) that is appropriate for the operations within their organization.

A high-level framework for conceptually viewing a process and the inputs-to-outputs conversion of a process is available in the SIPOC (Supplier–Input–Process–Output–Customer) chart (Figure 11.3).

In today's complex manufacturing environment, an internal process is often affected by elements outside the organization — from the supply side and customer side. To capture the relationships on both sides of the SIPOC, QFD helps to show

1-57444-300-3/02/$0.00+$1.50

Process Step	Process Input	Output #1 Dimensional Stability	Output #2 Uniform Density	Output #3 Smooth Finish	← Process Outputs
		10	8	6	← Importance
		Correlation of Input to Output			Total
1	Pre-Heat Temp.	0	3	0	24
2a	Barrel Temperature	9	3	0	114
2b	Auger Speed	3	9	0	102
2c	Gate Size/Config.	9	3	0	114
3a	Mold Temperature	3	9	3	120
3b	Gate Distribution	3	9	1	108
4a	Dwell Time	3	1	9	92
4b	Dwell Temperature	3	0	3	48
5a	Extraction Pressure	9	0	3	108
5b	Cool down slope	0	0	9	54

FIGURE 11.1 The cause-and-effect matrix. Correlation of Input to Output: 0, no relationship; 1, possible relationship; 3, medium; 9, strong relationship.

Control Plan No.:	Key Contacts/Phone	Date (Orig.)	Date (Rev.)	Effective:
Part No./Latest Chg. Level	Core Team		Customer Eng'g. Approval/Date	
Part Name/Description	Supplier/Plant Approval/Date		Customer Quality Approval/Date	
Supplier/Plant	Supplier Code	Other Approval/Date	Other Approval/Date	

Part/ Process Number	Process Name/ Operation Desc.	Machine, Device, Jig Tools for Mfg.	Characteristics			Special Class.	Methods					Reaction Plan
			No.	Product	Process		Product/Process Spec/Tolerance	Evaluation Measurement Technique	Sample Size	Sample Freq.	Control Method	

FIGURE 11.2 Process control plan. (Reprinted with permission from the *APQP Manual* (DaimlerChrysler, Ford, General Motors Supplier Quality Requirements Task Force).)

and integrate the supply chain management and customer relationship management elements with the internal SIPOC (Figure 11.4).

Managing this chain of events and relationships that extends from our suppliers through our own operations to our customers is essential for our success. QFD

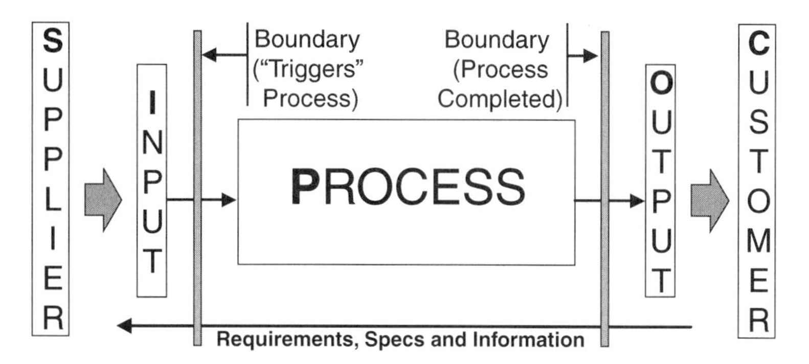

FIGURE 11.3 The SIPOC chart.

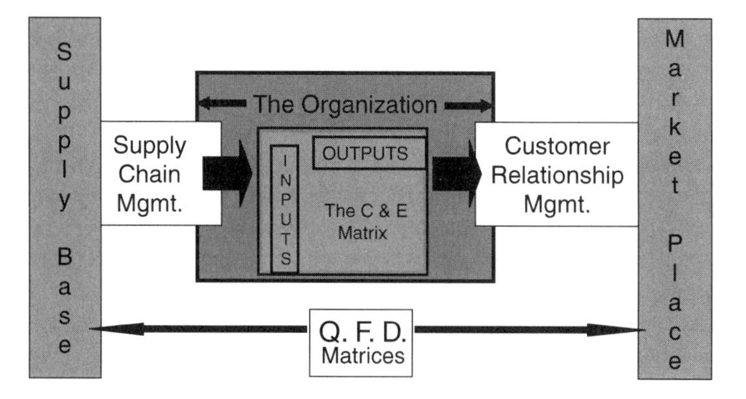

FIGURE 11.4 The span of quality function deployment vs. the cause and effect matrix.

concepts and tools assist in this by providing a structure to capture all the elements and prioritize them, enabling us to focus our limited resources in the most gainful way, i.e., from our customers' perspective.

Manufacturing can use QFD concepts and structure in three situations:

1. With the current product and process
2. With the current product and a new or redesigned process
3. With a new product and the current process

Because new product launches (situation 3) are rare compared with manufacturing operations' everyday need to address characterizing, monitoring, and improving the current processes, manufacturing's first use of a QFD tool is often the cause-and-effect matrix (for situation 1).

The cause-and-effect matrix shows how multiple inputs have varying levels of impact on the desired outputs sought from the process. For a process to consistently deliver satisfactory or even superlative output with no defects, it is essential to define the relationships among all of its inputs and outputs.

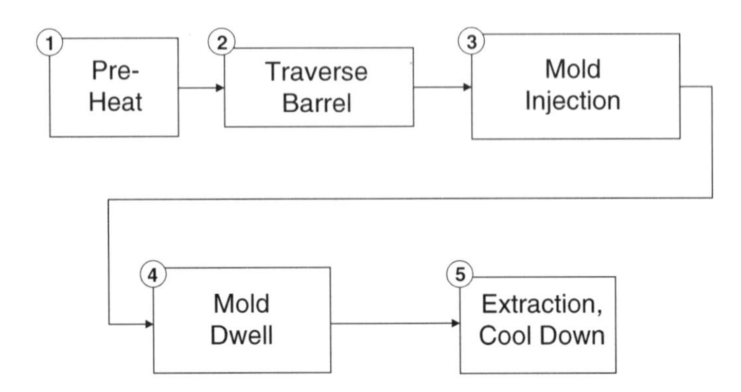

FIGURE 11.5 Process map example.

The cause-and-effect matrix does this most efficiently. Take a common manufacturing process: plastic molding. In a five-step plastic molding process (Figure 11.5), there are several inputs that affect the desired outputs.

As always when studying a process's input–output relationships, the desired outputs are determined first. The molding process's customers have indicated that dimensional stability, uniform density, and smooth finish are the most important output characteristics and have assigned weights of 10, 8, and 6, respectively, to those outputs. A group of plastic molding operators, supervisors, technicians, and engineers then review elements in all five steps and decide on ten that affect the desired outputs. These are entered into the cause-and-effect matrix and the degree of effect that each makes on the output is noted. A strong effect is rated 9, a medium effect 3, a slight effect 1, and no effect, 0. The strength of the effect of each relationship is then multiplied by the importance those effects are given by the customers to get a total value. The total value is used to guide the allocation of resources, monitoring, and improvement efforts. The five inputs — 2a, 2c, 3a, 3b, and 5a — are the most important out of the ten listed.

The completed cause-and-effect matrix is a valuable input for the process control plan. The latter defines the monitoring system to be used to maintain consistent production as well as the set of measures that will be used to highlight (1) the need for adjustments to production parameters and (2) opportunities for further process improvements.

With respect to situation 2, the expected outputs are well known and it is up to manufacturing to decide on how each of the outputs is to be met with the new or redesigned process. Again the cause-and-effect matrix is used. In this case, with the outputs defined, it is essential that the new process's inputs meet or exceed the performance of the old process.

The most complete use of QFD concepts and tools happens when situation 3 occurs. In many cases, there is an entirely new product to be manufactured by a series of known process steps. It is in this situation where there are many different steps using technologies of varying degrees of maturity that QFD can assist the manufacturing manager the most. This situation is the one of greatest complexity, but QFD helps to organize and overcome that. In fact, the initial application of QFD principles often shortens the concept-to-launch time up to 30%. In addition, if the same (but updated) matrices are used when launching the second generation or a follow-on product, there are additional time savings.

Because markets and the competitive environment are changing at a faster pace and innovation is causing technical obsolescence, many products now have a shorter life cycles. In addition, many organizations are decentralizing their activities and creating specialized approaches for interfacing with inputs (supply side) and outputs (customer side). The manufacturing function is experiencing more change, and the traditional functions, which were all in-house, may now be spread among several different entities, both inside and outside the organization. Given these changes, there is a greater need for faster and more accurate communications with a broader variety of groups than manufacturing experienced in the past.

11.2 RISK IDENTIFICATION

How can an organization increase the accuracy and timeliness of its responses to market demands and at the same time reduce the economic risk associated with the substantial investments necessary for new and reengineered products? Risks arise from

1. Products that are (a) more complex, (b) involve more technologies, materials, and processes due to increasing innovation, (c) come from more suppliers (a supply-chain management issue), and (d) involve more customers and modes of usage (a customer relationship management issue).
2. Products that have previously been "hardware" only, now have an electronic component incorporated or, along with the associated sensor and human inputs, have some limited monitoring or feedback capabilities. The adaptation of electronics to traditional products makes some really gee-whiz features possible, but with these newfound capabilities come further complexities in the form of additional software or firmware. These kinds of applications have only recently been seen in common commercial and consumer products, but they are a growing trend.
3. The introduction and support of products that are more complex.
4. The fact that (most importantly) the manufacture of products is continuing to become more complex.

Just based on sheer numbers, there is the possibility of missing the combination(s) of inputs that will give the greatest economic return to the organization. Using QFD principles reduces the risk that something will be overlooked. It also helps all areas of an organization understand what knowledge needs to be gathered and shared to assist the overall (both design and manufacturing) engineering effort on a new product.

11.3 THE SEVEN-STEP PROCESS

The QFD methodology is a structured way of capturing the spoken and unspoken needs of a product's various customer groups. It typically follows a seven-step process:

1. Define the product's customers, specifically their expectations and where they are in the product's life cycle.
2. Analyze (a) current industry offerings, (b) industry trends, and (c) the expectations of customers from three quality perspectives: normal, expected, and exciting to derive customer expectations. The three levels of quality are from the Kano model and they affect how firms choose the means used to capture customer input (more on the Kano model later).
3. Organize and prioritize these inputs: the voice of the market (2a and b, above) gathered through market research and the voice of the customer (2c, above) collected via *verbatims*.
4. Translate these "voices" into technical objectives. This is where QFD bridges a major gap between the users of the product and the designers and manufacturers. This is an extremely useful exercise because it gives the technologists (design and process engineers and technicians) specifics on which design or production efforts have the most value to the customer and which are less important.
5. Draw on the initial translation of the technical objectives to determine how each of the customers' expectations can best be satisfied. The technologists are in charge of this — the concept coming from design engineering, with input from process engineering. To the extent that the production and ultimate use of the concept are kept in mind during the design, ease of manufacture and early acceptance in the marketplace are assured.
6. Plan for production. The objectives focused on for the concept and design drive the manner of production. The QFD structure invites early consultation and input from the people involved in planning production. The collaboration of design and production activities is what makes the ramp up rapid and the initial and ongoing production smooth. If post-introduction demand increases, manufacturing operations have a much better chance of supplying consistent product quickly because of the guidance from QFD.
7. Update the original customer expectation QFD matrices as the product ages and the market changes. If the original QFD matrices are updated as new information becomes available, product launch time can be further reduced and new products can be introduced in progressively shorter cycles. This allows the organization more learning cycles and much greater flexibility, both in meeting market opportunities and in introducing innovation.

To maximize the synergy among marketing, design engineering, and production, a good project management structure is essential. Just as essential is a structure that provides clear information for the project team. There is always the chance of poor results from the GIGO (garbage in, garbage out) effect. To avoid these results, it is essential that two principles be stressed:

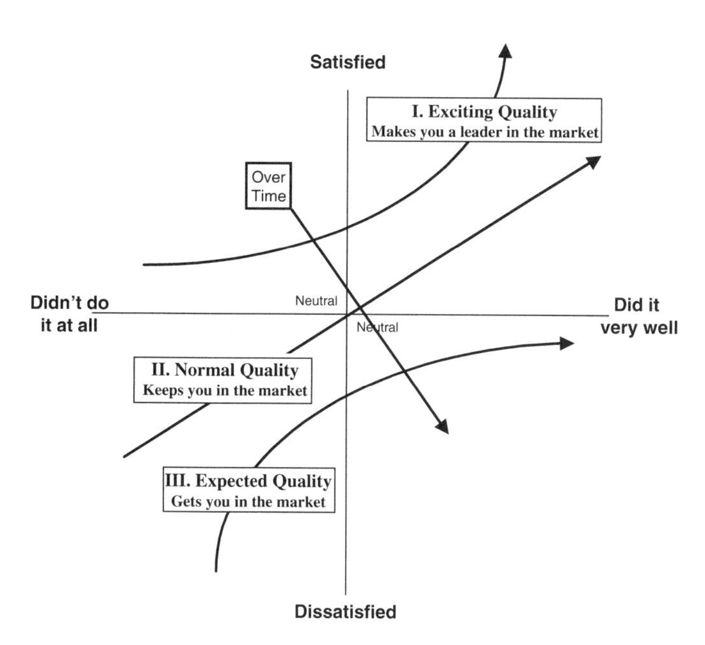

FIGURE 11.6 The kano model. (From Wm. Eureka and N. Ryan, The Customer Driven Company, American Supplier Institute, Livonia, MI, 1988. With permission.)

1. There is a well-defined process for gathering data and organizing it into information.
2. The people involved are the right (best) sources for the data or information.

Both are equally important if you are to capture the necessary knowledge from your different sets of customers and answer their spoken and unspoken needs.

11.4 KANO MODEL

The Kano model (Figure 11.6) helps define the process for gathering data and organizing it. The model stratifies each customer group's perceptions of the product into three types of quality: expected, normal, and exciting. Each of these types of quality requires a different approach for gathering data. The easiest to gather data about is normal quality — it is the basis of most of our conversations about a given product group and is usually the basis for advertising. The issues involved in normal quality are known by most customers. For example, in the case of tires, two major issues are price and length of warranty. Because the issues are well known, it is possible to gather information from the customer using simple surveys — telephone, mail-in, or in person. Satisfying customer expectations for normal quality keeps a firm competitive.

Expected quality issues are those that no one thinks about because everyone takes them for granted — until, that is, they are not met. An example would be tire treads that delaminate at high speed, a tire that does not hold air, or a tire with sidewalls that give out within 5000 miles of purchase. In the case of expected quality, the customers interviewed have knowledge and can provide answers, but it often takes some digging because these are not top-of-mind issues. Some approaches used to get this information are one-on-one interviews and focus groups. Being able to satisfy the expected quality issues only gets the firm into the marketplace.

The third type of quality, exciting quality, is the most difficult to obtain information on because, unlike expected and normal quality, the customer is not aware of exciting quality issues. In many cases, a new feature or function may be technologically feasible, but the technical persons are not aware of how much value the customer would place on the feature. The customer, on the other hand, is not technologically sophisticated enough to pursue innovation. For tires, an example of exciting quality would be a tire that never went flat, or one that could be driven several miles even though flat. To expose exciting quality issues, it is necessary to have multiple conversations with progressive customers and innovative designers. These conversations are best conducted in facilitated focus groups. Success in providing exciting quality helps make a firm worldclass.

Whereas the Kano model (Figure 11.6) offers the background for gathering data from different customer groups, general and specialized marketing information (including benchmarking) is used for competitive analysis. For both customer expectations and competitive market data gathering, it is necessary to define who will be consulted (the right persons to involve). You want them to be a good representative sample of the various groups.

The Kano model graphic shows how addressing expected quality issues only minimizes customer *dis*satisfaction; it never contributes to customer satisfaction. Normal quality issues can be either satisfying or dissatisfying although offering provision for greater customer satisfaction. Exciting quality issues can never be customer dissatisfiers (how can they be if the customer does not even know they exist?), but they can have a tremendous impact on customer satisfaction. Over time, items that are exciting quality become normal quality, and normal quality items become expected.

When gathering information from various groups of customers, it is important for the design team to realize the importance of the definition of the original design concept. It is also critical to check the concept against all the customer inputs and validate before proceeding. The cost of changing concepts is small at the concept stage, then rises exponentially. This is the reason it is so important that manufacturing be part of the QFD effort from the earliest phase (concept selection). The concept to production graph, Figure 11.7, shows how rapidly a company can become financially committed to a concept.

11.5 VOICE OF THE CUSTOMER TABLE

Once the groups of customers have been defined, there will be inputs from each group about the three different types of quality as defined by the Kano model. The resulting collection of *verbatims* from the customer groups will be entered into the Voice of the Customer Table (Figure 11.8). Then the *verbatims* are reworded to fit

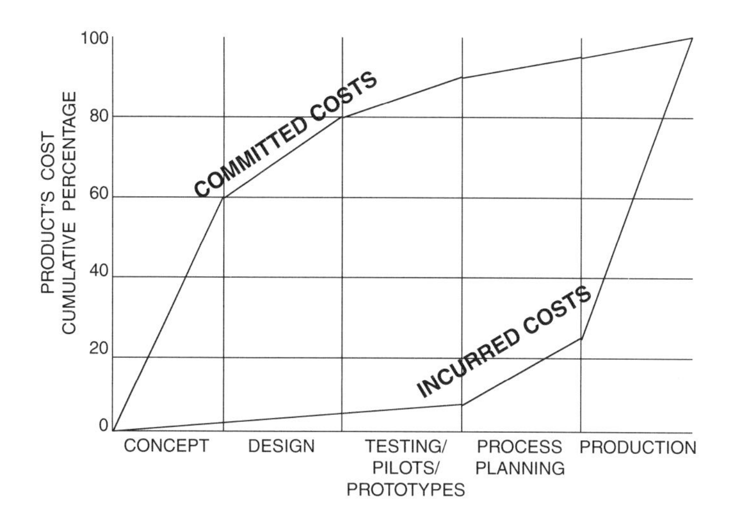

FIGURE 11.7 From concept to production. (From J. ReVelle, J. W. Moran, and C. Cox (Eds.), *The QFD Handbook,* J. Wiley & Sons, New York, 1998. With permission.)

Voice of the Customer Table 1

Customer Verbatim	Who	What	When	Where	Why	How

Voice of the Customer Table 2

Reworded Demands	Demanded Quality	Quality Characteristics	Function	Reliability	Other Issues

FIGURE 11.8 Voice of the customer tables. (From J. ReVelle, J. W. Moran, and C. Cox (Eds.), *The QFD Handbook,* J. Wiley & Sons, New York, 1998. With permission.)

into the categories in the Voice of the Customer Table 2. Figure 11.8 shows examples of VOCT 1 and 2 for a flashlight.

VOCT 1 categories are self-explanatory, but VOCT 2 categories are defined below:

- *Demanded quality* is a qualitative statement of the benefit the product gives the customer. These statements must be brief and phrased positively, for example, "can hold easily."

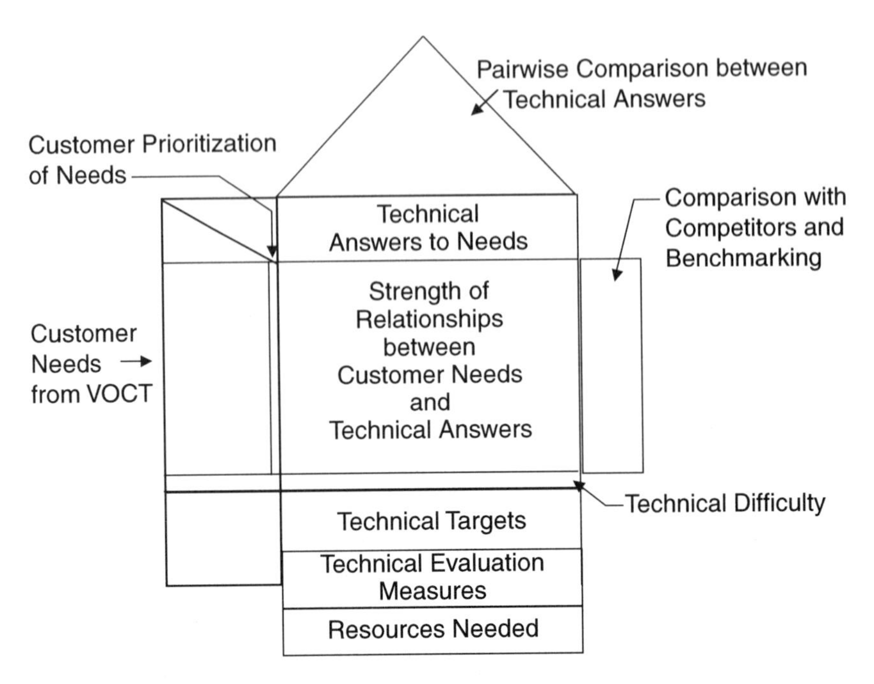

FIGURE 11.9A The house of quality (HOQ) matrix.

- *Quality characteristics* are quantitative — something that can be measured and that helps to attain demanded quality, for example, diameter.
- *Function* is the purpose of the product. Drawn from value engineering's standard practice, a function is stated as a verb plus an object, for example, "keeps aim."
- *Reliability* is the expected life of the product. Failure modes, typical warranty claims, or customer complaints can be included here. An example would be a complaint that the flashlight "won't light" or "won't turn on."
- *Other items* might be something emphasized in this particular design project, such as safety, environmental impact, price, or life-cycle cost.

11.6 HOUSE OF QUALITY (HOQ)

The results from the VOCT 2 are key inputs to the QFD beginning matrix shown in Figure 11.9A. Sometimes called the A-1 matrix, sometimes called the House of Quality (HOQ), this first matrix organizes the inputs from the various customer groups as well as marketplace intelligence, and has several elements or "rooms" that allow a tremendous amount of information to be organized.

The first "room" in the HOQ lists the wants of the various customer groups, which are referred to as WHATs. Each of them comes from the Voice of the Customer Table and has an importance rating (also from the customers) (see Figure 11.9B, #1).

The second room contains the HOWs and represents a technical, organizational response to explain how the WHATs will be achieved. It is possible that a single

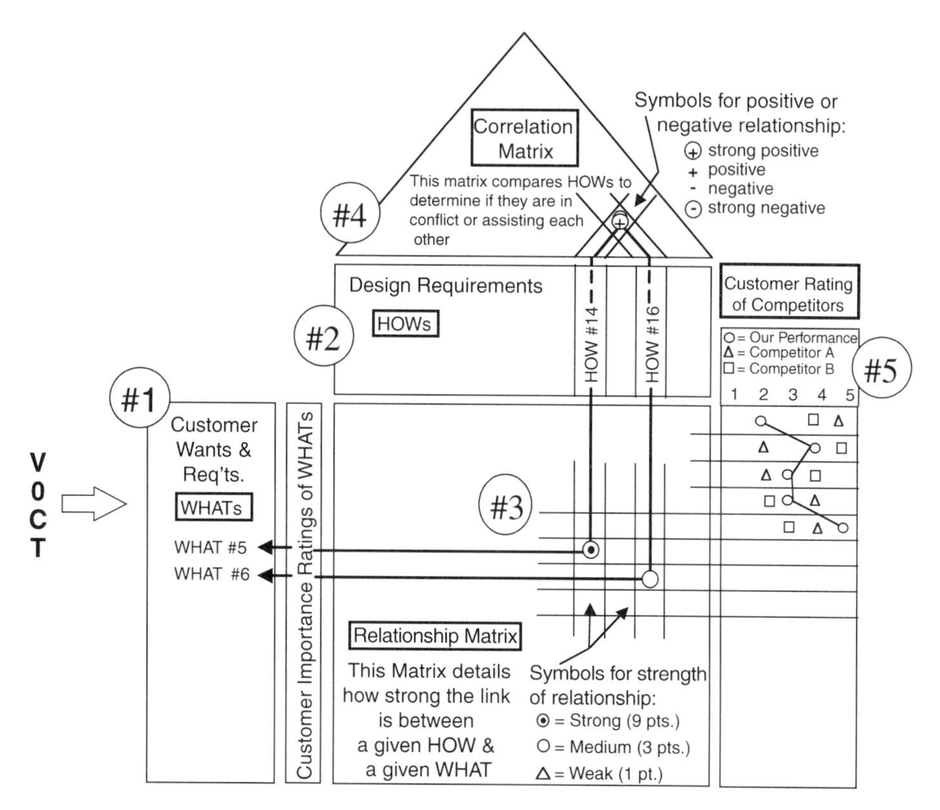

FIGURE 11.9B Five elements of the HOQ. (From J. ReVelle, J. W. Moran, and C. Cox (Eds.), *The QFD Handbook,* J. Wiley & Sons, New York, 1998. With permission.)

HOW can apply to several WHATs or that one WHAT may require several HOWs. (see Figure 11.9B, #2).

The third room is the relationship matrix located between the WHATs and the HOWs. It shows the extent to which the WHATs and HOWs are related and supplies a weight to the strength of the relationship. A strong relationship is rated 9, a medium relationship, 3, a weak relationship, 1, and no relationship is left blank (see Figure 11.9B, #3).

The fourth area of the HOQ is the "roof"(see Figure 11.9B #4). It is actually an L-shaped matrix which does a pair-wise comparison between each of the HOWs to seek out those pairs that are in conflict, but also notes those pairs which leverage each other. Again there is a multilevel rating system, in this case four leverage levels of relationships: strong positive, positive, negative, and strong negative. It is the strong positive and negative relationships that need to be noted and addressed. For strong negative relationships, the design team can look for ways to compromise, or the team can apply either TRIZ (Russian acronym for Theory of Inventive Problem Solving) or robust design. TRIZ refers to a technique, based on the study of thousands of patents, that allows these conflicts to be overcome without compromise. Robust design, on the other hand, is a methodology employed to make both product and processes robust, i.e., insensitive to conditions of use or manufacture.

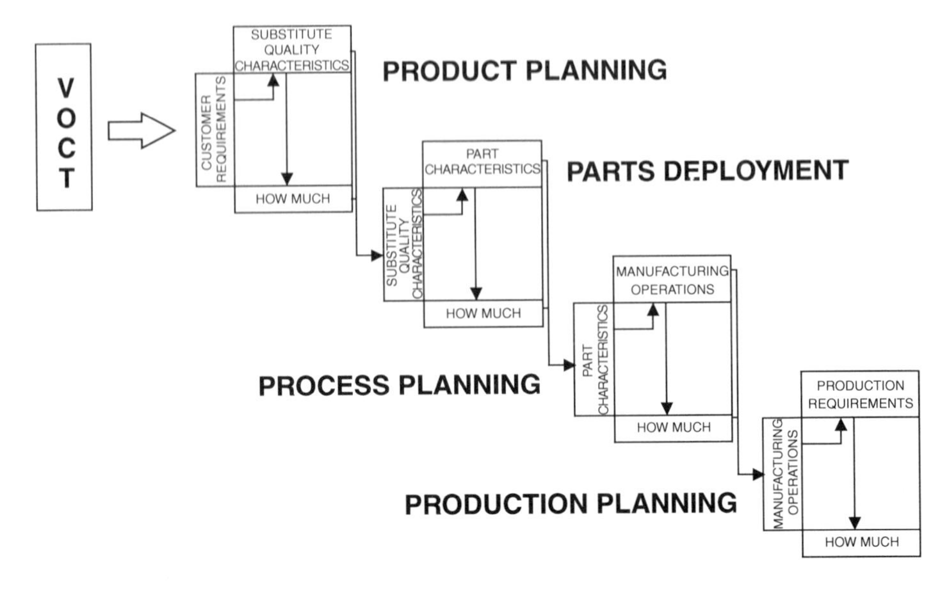

FIGURE 11.10 ASI four matrix approach — linking customer requirement to the production process(es) requirements. (From Wm. Eureka and N. Ryan, The Customer Driven Company, American Supplier Institute, Livonia, MI, 1988. With permission.)

The fifth room captures and presents the competitive intelligence, comparing our new product's features and functions with those of our competitors, and indicating the marketplace's perception on a feature-by-feature basis (see Figure 11.9B, #5).

11.7 FOUR-PHASE APPROACH

One series of matrices popularized by the American Supplier Institute (ASI) consists of four matrices (Figure 11.10). These start with high-level customer wants and requirements and finish with well-defined production requirements for manufacturing operations. The output from the voice of the customer tables feeds the first matrix, called the product planning matrix. Product planning changes the customer-defined requirements into substitute quality characteristics, which quantify the customer requirements and enable engineers and technicians to have design targets. The second matrix takes the high-level quantified concept and defines the components or parts of the system. The third matrix details the production process layout and the fourth matrix gives the measures and monitoring needed to assure consistent production.

An example of using the ASI approach might be in the design of a passenger vehicle. Among other wants, a potential buyer might say, "I want low cost of ownership," or "I want low fuel consumption." In the first matrix, these generalized wants, low cost and low fuel consumption, are quantified. The result would be agreement on a concept that included specifics on the coefficient of drag (the aerodynamics of the vehicle's movement through air at high speeds), targets for the mass of the vehicle, nature of the transmission (manual shift) and cubic displacement, breathing and fuel delivery configuration of the engine (multivalve, overhead cam, naturally aspirated or turbocharged, throttle-body or manifold fuel injection, etc.).

The results of the first matrix, product planning, would then feed the second matrix, parts deployment. In the example, if we focus on the mass-of-the-vehicle part of the vehicle's design in the parts deployment matrix, conclusions about the nature of the vehicle's structure (frame and body vs. unibody) and materials to be used can be decided. It may happen that a frame and body structure with a fiberglass skin is selected.

The output of the second matrix, parts deployment, serves as input to the third matrix, process planning. Knowing the type of vehicle structure (hence the sequence of production steps) will limit the options available for laying out the actual production operations. Once these decisions have been made, the results are transferred to the final matrix, production planning. Production planning addresses all the measuring and monitoring necessary to ensure that basic items (such as the fiberglass) are produced correctly. As a result of the last matrix, for example, there might have to be a very specific manufacturing procedure for mixing the resins that go into the fiberglass. Any requirement on the manufacturing floor would be directly traceable all the way back to some customer requirement (such as low fuel consumption).

11.8 MATRIX OF MATRICES APPROACH

The ASI four-phase approach can demonstrate a commonly used subset of a larger set of matrices, the matrix of matrices (popularized by GOAL/QPC), Figure 11.11. This larger set of matrices includes those that might be used when doing other types of analysis, such as value engineering, reliability planning, quality control, or cost analysis (all analyses that would also have an impact on manufacturing operations).

11.9 RECOMMENDATIONS

11.9.1 SOFTWARE

- A longtime major software package for assisting the QFD process is *QFD Capture* from International TechneGroup, Inc., Milford, OH. 513-576-3900. <http://www.iti-oh.com>
- Another software package that is available is *QFD Designer* from QualiSoft Corp., West Bloomfield, MI. 248-357-4300. <http://www.qualisoft.com>

11.9.2 BOOKS

- Cohen, L., *Quality Function Deployment: How to Make QFD Work for You*, Addison-Wesley, Reading, MA, 1995.
- Day, R. G., *Quality Function Deployment: Linking a Company with Its Customers,* ASQC Quality Press, Milwaukee, WI, 1993.
- King, B., *Better Designs in Half the Time: Implementing QFD in America,* GOAL/QPC, Methuen, MA, 1989.

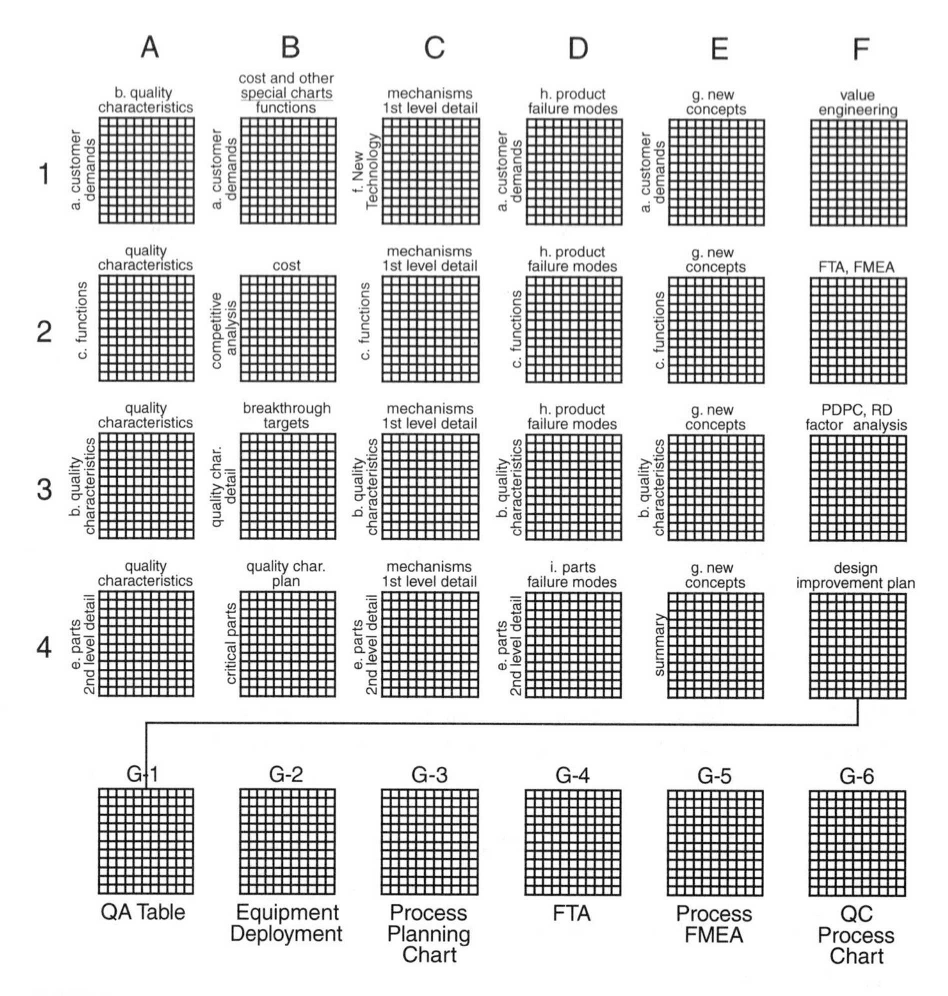

FIGURE 11.11 The matrix of matrices approach. Use for value engineering, reliability/durability, or other focuses in the product/service development process. (From B. King, *Better Designs in Half the Time,* Goal/OPS, Salem, NH, 1989. With permission.)

- ReVelle, J., Moran, J., and Cox, C., *The QFD Handbook*, Wiley, New York, 1998.
- Terninko, J., *Step by Step QFD: Customer-Driven Product Design*, CRC/St. Lucie Press, Boca Raton, FL, 1997.

11.9.3 Web Sites

Note: These references are listed in order of complexity.

- A quick 26-slide overview of QFD is available at (http://www.mines.edu/Academic/courses/eng/EGGN491/lecture/qfd/>

- A well-thought-out three-exercise tutorial from the Software Engineering Research Network at the University of Calgary is available at <http://sern.ucalgary.ca/~dweening/SENG613/Exercises/Exercise1.html#HouseOfQuality>
- A detailed write-up is available at <http://www.proactdev.com/pages/ehoq.htm>
- A *very* detailed write-up, which includes how the various features of the *QFD Capture* software can be utilized, is available at <http://www.iti-oh.com/cppd/qfd/qfd_basics.htm>
- An overview and commentary (part of the E. B. Dean/NASA series on Design for Competitive Advantage) on other good sources of information are available at <http://mijuno.larc.nasa.gov/dfc/qfd.html>
- A listing of many varied QFD resources and multiple bibliographies is available at <http://www.postech.ac.kr/ie/qelab/QFD_resource.html>

12 Manufacturing Controls Integration

R.T. "Chris" Christensen

12.1 THE BASIC PREMISE OF INVENTORY

Ever since the pharaohs built the pyramids, humans have been faced with the problem in production management of how inventory should be used to maintain, balance, and level load production. In the case of the pharaohs, they needed to have a big pile of big rocks on hand to maintain a continuous production schedule. And since the time of the pharaohs, we hadn't made any significant inroads into the pile-of-rocks theory of manufacturing and inventory control until 1959. That was when Joe Orlicky of IBM developed the matched sets of parts relationship required to get the right parts to the right job at the right time. He called it materials requirements planning (MRP).

Although we had the tool, we had only a very limited application of MRP. Although the work required for processing information in an MRP environment is ideally suited for computer processing, the limiting factor in the early 1960s was our limited and expensive computer power. The repetitive work required to process the information and do the calculations was cost prohibitive. This left us with finding the cheapest way to balance the matched sets of parts. We found the method necessary to minimize our manufacturing cost and called that tool *inventory.* Like the pharaohs, we now have our "pile of rocks" — the cheapest way to do it. From this point on in the development of manufacturing theory, all we have really done is add tools to accomplish the task of controlling the matched sets of parts. The primary tool that we use is the computer, so we can do the calculations needed to control our operations. As we continue to increase the level of computer involvement as our tool, our processing time becomes cheaper than that pile of rocks. When computer power becomes cheaper than inventory, we reduce inventory and add power.

This new and cheaper information processing power has brought us to today, where our goals are to run a quicker and leaner operation using the information we have gained from tools such as the theory of constraints (TOC), takt time, and advanced planning systems. Today's quick-response manufacturing facility is an internal cog inside the supply chain, bringing the goods and services to the industrial or retail consumer at the right time and the right place with the right product. To do this we must use these tools so that we can be "just-in-time" to meet our customers' needs. Though holding inventory in the past helped to speed up the delivery cycle to get product to our customers, we must remember that inventory adds no value in itself.

This chapter shows how we can eliminate inventory and at the same time meet the customer's rapid demands — essentially having only one Big Mac ready just as

1-57444-300-3/02/$0.00+$1.50

you open the restaurant door. This chapter identifies advanced and economically viable techniques that now involve the use of e-manufacturing, Web-based information systems, and integrated control systems.

12.2 NEED FOR INVENTORY IDENTIFIED BY DEFINITION

The following different definitions of types of inventory will help you get an idea of why you have inventory and what it really is. Once you understand why you have inventory, you can determine what you need to keep on your shelves. The reason that we define the different inventory groups is so that we can apply various tools to control and manage that inventory based on the reasons that cause you to have inventory. The definition of inventory is

Material: In the traditional sense, inventory is the parts and material stocked to meet your short-term and long-term sourcing requirements.

A decoupling activity: Inventory is the tool that decouples the customer's demand from production capacity to enable the organization to flat load the plant.

A fixed investment: If you have $2 million in inventory now, you'll always have $2 million in inventory. You use parts and materials from your inventory supply, but you immediately replace them with new stock upon consumption.

Insurance: What is insurance? It's being reimbursed for an incurred loss. Insurance minimizes your loss if disaster strikes. So, isn't inventory just that — insurance against an inability to get the parts needed to meet the production order?

A bet: Similar to insurance. When you carry auto insurance for your teenager, you're placing a bet that he or she will wreck the family car. The insurance company is giving 10-to-1 odds that it won't happen. As a manager concerned about inventory, you're like that insurance company. You bet there will be no downtime, and you stack the odds in your favor by the amount of inventory you carry.

A buffer stock against use: Inventory is a hedge against the unknown. If you knew exactly when a part was required, you wouldn't need to carry it in stock. You'd buy the part and have it arrive exactly when needed. This sounds good in theory, but because you don't know exactly when you'll need that part, you carry it.

A buffer stock against delivery: Inventory also protects you from the uncertainties of delivery. If you knew exactly when a supplier would deliver your order, you'd never need inventory to cover for erratic delivery schedules. Hey, suppliers have problems, too.

Safety stock: How big a risk taker are you? What are you willing to risk by not having parts on hand? We're always being asked to reduce inventory and we come up with excuses for not meeting the reduction goals. The flip side is, if you reduce inventory and then run out, you are past the excuses point in defending your inventory policy. That's when you get yelled at.

CYA stocks: We all know what "cover your a--" inventory is and why we have it. See above.

A quantitative measure of your inability to control yourself: I can always tell how well a person is able to run his or her operation by looking at the amount of inventory. The better you manage your operation, the better you control your inventory level.

"Unobtainium": There is a layer of parts that fall into the category, "must have, can't find." These are rare, almost impossible-to-obtain parts; or the lead-time to acquire them is so long, it just *seems* like you can't get them. These sit on your shelf awaiting your need, and there is little you can do about it.

Hidden stock: This is the inventory your production people stash under conveyors, under stairwells, inside parts cabinets, or in their lockers and toolboxes. This is the stuff you call "lost" each year when you do physical inventory. It's a real problem because you don't know the condition of those parts. This happens a lot in an incentive environment that allows the worker to turn in work for pay while the machine or line is down. The operators make this material during breaks, at lunchtime, between shifts, and at other times when they are present and can't get paid for their time. This not only presents raw material and finished goods problems but also is a serious safety and quality issue.

Rogue parts: These are the parts you don't list in your system. You have errors in your bill of materials that your schedulers know about, which forces the scheduler to make manual inputs whenever the problem arises. These parts may be good and they may be useful, but many times you can't find them when you need them. The mechanic has misplaced them, is on vacation, or has quit or retired. The parts are out there somewhere.

Anticipation inventory: This inventory allows an organization to cope with the anticipated changes in demand. Vacations, shutdowns, peak sales periods, sales promotions, or strikes are situations that can lead an organization to produce or purchase additional inventory.

"Cheapest way to do it" inventory: There are many ways to get the parts you need, but what it really comes down to is, "What is the cheapest way to get those parts when you want them?" However, you get these parts, there is a cost. There is a balance between the cost of acquiring and keeping parts in inventory, and your ability to plan or forecast needs. But somewhere along the line it will become clear that the overall cheapest way to get parts is to just carry them in inventory. This won't apply to all your parts needs, but you'll find a group of parts here that falls into this category.

Lot size inventories: This inventory comes about when it becomes inefficient to produce or purchase goods at the same rate at which they are consumed.

Fluctuation inventories: These inventories are used to provide a buffer for both demand fluctuations and supply fluctuations. These inventories help smooth out the production cycle.

Transportation inventories: These inventories are used when stages of the production cycle are not always adjacent to each other. This is true for multiplant operations; the general rule is that the farther apart the plants are, the more inventory will be required to keep the system running.

Reasons for inventory definitions and answers. We must know why we name these different groups. If you look at the goals and objectives top management gives you each year, you will invariably see items such as

- Reduce inventory
- Lower inventory costs
- Improve on-hand availability of parts

- Reduce annual parts costs
- Shorten the delivery cycle

Those items are actually the savings that management wants to realize from your inventory. What management fails to do is give you the tools or a road map to achieve those lofty goals. So just add the words *how to* in front of the five bullet points and you'll see an outline for categorizing your inventory for cost reduction. Once you have the reason for each type of inventory identified from the definitions given, you can then work on eliminating that inventory. If you can do this, then you can get the savings you are looking for. You can work on the tools needed to answer these concerns:

- How to reduce inventory
- How to lower inventory costs
- How to improve on-hand parts availability
- How to reduce annual parts costs
- How to shorten the delivery cycle

Now you know why we spent the time defining inventory. Before you can work on the five "how to's," you must define the reason you have inventory in the first place.

12.3 MANUFACTURING IS REALLY JUST A BALANCING ACT

In order to understand the different elements in a manufacturing operation, we must begin by understanding the interrelations among the functions that make an operation run. The best way to visualize this is to imagine an operation as being a balancing act between the various components of the operation, the systems, and the manufacturing capabilities. Upset this balance, and problems arise. Keep the elements in balance, and all should run fine. Understanding how these functions work will help you to understand solutions to the problems we face in operations and how the solutions affect the outcomes.

12.3.1 THE BALANCE

Take a look at Figure 12.1. We have a balance beam that represents the operation. It is a simple balance beam, not that different from a balance scale. On the left side we have the system capabilities. These are the tools that are used to run the manufacturing operation. These are the sales plans, the computer system, the suppliers' capabilities, the forecast system, the customers' requirements, and transportation issues. All the items in the systems box on the beam are the issues or constraints to be dealt with from a planning point of view.

On the other side of the balance beam is the box representing the operations in which the manufacturing capabilities reside. This box contains the production capabilities, available capacity, throughput processing capabilities, manufacturing lead time, capacity constraints, inventory record accuracy, the accuracy and completeness of the bills of materials, and the route sheets.

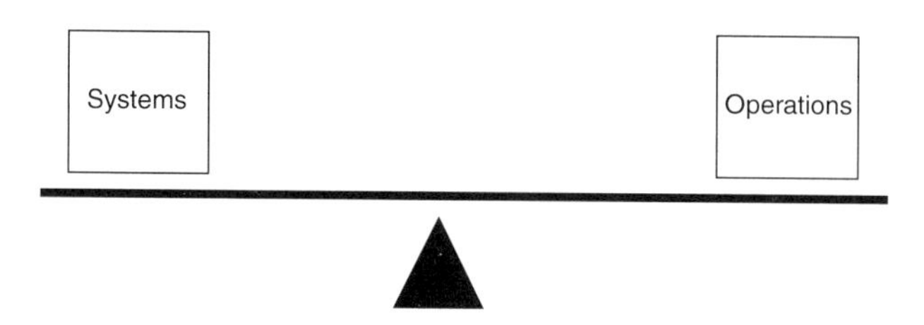

FIGURE 12.1 The balance beam of manufacturing — the basic components.

Now, looking at the manufacturing system, we will begin with the premise that when each side is equal and in balance, all is fine — sort of like the teeter-totter we played on at the park when we were kids. When your weight was the same as the person on the other end of the beam, the beam remained stationary in a horizontal position and you were in balance. If your friend was larger than you, then that end of the beam went down while you went up and were trapped high in the air. The beam was out of balance and no longer functional. If you had a really big friend, there was no way for you to rock the beam up and down, as the beam was way out of balance. To solve this problem, you could have had another friend of yours climb on the beam at your end so your combined weight could bring the teeter-totter back into balance.

Applying this analogy to the manufacturing operations, if the systems and operational capabilities are in sync, the beam is in balance. But if the system capabilities are not in balance with the operational capabilities, then the beam tips. When a manufacturing system is not in balance with operations, we can easily see what the effects are — longer lead times, stock-outs, missed shipments, or worse, lost customers. The system is out of balance and experiencing problems.

To get back in balance, we again turn to the playground example. When we were on the light side of the balance beam and up in the air, we had a friend climb on the beam with us for weight to get us back in balance, and all was working well. In the manufacturing arena we also have a friend we can add to the light side to get us back into balance. That friend is called inventory. Inventory is the weight that we add to an operation to bring it back into balance so everything is back in sync again. It can be placed on either side of the beam as necessary to regain balance. It can be used to add weight to weak systems and weak operational capabilities. In essence, the inventory box can be moved to wherever it is needed, anywhere on the beam. If placement of the box cannot add enough leverage to balance the beam, then we can add a bigger box for more inventory. This now begins to explain the quantity of inventory we have in our operations and why we even have inventory. Inventory is a universal equalizer. Inventory supports the areas of operations that are weak, and it is essential for keeping us in balance.

Look at Figure 12.2 to see how we have added weight to the beam in the form of inventory. Let's assume that our customer requires us to produce and ship an order in 5 working days. If we can do it in 5 days, everything is fine and the system is in balance. But if the customer wants the order in 3 days and we still need 5 days to deliver, then we are out of balance and cannot make the delivery. If we cannot produce and ship in the time required, we have only two options. The first is to turn down the order. The second option is to add inventory to meet the customer's

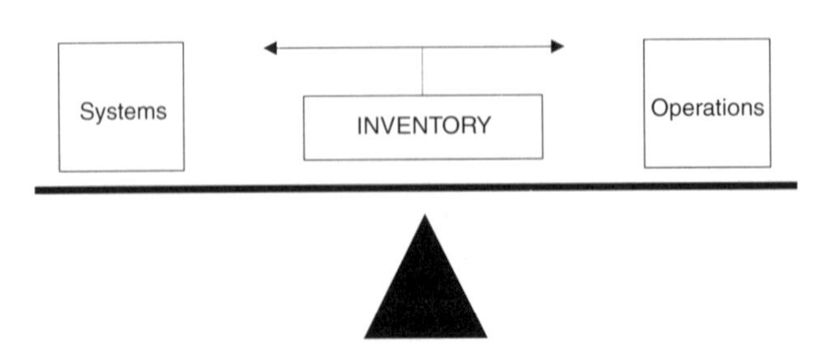

FIGURE 12.2 The balance beam — positioning inventory.

shipping demand of 3 days by shipping from that inventory. Because we are unable to meet the customer's demand for 3-day delivery with our present manufacturing and inventory policy, we must increase our inventory as a short-term solution to the problem. We ship from that inventory, and the need to stay in balance begins to determine the amount of inventory necessary to meet requirements of the customer. The size of the gap between what our customer wants and our ability to deliver dictates the amount of inventory we must keep on hand. The long-term solution is to do something about the size of the systems or operations box to enhance the robustness of the weak link in the delivery chain to meet the 3-day delivery window, but that is a long-term fix and could be costly.

If you remember your physics, you will recall that length times weight equals mass or, in plain English, multiplying the weight of the box by the distance from the box to the balance point determines how much weight is acting on the beam. This tells us how much weight is necessary and where to place the weight on the other side of the beam to keep it in balance. From this, we can see that a weak aspect of our operation can be brought back into balance just by moving the weak box farther away from the balance point and lengthening the beam until we are in balance again. Although this strategy works in theory, in reality we have a name for the length of the beam — lead time. If we move the weak link farther from the balance point on the beam by increasing lead times, we do, in fact, bring the beam back into balance. But we do so at a cost, and that cost is the amount of lead time necessary to deliver.

If business requirements have weakened a link in the system, we could bring the beam back into balance by adding length to the beam and moving one of the boxes farther from the balance point until the system is balanced again, but our lead time has now increased. Now that lead time has been added to the balance beam, we have a balance beam that looks like Figure 12.3.

We have now identified all the components of the beam that represents our operational capabilities. Now we can clearly see what happens to the system when we try to meet the customer's 3-day delivery requirement with a 5-day delivery operation. We can approach the delivery requirement in two ways over the short term — lengthen the beam and keep the 5-day window, or add inventory and make the 3-day window. In each case, we now notice that if we change one of the parameters of the balance beam we will need to change another parameter to keep the beam in balance and meet our objectives. What we see now is that there is a cause-and-effect relationship to consider when working at keeping the beam in balance. That cause-and-effect relationship means

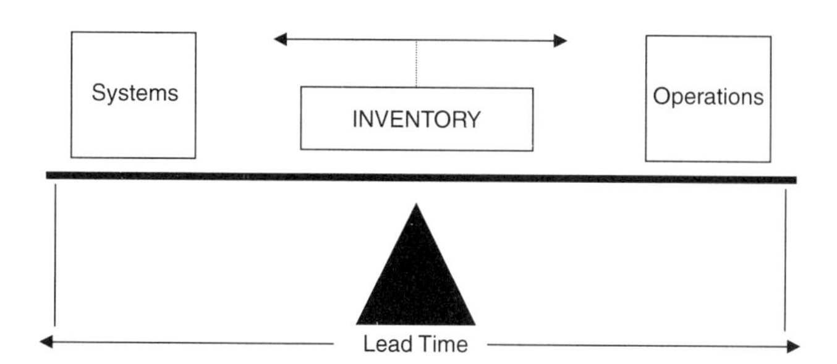

FIGURE 12.3 The balance beam — understanding the role of lead time.

that if we change one of the elements of the beam, another element on the beam must also change to keep the beam, and our operation, in balance and in sync.

These are short-term considerations. In the long term, we need to assess what our needs are and the delivery window necessary to meet our customers' requirements, and then make the changes necessary to meet the new demands and keep our system in balance. Inventory is the tool we use to keep the system in balance. If our goal is to reduce inventory and we install a new order-processing system that makes the system box more robust, we could then reduce inventory and maintain balance in the system. But, if our sales force starts to promise shorter delivery lead times to their customers based on the efficiency of the new system, we have really traded a more efficient system for a shorter lead time, and our goal of inventory reduction is in jeopardy, because inventory is now required to maintain balance to meet the new customer needs.

According to a fundamental law of physics, for every action there is an equal and opposite reaction. That is true here, too. For every change that you make to one of the elements in a manufacturing system, as represented on the balance beam, there is another element in the system that must also change to keep the operation in balance. There is a cause-and-effect relationship to everything that you do. When you want to establish a goal of reducing inventory, remember that there is also another change that must be made to the balance in your operation to attain that goal. The balance beam represents this concept clearly.

12.4 THE PRIMARY CONTROLS FOR INVENTORY

You cannot achieve manufacturing excellence by starting out with poor records. Remember that the first question you ask yourself when you receive an order is, "Do I have any of this on hand?" The answer to this question comes from your inventory records. That is the place where you go to see if you have any finished goods or components in stock to fill the order. If you do not have good inventory records, one of two things will happen, both bad. You will think that you have inventory when you don't and will make a promise to fill the order that you can't meet, or you will think that you don't have product, thus order or produce more. Now you have too much inventory. One of the things a lot of people do is called "sneaker net." You put on your sneakers, go out into the warehouse, and look for yourself. In the meantime, your buddy in the office is promising the same inventory to another customer. And so the story goes. What you need is an accurate inventory record program so that you can easily and instantaneously answer the question, "Do I have any of this on hand?"

The first thing you need to do to control your inventory is to stop its continual outward movement through "unofficial" channels. Lock it up and put someone in charge. Then, to get an accurate input of data in your inventory records system, bar code your inventory system. This will give you the 99.99(+)% record capture accuracy that you need. This will take care of the "midnight acquisition" problem and give you a tool to minimize data-transfer errors. The best tool we have seen to ensure that you will reach and maintain a high level of inventory accuracy is the old tool of cycle counting. Let's look at the tool that will help you find, control, and eliminate the human error side of the problem.

Federal law requires us to take at least one inventory each year. The tax man is waiting for this. But more important, we need to understand what we have in our inventory. The physical inventory is the most inaccurate way of determining what we have in inventory. It is basically an accounting procedure designed for tax purposes, and it does nothing for the inventory records necessary for manufacturing. As long as the numbers come close, the accountants are happy and we can all go home. But the big problem from a manufacturing point of view is that the taking of the physical inventory does nothing to correct the cause of the problem that created the inventory errors in the first place. So next year when you take the physical, you will find the same errors, and make the same adjustments to the inventory record, but you are still stuck with the problem. You have gained nothing.

One of the biggest abuses we find with cycle counting is that the name is used without understanding the technique. The abuse? Calling the taking of a monthly physical inventory "cycle counting." We find people who recognize the need for having excellent record accuracy, but all they do is count it over and over again. As we have said, this physical approach does nothing to correct the cause of the problem.

What you want is a tool that will not only give you a higher level of inventory record accuracy, but also lower the cost of maintaining that level of accuracy, while still keeping your operation in business. Remember that you shut down your operations to take the physical and you lost all that production time. With cycle counting, you keep right on going while you're doing the count. And do you know who handles the cycle count? The people in your stockroom, that's who. And do you know why? Because they are the ones who are the most knowledgeable in your operation as to what your materials look like and what the part numbers are, and they are the ones whose lives will be made the easier by having your inventories under control. All is not free in this world and cycle counting does come at a cost, but the cost savings can be immeasurable.

Let's take a look at both the physical and the cycle counting methods of checking your inventories. The following is a list of disadvantages of taking the physical.

- No correction of causes of errors
- Many mistakes in part identification
- Plant and warehouse shut down for inventory
- No improvement in record accuracy

Now let's look at the advantages you can gain by using the cycle counting approach, in terms of the same items:

- Timely detection and correction of causes of error
- Fewer mistakes in part identification
- Minimal loss of production time
- Systematic improvement of record accuracy

Cycle counting is basically very easy. Every morning you come in and count a portion of your inventory. The cycle counter is given a list of parts to count and given all the information about the part that is available except one. You never give the cycle counter the number of parts your records show that you have in inventory. Why? Because if you send someone out to find 1675 "unicroms" in your inventory bin, guess how many unicroms he or she will find: 1675, that's how many. Of course, it is easier to count an inventory location when the bin is near empty, as there will be fewer parts to count. So this is when you do your cycle count, when it is time to reorder. Because cycle counting is a daily activity, you can then choose when to make the count, so you do it when the bin is empty. This minimizes the workload.

After the count is complete, you check the record, looking for a count match. If the counts don't match, this is the list of items to do in sequence:

- Total counts for all locations
- Perform location audits
- Recount
- Check to account for all documents not processed
- Check for item identity
    - Part number
    - Description
    - Unit of measure
- Recount again if needed
- Investigate error factors
    - Recording error
    - Quantity control error
    - Physical control problem
    - Positive and negative counting errors

Now that the count is complete and you know the reason for the errors in the records, you then just change the records, right? Wrong! Now that you know the reason for the error, you correct the cause of the error so that it won't happen again. But this sure sounds like a lot of work that we don't do now. And by the way, how many times a year do you count your inventory items?

Generally speaking, to meet IRS standards, you must count your entire inventory at least once a year. So that is what you do to the C stock items only. You count your B stock twice a year and your fast-moving, high-dollar items at least six times each year. Sounds like we have added a lot of work, but we really haven't. Let's look at the workload for the people in your stockroom. As an example, we'll

TABLE 12.1
Inventory Counts

		Work Load		
Inventory Class	Number of Items	C/C Counts per Year	Total Count C/C	Total Count Physical
C	8000	Once	8000	8000
B	2000	Twice	4000	2000
A	500	Six	3000	500
Total inventory counts per year			15,000	10,500

assume you have 10,500 stockroom parts. Table 12.1 compares cycle counting with physical inventory and shows the relative workload of each method of taking inventory.

And, yes, you would be right. The workload did go up by requiring an additional count of 4500 parts per year. But look at the workload. Cycle counting should be done every day. If your operation works 5 days a week, 52 weeks a year, then you would work 260 days a year. If you cycle counted 15,000 parts per year divided by 260 days, then you would have to count only 58 parts per day. Is that a lot? Not really. First, if you have this size stockroom you probably have more than one attendant, probably three, one for each shift. Now you are looking at 19 or 20 parts per person per day. Not much of a workload here. And the workload gets even less.

When you do a physical inventory, you must count all the items at the same time. As such, some of the bins are full and some are empty and on the average they are half full. So you are counting an average volume of inventory items. But when you cycle count, even though you count some items more than once a year, you can choose when in the year you will do the count. How about when the bin is at or near empty? Count accuracy goes up and the workload goes way down. And think of this: once you have completed a count, the cause of the error has to be resolved so that subsequent counts will be simple. The workload is going down.

Now consider this additional information. Realizing that the operations will be ongoing when the count is made, you will save the production time lost for inventory record purposes. Here is a list of how to determine when to cycle count to save you time and money and to minimize the inconvenience to the operation.

- Count when the bin record is near empty.
- Count at reorder point (also verifies the need for the order).
- Count during off shifts when no receipts are processed.
- Count early in the morning just after the MRP has been updated and parts have not been pulled for the day's operations.
- Count when a bin record shows less than needed for an upcoming job.
- Count C items at the slow point of the year. And take a look at this one:
- Do a cycle count on the empty bins.

Why on earth would you want to count a bin that your inventory records show has nothing in it? Because that is where you misplaced that last shipment of gold bricks you haven't been able to find. You put them in a bin that your records show is empty. And if you think it is empty and never look at that bin location, you will never find that pile of gold bricks!

After you have gone through all this and have discovered the errors of your ways, it is finally time to correct the inventory bin record. Accountants may not like this because you are always changing the value of your assets, but this can easily be handled with a variance account. Do you need to continue with the physical inventory? Generally speaking, you will need to verify to your auditors that the cycle-counting procedures are better than the physical. It usually takes two physical cycles to establish credibility and stop taking the physical.

So are there savings in the cycle count? Yes, and here they are

- Elimination of the physical inventory
- Timely detection and correction of inventory errors
- Concentration on problem solving
- Development of inventory management specialists in your stockroom
- Fewer mistakes
- Maintenance of accurate inventories
- Reinforcement of a valid materials plan
- Less obsolescence
- Elimination of inventory value write-downs
- Correct statement of assets

And now a final word about the physical inventory. When is your inventory most accurate using the physical count? The record is at its best the day after the physical count has been completed and goes downhill from there for the rest of the year. And the list of all the benefits of the physical inventory procedure is very short. The savings generated by taking the physical inventory are

- None

12.5 THE TOOLS FOR INVENTORY CONTROL

Take a real hard look at this. This is where the theory meets the road. We hope you are beginning to understand that, throughout this book, the concepts that are talked about are not just concepts. They are things that you can and should do in your corporation that will generate real savings for you. The following concept is one of the best revenue generators and cost-reduction approaches you will discover. Taking something as simple as the ABC inventory concept and applying it to your operation is something that can generate both profits and savings for your company.

First, you have to understand the nature of inventory in meeting your customers' needs (either an inside or outside customer). If you cannot produce within the demand window that the customer requires, then of course you must ship from inventory. Inventory is then the medium that you use to meet the needs of your customer in both time and quantity. Having said that, are we saying that you need to own the world's supply of everything that you sell? That would ensure that you could meet any demand the customer would require of you, wouldn't it?

Yes, but look at the expense. The notion of having an inventory huge enough to meet any and all customer demands is obviously cost prohibitive. But what if we could "own the world's supply" of inventory of our products? We would never have a late, partial, or missed shipment, period. But how can we do this? How can we accomplish each of the following two conflicting objectives? On the one hand, we want to reduce cost by minimizing the amount of inventory that we have on hand. Balanced against this we also want to meet every customer request by carrying all the inventory needed to meet their requirements. The tool to accomplish both these objectives and give you the best of both worlds is the ABC method of evaluating inventory.

12.5.1 The ABC Inventory System

The first thing to do is to stratify your inventory by an ABC classification. This is the starting point to begin to understand the concept of your inventory. The Pareto principle stated that 80% of your sales come from 20% of your part numbers and, conversely, the bottom 20% of your sales must come from the remaining 80% of your part numbers. This is the significant few and the trivial many. But rather than use only two categories, we use three, A, B, and C, and then apply different approaches to managing the inventory based on its category. You need to apply different thinking and tools to manage inventory based on the classifications that you determined.

Once you have your inventory categorized and displayed in descending order by annual usage value, you will begin to see the cost value by classification. You will see that the A-classified items will represent about 70% of your total dollar value of inventory and probably the same amount of your revenue from sales. Your B items will be another 15% of your value and the remaining C stock will be the remaining 15% of value. But the number of parts or stock keeping units (SKUs) represented by the A items, while being 70% or so of value, will be only about 15% of your part matrix. The B stock will be about 15% of both value and SKUs while the C stock will be the remaining 15% of value but a whopping 70% of your part numbers. So if this is true, why do you insist on using the same inventory tactics across the board. It takes different strokes for different folks, or different approaches for different inventories.

First, we need to understand this basic fact of C items in stock. Volume is low and customer demand is erratic at best. This means that you can't forecast it to begin with, but not having a C part in stock can cause a missed shipment. Remember that the customer doesn't care what your problems are. The customer just wants what it ordered and doesn't care how you go about meeting the demand. So, if you can't forecast it, then you can't manage it. You can't develop an inventory plan, so no matter what you do, you can't ship it when the customer wants it. You have a stock-out. So, what do you do?

Give someone else ownership of the inventory and the responsibility of maintaining it. This tool is called the *vendor-stocking program*. The materials are still on your floor and available to ship, but the management and ownership of those materials belong to someone else. These are the parts that are a small volume of your business, so why not turn over the management of them to someone who takes ownership of them? It may be a small portion of your business and is therefore not significant. And the management of this C stock inventory is not one of your core competencies. But to your suppliers this part of your inventory is a major portion of their businesses and a core competency for them. You have other fish to fry and cannot manage this inventory as well as your vendors can. Usually a company can turn over the management of about half of its C stock to others. This means that by giving up control of the dollars on about 7.5% of your stock, you have relieved yourself of doing the work of managing about half of your part numbers associated with C stock. You no longer need to control about 35% of your part numbers and you have retained and improved the delivery rate on these items at the same time.

Now let's take a look at the other half of your C stock. This is another 7.5% of your inventory or sales value of unmanageable stock. What we suggest you do here is to buy a 1-year supply and put it on the shelf. Don't choke on this. How many inventory turns do get on your C stock now? Two? Three? And how many stock-outs do you get on this inventory a year? And what is the cost of air freight on this portion of your stock? And what is the cost of all the expediting done by your staff? And what did it cost you to place two or three orders for this stock each year? And what did it cost you to manage this stock last year? How many sales or customers did you lose last year because you were not able to fill the order? OK. Got the answers to those questions?

Now take a look at the savings. By having a 1-year supply on hand, you can have this happen only once a year, period. Yes, you've doubled or tripled the investment in your inventory. But assuming that a storage bin is always half full or half empty (your point of view), the worst case is that you have taken 7.5% of the annual usage rate divided by two (half empty), which is 3.75% of your total usage rate and tripled it. How big a number in real dollars is it for you to triple 3.75% of your inventory's annual usage rate? Not much. Then compare the costs of managing two or three turns and it will quickly become apparent that you are much better off with a year's supply of this kind of inventory.

The way to handle the B stock is to place it on an automatic reorder system, such as a min/max. There is sufficient volume in this inventory to begin to forecast and manage it accurately. But it still is not worth your time to oversee it. So, let the system automatically reorder inventory but with limits set on how much you will allow the use rate and order interval to change from the forecast before you intervene. You, not the system, manage the exceptions in this case. Most people feel comfortable in allowing the order volumes to fluctuate up to 20% of the forecast before intervention is needed.

So now we have given you some tools to relieve you of the work of managing 85% of your part numbers and 30% of your costs. And having so many part numbers in the B and C groups and so few dollars tied up in them, just think how extensive a cost-reduction program would be needed to get any savings out of this inventory.

Now we get to work on the part of the inventory that you can affect: the big movers, your A-category inventory. Now you have something that you can accurately forecast on both the supply side and the demand side of the equation. This is where the forecasting tools come into play. This is where you apply JIT to the inventory. This is where you receive the materials weekly. Or daily. Or even hourly. It is now cost effective for you to do this. If you now have three turns in your A stock, you have in stock half of 70% of your cost, or 35% of your annual usage cost, divided by three, or about 12% of your inventory cost would be with you all the time. If you managed this inventory on a weekly basis you would have in stock 35% of your cost not divided by three, but divided by 52. You would have only 0.5% of the annual usage value in stock, for an inventory reduction of 34.5%. Real money.

Let's add up the potential savings:

- Bottom of the C stock now vendor managed. *Savings: 7.5% of inventory value.*
- Top A items managed weekly. *Savings: 34.5% of inventory value.*

Let's add up the break-even trade-off:

- All B stock that is now on a min/max system. *No change in value.*

Let's add up the costs:

- Top of the C stock now an annual buy. *Added costs: 9.75% of inventory value.*

The bottom line is that you have reduced your on-hand inventory by 31.25%. Think about the accuracy savings. Now you have something.

What we are really saying is that the money invested in inventory is in the high-volume parts. And by stratifying inventory by ABC you now have the opportunity to apply some tools that will affect the overall investment in inventory. A basic fact to remember is that the most important part at the time of shipment is the part that you do not have. It doesn't matter if it is a high-volume, expensive part or the lowest cost, lowest volume part in your part matrix, if you don't have the part you will not ship, period. If you use a min/max system to manage your inventory, you set an order quantity and determine an inventory level such that when your inventory gets down to that point, you reorder and hopefully receive new materials just as the bin becomes empty. This is a system that reacts to past history and assumes a level demand and a fixed constant replenishment point. If you use this type of system you will have an inventory valuations curve that will look like Figure 12.4. You will tend to have a constant investment in inventory at all levels and generally a higher than necessary investment in inventory.

If you now include safety stock on your inventory investment curve, you have what is called a permanent investment in fixed inventory. You never sell your safety stock. Figure 12.5 outlines a typical safety stock fixed curve in inventory. Notice how the curve increases in the C stock because your forecast accuracy decreases in this area.

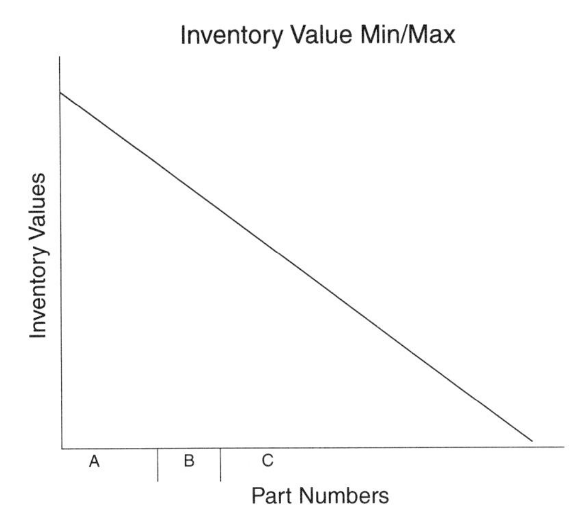

FIGURE 12.4 Inventory value min/max.

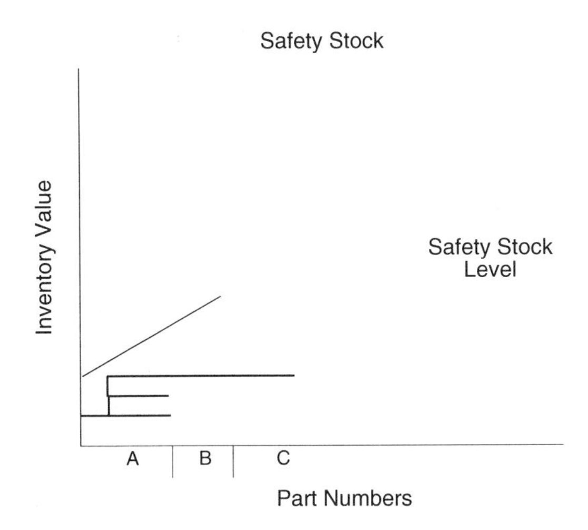

FIGURE 12.5 Safety stock impact on inventory value.

This is due to less usage and, therefore, less accuracy in the forecast that requires more safety stock to protect against the unknown. Figure 12.5 shows that increase.

If you combine the two curves from Figures 12.4 and 12.5 you will get something like Figure 12.6. High inventories are needed to meet your customers' requirements. You will have a lot of money invested in inventory and, as we said earlier, inventory does not add value to the product. When you go to McDonald's and order one Big Mac, the fact that there are ten other Big Macs on the shelf does not mean that they can charge you more. Because you want only one Big Mac, the other nine on the shelf add no value to your purchase. The other nine are on the shelf to level load

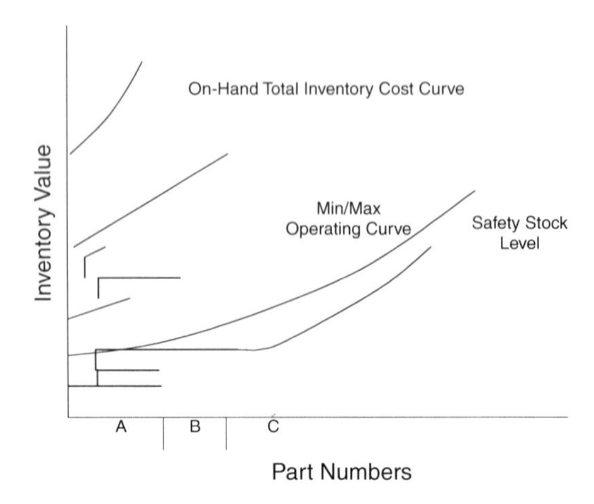

FIGURE 12.6 Total inventory value by classification using min/max.

the production cycle and meet your random demands. But think about how much cheaper it would be if they could meet your demands with no inventory. What we want to do is achieve two conflicting objectives. We want to meet the customer's delivery requirements by keeping the balance beam that we discussed in Figure 12.3 in balance. At the same time, we want to reduce the fixed cost that we have in inventory. We cannot reduce inventory and at the same time improve our delivery capability to our customers. But if we stratify our inventory and apply some of these tools, we can achieve both goals and keep the operation in balance.

The first thing that you want to do is turn over your bottom C stock to a third party. This is vendor-managed and vendor-controlled inventory and it is in your operation on consignment. The management of this inventory is a core competency of the supplier and not yours, so this is something that you want to give to someone else to do. In almost all situations that we have seen, there are only rare instances of a missed shipment because of a stock-out. You have achieved the best of both worlds, reducing the value of your inventory to zero and at the same time virtually eliminating the possibility of a stock-out. And in most cases, the part cost has gone down, too. Figure 12.7 shows how your inventory cost curve looks with consignment C stock.

The next stratum of inventory, the higher C stock, is the oddballs in your product matrix, with low annual volumes or stock that no vendor can manage or wants to. Because of the volume, you virtually cannot forecast the use rate. This is the nonforecastable and unmanageable portion of your inventory and there is nothing that you can do about it. So, you buy a 1-year supply of parts, put them on the shelf, and forget about them. Figure 12.8 shows this inventory. Worst case dictates that you would have only one stock-out a year. Leaving a 1-year supply, you are now fairly certain that the materials will be there when you need them. And though you have an increased investment in inventory, you have decreased order-processing costs because you have to process only one order per year. And think about all the air freight costs you can save that you spent on getting this stuff into your operations when you've run out. So buy it, put it on the shelf, and forget about it. One year is

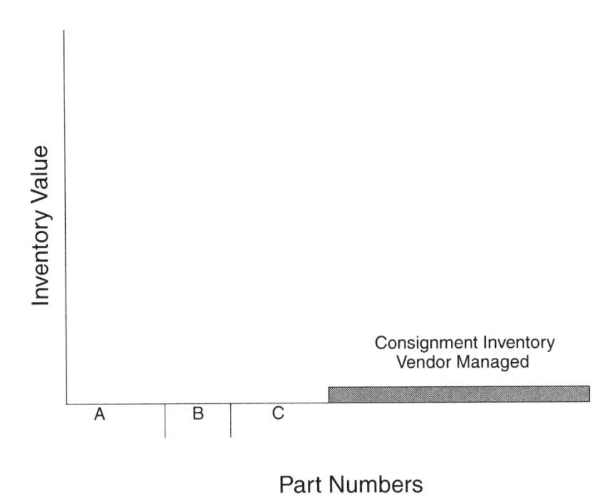

FIGURE 12.7 Inventory component cost as a result of consignment inventory.

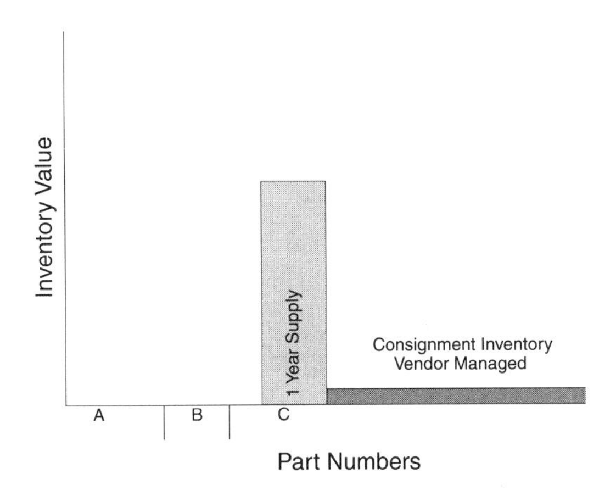

FIGURE 12.8 Inventory component cost as a result of a year's supply of additional inventory.

a large enough supply; things change over time, so cut off the supply at a year in most cases.

You are still going to use the min/max system, but only for the high C and low B categories. In Figure 12.9 you have some volume and a steadier demand for the part but not a high investment based on the use rate. So you let a mechanical system such as min/max reorder these parts, and though you may carry more inventory under a min/max than if you managed these parts individually, the cost to manage far exceeds the cost of carrying the extra inventory. This system will do a relatively good job of batch-ordering materials. At the same time, you establish parameters within which the computer can generate the order. This means that you get involved only when fluctuations vary beyond the set parameters, usually not more than plus or minus 10% of the projected demand, saving you considerable time. If the use

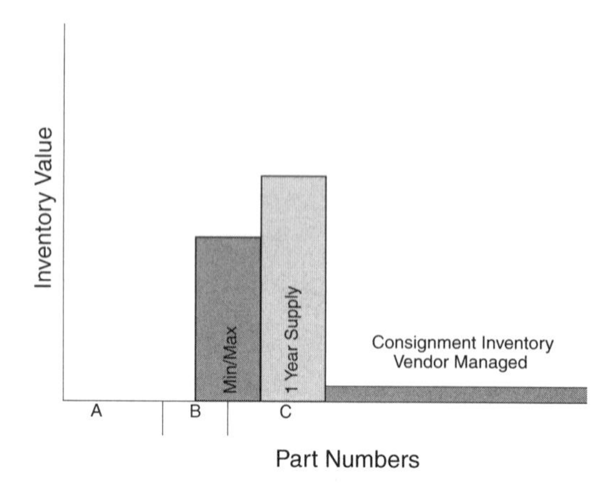

FIGURE 12.9 Inventory component cost as a result min/max inventory.

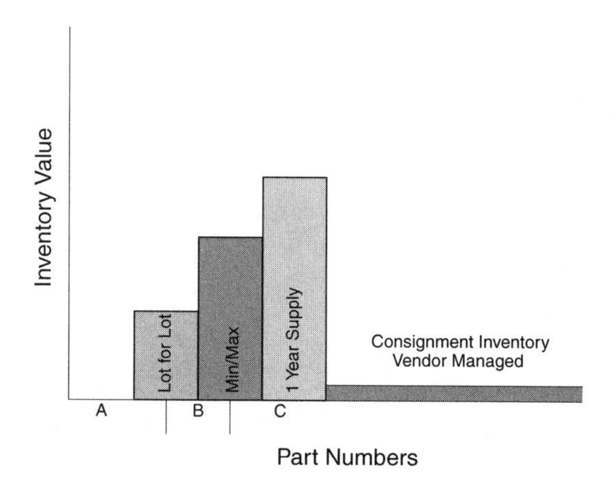

FIGURE 12.10 Inventory component cost as a result of lot-for-lot inventory.

rate or delivery rate exceeds the tolerance, the computer generates an error message and you would then manually get involved. This manual intervention protects you from an out-of-control ordering system.

What we have done up until this point is to either eliminate or greatly minimize the need for direct management control over the status of the inventory. While you can spend as much time managing a low-volume part as a high-volume one, it's not a good use of your valuable time. Either let someone else do it or have systems in place that require a minimal amount of your time. You want to focus all your time on managing the high-volume dollar amounts that you have in stock.

The next level of inventory is shown in Figure 12.10. This is a high-budget area involving only a small number of parts. In this segment of your inventory the best approach is to apply the lot-for-lot technique. It is a batch buy, but you buy only what you actually need to satisfy the near-current demand or possibly what is in the near-term forecast.

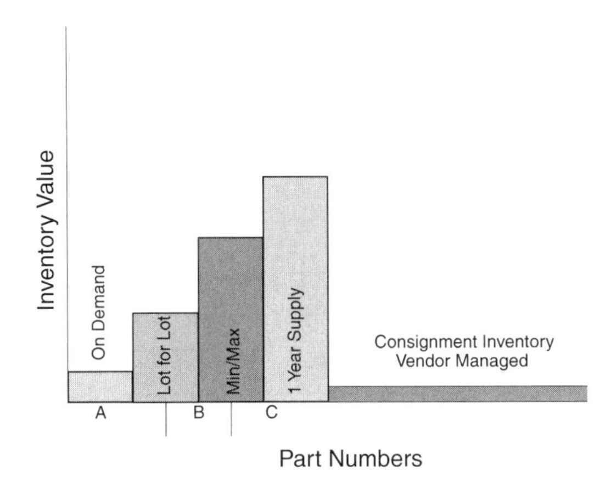

FIGURE 12.11 Inventory component cost as a result of on-demand inventory.

This now leaves the final portion of your inventory, the top-level parts. Here you buy exactly what you need with a just-in-time approach. These are high-volume parts that are easily forecast. You buy only to a firm production schedule. You might need some inventory of this type of part if it has a long lead time for delivery, but this is rare. Usually these are common parts that can be easily sourced and should not give you delivery problems. In the rare situation wherein there is a long lead time on one of these parts, the vendor may be making a lot of money selling these parts to you. This occurs when you ask him or her to stock the parts for you, and you give the supplier your production requirements and have them shipped to you on demand. In that way, there is no inventory and you are again ordering and paying for the materials on a JIT basis. Most suppliers will work with you on these high-volume components. This situation is displayed in Figure 12.11.

Figure 12.12 shows the total cost curve connected. Compare this with Figure 12.6 and see how much inventory you need to run your operation. And look at your workload. This is how the ABC inventory stratification process manages your inventory.

12.5.2 Capacity Capability and the Effect on Inventory

As inventory is a measure of how well you manage your operation, there are several tools that allow us to manage our operation to meet both customer and inventory requirements. Takt time is the building block of lean manufacturing. The question is, "How can I run my business using the minimum of effort, machinery, and inventory?" The easy answer to this is to properly use your operation's *demonstrated capacity* — the capacity that you have actually demonstrated you can attain. This is not what the machine salesman said the machine would produce when he sold it to you. It is not what the machine can produce when it is running or what it can produce when it is running at its best. Demonstrated capacity is what you actually can get from the machine. In the same way, demonstrated capacity for your operation is what you can actually get from the operation, given real machine capabilities, schedule requirements, etc. It is the maximum that you can load the operation.

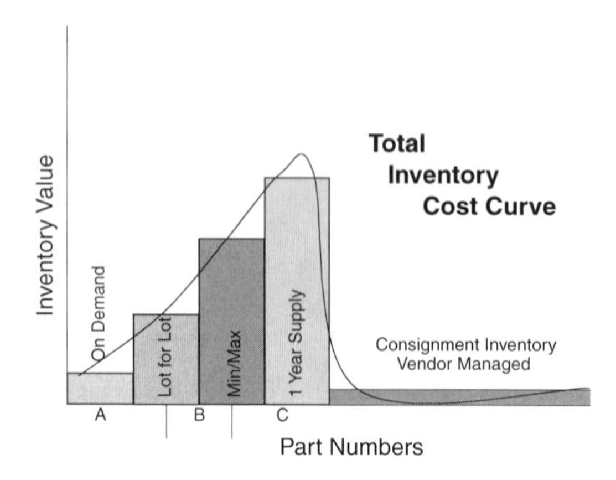

FIGURE 12.12 Resulting cost curve developed by applying specific tools to ABC-segmented inventory.

This is what your sales people sell: the next available minute of capacity. If your salespeople want a ship date sooner than the next available minute, your answer goes something like this. "We can give you the date the salesperson wants, assuming that the parts are here, but to do this, what in the current schedule would you like delayed so that we can build this part?" The delayed part must then move to the next available minute of capacity. Ask your salespeople this. See what kind of an answer you get. If neither option suits your salespeople, then ask your finance officer to either authorize overtime or make an investment in inventory. If your salespeople cannot work within your capacity, then you must either ship from inventory or authorize overtime (if available) to make the ship date. There is no other way.

12.5.3 Production Constraints

There is always some part of your operation that is constantly loaded to its demonstrated capacity. Some call this a bottleneck and treat it as if it were a negative to your operation. But a bottleneck is something that is good to have. It means that you are utilizing your operation to the utmost of its capability. Eli Goldratt has developed the theory of constraints (TOC), and it is relatively simple to understand. First, determine which piece of equipment in the operation is loaded to the fullest, which is fairly straightforward. Then schedule your operation so that the constraint is given top priority. In other words, schedule the bottleneck first, and then schedule all the other parts of your operation. In reality, you can only have one bottleneck. All other components of your operation are loaded to somewhat less than their demonstrated capacity. Understanding the bottleneck is like understanding the orange juice squeezer in Figure 12.13.

Think of your operation as a manual juice squeezer and that this is the bottleneck. I have work-in-process inventory waiting to be squeezed. I have containers waiting for the finished product, but I just can't get enough juice through the squeezer. Production is the forcing of the handle against the orange and no matter how much

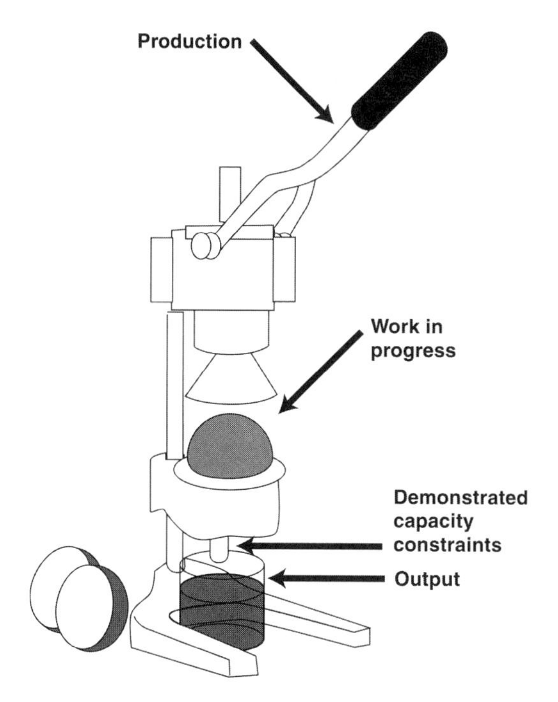

FIGURE 12.13 Manufacturing constraints as they would look if your process was a juice squeezer.

force we exert, the output nozzle will allow only a fixed amount of juice to pass. There are two nonproduction times in the cycle which little can be done to eliminate: the time required to load the part and the actual production time. No matter what you do with this piece of equipment, the capacity is fixed and there is no more output to get from it. There are only two solutions to gaining increased capacity: buy a second squeezer or make a larger hole in this one.

Managing the point of constraint is key to managing your operation. The only tool necessary to manage the constraint bottleneck is that you must load it first. When dispatching orders to the shop floor, you must schedule, load, and sequence the constraint first. Then schedule the next highest loaded machine, then the next and the next and so on until the operation is loaded.

Case Study on Constraint Management. I had a project at a company located in Montreal that had problems getting product shipped on time. They were a manufacturer of custom one-of-a-kind equipment with a 3-month average delivery lead time. They had design engineering scheduled into the delivery lead time. They had a master schedule for the production. They sequenced the shop floor and understood demonstrated capacity and constraints. But they were always late. I worked with them for 6 months from the corporate office to solve the problem. The product managers were arguing in the president's office at the end of every month trying to expedite their products to keep some of the customers happy. When I studied the shop floor, it became obvious that there was no machine or piece of equipment that was utilized to the limit of its demonstrated capacity. Therefore, no bottlenecks. But the master schedule showed that there was a bottleneck. Everything was late. Where was the problem? When I got into the operation, I discovered that because of the product mix in the operation, the loads on the equipment varied from month to month based on what was to be produced. The bottleneck existed. But it moved. Each month there was a new bottleneck to the operation and no one knew it. The problem was that the plant continued to allocate production to what they thought was the bottleneck when in fact there was a new bottleneck. By the time they finally realized that the bottleneck had moved, it was the middle of the month. They discovered that the old bottleneck was no longer their production constraint when they ran out of jobs to process on that equipment.

Because of the long production time required for each special product, the dispatch of jobs to the floor was in monthly increments. It could be as late as the second or even third week in the month before they realized that there was a shift in the plant loading. Of course, when they discovered this, 2 to 3 weeks of production on the new constraint had been lost. This was the cause of the end of the month hassle to get product out the door. It took 6 months to figure this out.

The solution was simple. All the information we needed was available, but it wasn't used. All we did was to take the available orders and develop the outgoing plant-load report enough ahead of time to include all open and planned orders. Then we knew what the planned and proposed plant load was. Then, by performing a plant machine load report, we could see where the bottlenecks were. Knowing where the bottlenecks were, we could then sequence the jobs to use the plant load. It took 2 months to develop the reports; then, with a schedule report in the foreperson's office, we could monitor progress daily and load each of the bottlenecks. The problem wasn't completely eliminated. The arguments continued. But there was something new. The arguments were no longer on the shop floor; the fighting was now in the president's office as each project manager tried to get on the schedule. It was a nice problem to have because the situation moved to a control point, and the due dates began to be met. Think about your operation and understand that constraints (bottlenecks) can, in fact, move. Planning is the solution to solving the problem.

As tools to manage operations become more robust, substantial savings can be generated through better control of operations. But there can be some serious penalties with these new systems and philosophies of operation. These penalties arise because companies are living very close to the edge of disaster. In the past we could use inventory as a cushion to keep our outbound shipments flowing to our customers when we were faced with major disasters. On-hand inventory is what we used to keep materials flowing when faced with problems such as equipment breakdowns, quality problems, late shipments from our suppliers, and all the other difficulties of everyday manufacturing. With the advent of E-commerce and other high-speed tools to receive orders, and high-powered software systems to control operations, savings have emerged in the form of reduced inventories. However, we still have problems in our manufacturing processes and we have eliminated the cushion or insurance inventory from our operations. So now when we have a problem, our customers will have a problem, too. Be advised that one of the ways your customer can resolve your delivery problems is to get rid of you as a supplier. This is a very effective tool, and in all cases it works very well. Obviously, this is not the solution you want.

The tool that we use today is called lean manufacturing. Lean manufacturing is a technique that allows you to effectively meet your customers' requirements with a minimal amount of inventory, a minimum of planning and scheduling effort, and a high machine utilization for your operation. The key to accomplishing these objectives is to match your production output to your customers' order requirements. Sounds easy but it's really hard to do. The tool to accomplish this is called *takt* time. Takt time is at the core of lean manufacturing. In effect, takt time is an algorithm that allows you to flat load your facility, which reduces the day-to-day variation in production load experienced because of fluctuating sales requirements. Takt time is a tool that matches your capacity with the customers' requirements, allowing you to reduce inventory and manufacturing costs while still meeting the customers' demands. Takt time breaks the customer forecast into repetitive units of time, creating a basic repetitive schedule per unit that, in multiples, allows you to produce the same quantity of product day in and day out. You have now flat loaded your plant. However, there are a few requirements you need to understand to allow you to do this.

First, you must be able to forecast your customers' requirements well enough to schedule your production over a fairly long period of time, usually 6 months or more. Next, you must know what your demonstrated capacity is to ensure that you can meet the production requirements on a daily basis.

Your customers' requirements should be fairly constant over the planning period. But be careful with this one. We were talking to a first-line supplier of steering wheels for an automotive manufacturer and at first meeting the demand sounded simple. The automaker's production rate was 60 cars per hour on two assembly lines, for a total of 120 cars per hour. Obviously, every car gets exactly one steering wheel, which means you need to flat rate your plant for 120 steering wheels per hour. The question is, which steering wheel do you need to produce? What color is the wheel? Are the radio controls mounted on the wheel? And which radio does the car have? Is there a wheel-mounted cruise control? So now what appears to be a standard part for your customer, the automaker, is really a custom mixed order for you. Generally, automakers tell you which steering wheel will be required 8 hours before you need to have that wheel in its inventory in the assembly sequence. Can you meet this delivery window with what is now a variable product? How much lead time do you need to produce this product? How much WIP inventory will you need to have the parts available when the final sequence is made available to you? Take this into consideration when you are scheduling. If you are unable to meet the product variation inside the 8-hour window, you have one solution: inventory.

But say you have a solid useable forecast and can meet the demand window with your capacity, and you can handle product variation but your customer has an erratic schedule. Your takt time will average the demand for the flat load but cannot manage the daily variations. This is where inventory comes in. All you need to do is carry enough inventory to cover the worst possible scheduling variation. In that way, you can flat load the facility based on the long-term forecast and ship on a variable basis to the customer. Schedule variations for higher than flat-rate production loadings are then made up from the product in inventory. When schedule variations are lower than your flat rate, you meet the demand with what you produce and put the excess back into inventory.

But what do you put into inventory? You place your standard high-volume parts in inventory — not your customers' high-volume parts, but your high-volume parts. Because the high-volume parts turn quickly in your inventory and are easily forecast, it is rare that any of these parts become obsolete because you are always monitoring them and they move off your shelf long before becoming obsolete.

When your flat rate meets customers' demand, you then produce all parts, both high volume and low volume, to meet your customers' demand. (You might want to rotate inventory here just to keep a fresher stock of parts.)

At periods of high demand you produce the odds-and-ends, low-volume parts of your product mix first and then all the standard high-volume parts up to the flat rate determined by takt time. You then ship the rest of the order above takt time from inventory.

How much do you keep in inventory as a buffer against the variability of demand? Simple answer. You carry enough inventory to cover the worst possible case of excess demand that your customers could ever require. That way you have flat loaded your plant and kept a constant schedule to most efficiently run your operation. At the same time, you are sure of meeting any and all demand variations from your customers.

Inventory is your tool to control variation in demand.

13 Robust Design

John W. Hidahl

Robust design is a methodology for improving product quality and reducing cost. It is generally recognized as being Dr. Genichi Taguchi's approach for determining an optimum set of design parameters that maximize quality, maximize performance, and minimize cost. Robust design techniques are applicable to all mechanical, electrical, and electronic hardware configurations. This well-proven methodology provides an efficient and effective disciplined approach to developing optimized designs in a design-to-cost (DTC) or cost-as-an-independent variable (CAIV) environment.

Today most U.S. engineering organizations focus on system engineering design and system tolerance design to achieve their performance requirements. This often leads to excessive product manufacturing costs and product delivery-cycle times. By forcing the system tolerance design process to minimize or eliminate the performance parameter variability that can have a large negative impact on the system, operability and functionality, higher costs, and longer cycle times are inadvertently imposed upon manufacturing. The higher costs arise from added inspections and higher scrap, rework, and repair of the product, due to the establishment of tight design tolerances. The longer cycle times result from all the added manufacturing process steps that must be performed to deliver quality products. The proper use of Taguchi's parameter design techniques to optimize performance while reducing sensitivity to noise factors is a preferred method that minimizes or eliminates the requirement for tight system design tolerancing.

Beginning in the 1950s, Dr. Taguchi developed several new statistical tools and quality improvement concepts based on statistical theory and design of experiments. The robust design method provides a systematic and efficient approach for finding a near-optimum combination of design parameters, producing a product that is functional, exhibits a high level of performance, and is insensitive or "robust" to noise factors. *Noise factors* are simply the set of variables or parameters in a process that are relatively uncontrollable, but can have a significant impact upon product quality and performance.

There are three primary advantages to a robust design. First, robustness reduces variation in parts and processes by reducing the effects of uncontrollable variation. More consistent parts mean better quality parts, and thus better quality products. Similarly, a process that does not exhibit a large degree of variation will produce more repeatable, higher quality parts. Second, a robust design enables the use of nonprecision, commercial off-the-shelf (COTS) parts, which saves development and production time and money. Finally, a robust design has more customer appeal and acceptance. Customers expect purchased products to be robust and, therefore, tolerant to the severe exposures and applications for which they were designed.

1-57444-300-3/02/$0.00+$1.50

13.1 THE SIGNIFICANCE OF ROBUST DESIGN

Many studies have been performed demonstrating that the early design phase of a product or process has the greatest impact on life-cycle cost and quality. These studies showed that the use of robust design techniques enables substantial product development and production cost savings, as well as cycle time reduction, when compared with more traditional design–build–test–redesign iterative approaches. Significant improvements in product quality can also be realized by optimizing product designs.

To optimize the performance of a product or process, it is necessary to consider three essential system design elements: system engineering design, system parameter design, and system tolerance design.

System engineering design is the process of applying scientific and engineering knowledge to produce a basic functional design that meets all customer-imposed and internally derived requirements. A prototype model of the design is typically created and tested to define the configuration and attributes of the product undergoing analysis or development. The initial design is often functional, but may be far from optimum in terms of quality and cost.

System parameter design is the process of identifying the set of independent variables that greatly influences and thus controls the quality and performance of a product. In the design phase, a set of design parameters is investigated to identify the settings of the various design features that optimize the performance characteristics and reduce the sensitivity of engineering designs to sources of variation (noise).

The third element, *System tolerance design*, is the process of determining tolerances around the nominal settings identified in the parameter design process. Tolerance design is required if robust design cannot produce the required performance without costly special components or high-process accuracy. It involves tightening tolerances on parameters where their variability could have a large negative effect on the final system. However, tightening tolerances almost always leads to higher costs. Robust design focuses on the middle process, defining an optimum set of parametric control-factor settings.

Robust design, which is also known as parameter design, involves some form of experimentation for evaluating the effect of noise factors on the performance characteristic of the product defined by a given set of values for the design parameters. This experimentation seeks to select the optimum levels for the controllable design parameters such that the system is functional, exhibits a high level of performance under a wide range of conditions, and is robust to noise factors.

Varying the design parameters one at a time as individual changes while attempting to hold all the other variables constant is a common approach to design optimization. Trial-and-error testing using intuitive and visceral interpretations of results is another common method used. Both of these approaches can lead to either very long and expensive time spans to verify the design or a termination of the design process due to budget and schedule pressures. The result in most cases is a product design that is far from optimal. For example, if the designer studied six design parameters at three levels each (high, medium, and low), varying one factor at a time would require studying 729 experimental configurations (3^6). This is referred

TABLE 13.1
L8 (2^7) Orthogonal Array

Column Experiment #	1 A	2 B	3 C	4 D	5 E	6 F	7 G	Outcome Being Measured
1	1	1	1	1	1	1	1	X
2	1	1	1	2	2	2	2	X
3	1	2	2	1	1	2	2	X
4	1	2	2	2	2	1	1	X
5	2	1	2	1	2	1	2	X
6	2	1	2	2	1	2	1	X
7	2	2	1	1	2	2	1	X
8	2	2	1	2	1	1	2	X

to as a "full factorial" approach, wherein all possible combinations of parametric values are tested. The project team's ability to commit the necessary time and funding involved in conducting this type of a detailed study as part of the normal design development process is very unlikely.

In contrast, Taguchi's robust design method provides the design team with a systematic and efficient approach for conducting experimentation to determine near-optimum settings of design parameters for performance, development cycle time, and cost. The robust design method uses *orthogonal arrays* (OAs) to study the design parameter space, containing a large number of decision variables, which are evaluated in a small number of experiments. Based on design of experiments theory, Taguchi's orthogonal arrays provide a method for selecting an intelligent subset of the parameter space. Using orthogonal arrays significantly reduces the number of experimental configurations. Taguchi simplified the use of previously described orthogonal arrays in parametric studies by providing tabulated sets of standard orthogonal arrays and corresponding linear graphs to fit a specific project. A typical tabulation is shown in Table 13.1.

In this array, the columns are mutually orthogonal. That is, for any pair of columns, all combinations of factor levels occur, and they occur an equal number of times. Here, there are seven factors — A, B, C, D, E, F, and G, each at two levels. This is called an L8 design, the 8 indicating the eight rows, configurations, or prototypes to be tested, with test characteristics defined by the row of the table.

The number of columns of an OA represents the maximum number of factors that can be studied using that array. Note that this design reduces 128 (2^7) configurations to 8. Some of the commonly used orthogonal arrays are shown in Table 13.2. As Table 13.2 depicts, there are greater savings in testing for the larger arrays.

Using an L8 OA means that 8 experiments are carried out in search of the 128 control factor combinations that give the near-optimal mean, and also the near-minimum variation away from this mean. To achieve this, the robust design method uses a statistical measure of performance called *signal-to-noise* (S/N) *ratio* borrowed from electrical control theory. The S/N ratio developed by Dr. Taguchi is a performance measure to select control levels that best cope with noise. The S/N ratio takes

TABLE 13.2
Common Orthogonal Arrays with Number of Equivalent Full Factorials

Orthogonal Array	Factors and Levels	No. of Experiments
L4	3 Factors at 2 levels	8
L8	7 Factors at 2 levels	128
L9	4 Factors at 3 levels	81
L16	15 Factors at 2 levels	32,768
L27	13 Factors at 3 levels	1,594,323
L64	21 Factors at 4 levels	4.4×10^{12}

both the mean and the variation into account. In its simplest form, the S/N ratio is the ratio of the mean (signal) to the variability or standard deviation (noise). The S/N equation depends on the criterion for the quality characteristic that is to be optimized. Although there are many different possible S/N ratios, there are three that are considered to be standard and are therefore generally applicable in most situations:

- Biggest-is-best quality characteristic (strength, yield)
- Smallest-is-best quality characteristic (contamination)
- Nominal-is-best quality characteristic (dimension)

Whatever the type of quality or cost characteristic being used, the transformations are such that the S/N ratio is always interpreted in the same way: the larger the S/N ratio, the more robust the design. This simply implies that the variation in signal is small compared with the magnitude of the main signal.

By making use of orthogonal arrays, the robust design approach improves the efficiency of generating the information that is necessary to design systems that are robust to variations in manufacturing processes and operating conditions. As a result, development cycle time is shortened and development costs are reduced. An added benefit is the fact that a near-optimum choice of parameters may result in wider tolerances such that lower-cost components and less-demanding production processes can be used.

Engineers usually focus on system engineering design and system tolerance design to achieve needed product performance. The common practice in product and process design is to base an initial prototype on the first feasible design. The reliability and stability against noise factors are then studied and any problems are remedied by using costlier components with tighter tolerances. In other words, system parameter design is largely ignored, or overlooked. As a result, the opportunity to improve the design (and thus product) quality is usually averted, resulting in more expensive products, which are often difficult to manufacture. These products lack robustness, and thus are oftentimes very limited in their potential for future, more demanding applications.

The use of Taguchi's quality engineering methods has been steadily increasing in many companies over the past decade; however, new survival tactics and the increasingly competitive worldclass market are dictating new tools. Robust design practices are becoming increasingly more common in engineering as low life-cycle cost, operability, and quality issues replace performance as the driving design criteria.

13.2 FUNDAMENTAL PRINCIPLES OF ROBUST DESIGN — THE TAGUCHI METHOD

There are nine fundamental principles of robust design, as outlined below:

1. The functioning of a product or process is characterized by *signal factors* (SFs), or input variables, and *response factors* (RFs), or output variables. These, in turn, are influenced by *control factors* (CFs), or controlled elements, and *noise factors* (NFs), or environmental and other variations.
2. In a robust product or process, the response factors are accurately meeting their target values as functions of the signal factors, while being under the constraint of the control factors, but subject to the noise factors.
3. The robustness of a product or process can be increased through the choice of operating values for the signal factors and the control factors (parameter design) or additional design parameters. This improves the accuracy of the response factor values in relation to the target values (system tolerance design).
4. A quality loss function is defined in order to be able to quantify the penalties associated with deviation of the response factors from their target values.
5. The combined principles of system parameter design and system tolerance design form the principles of robust design. System parameter design is the primary principle and is not associated with any additional cost. System tolerance design implies the addition of extra design and associated extra cost. System tolerance design is needed only if parameter design is not sufficient to improve the accuracy of the target values of the response factors. The cost of tolerance design is balanced against the decrease in quality costs according to the quality-loss function.
6. System parameter design uses nonlinearities in the signal factors and control factors to set their values such that the influence of noise factors on their values is insignificant.
7. In order to define meaningful values for the signal factors and the control factors, tests with different values for the actors have to be conducted. The tests are either performed on the product or process directly or are approximated by simulation. For each factor, two or three values are typically tested. To find useful nonlinearities, three or more values must be used. In order to limit the number of tests, and also to limit interdependencies between the factors to be tested, a set of Taguchi orthogonal arrays have been designed and these are recommended for planning and conducting the tests.

8. Statistical analysis of the test results provides the basis for deciding the set-point values for the signal factors and the control factors, leading to a more robust design. If this is not enough to provide the targeted result, then system-tolerance design principles must also be invoked.
9. The experimental tests must be conducted in the normal operating environment of the product or process to ensure that an accurate exposure to realistic noise factors and levels has been achieved.

13.3 THE ROBUST DESIGN CYCLE

Optimizing a product or process design means determining the best system architecture by using optimum settings of control factors and tolerances. Robust design is Taguchi's approach for finding near-optimum settings of the control factors to make the product insensitive to noise factors. There are eight basic steps of robust design:

1. Identify the main function
2. Identify the noise factors and testing conditions
3. Identify the quality characteristics to be observed and the objective function to be optimized
4. Identify the control factors and their alternative levels
5. Design the matrix experiment and define the data analysis procedure
6. Conduct the matrix experiment
7. Analyze the data and determine near-optimum levels for the control factors
8. Predict the performance at these levels

These eight steps constitute the robust design cycle. The first five steps are used to plan the experiment. The experiment is conducted in step 6, and in steps 7 and 8, the experimental results are analyzed and verified.

13.3.1 A Robust Design Example: An Experimental Design to Improve Golf Scores

The details of the eight steps in robust design are described in the following simple, yet illustrative example. The approach is applicable to any quality characteristic that is to be optimized, such as performance, cost, weight, yield, processing time, or durability.

13.3.1.1 Identify the Main Function

The main function of the game of golf is to obtain the lowest score in a competition with other players, or against the course par value. A point is scored for each stroke taken to sink the golf ball in a progressive series of holes (usually 9 or 18).

13.3.1.2 Identify the Noise Factors

Noise factors are those that cannot be controlled or are too expensive to control. Examples of noise factors are variations in operating environments or materials, and manufacturing imperfections. Noise factors cause variability and loss of quality. The overall aim is to design and produce a system that is insensitive to noise factors. The designer should identify as many noise factors as possible, then use engineering judgment to decide the more important ones to be considered in the analysis and how to minimize their influence.

Various noise factors (*N*s) that can exist in a golf game, and methods of minimizing their influence are

N1 = Wind — play on a calm day.
N2 = Humidity — play on a clear, dry day.
N3 = Temperature — play in a temperate climate.
N4 = Mental attitude — play only on good days!
N5 = Distractions — maintain concentration and composure at all times.

13.3.1.3 Identify the Quality Characteristic to be Observed and the Objective Function to be Optimized

In this example, obtaining a winning golf score is the objective. Therefore, the total score will be taken to be the quality characteristic to be observed. The objective function to be optimized is the total score (TS), which is the cumulative score resulting from each of 18 holes (*X*s) of play:

$$\text{Minimize TS} = X1 + X2 + X3 + \ldots X18$$

The objective now is to find the approach that minimizes the total score, considering the uncertainty due to the noise factors cited above.

13.3.1.4 Identify the Control Factors and Alternative Levels

In this example, the control factors (*CF*s) to be considered are

CF1 = Age of clubs
CF2 = Time of day
CF3 = Driving range practice
CF4 = Use of a golf cart
CF5 = Drinks
CF6 = Type of ball used
CF7 = Use of a caddy

For this example, two levels will be considered for each of the control factors to be studied.

13.3.1.5 Design the Matrix Experiment and Define the Data Analysis Procedure

The objective now is to determine the optimum levels of the control factors so that the system is robust to the noise factors. Robust design methodology uses orthogonal arrays, based on the design of experiments theory, to study a large number of decision variables with a small number of experiments. Using orthogonal arrays significantly reduces the number of experimental configurations. Table 13.3 identifies the control factor levels, and Table 13.4 displays the resultant experiment orthogonal array.

TABLE 13.3
Control Factor Levels

Factors	Level 1	Level 2
• Age of clubs	• Old	• New
• Time of day	• A.M.	• P.M.
• Driving range practice	• Yes	• No
• Use of a golf cart	• Yes	• No
• Drinks	• Yes	• No
• Type of ball used	• Titleist	• Wilson
• Use of a caddy	• Yes	• No

TABLE 13.4
L8 (2^7) Experiment Orthogonal Array

Construction of the Orthogonal Array

EXP #	Club Age	Time of Day	Driving Range	Golf Cart	Drinks	Ball Type	Caddy	Score
1	Old	A.M.	Yes	Yes	Yes	Titleist	Yes	TBD
2	Old	A.M.	Yes	No	No	Wilson	No	TBD
3	Old	P.M.	No	Yes	Yes	Wilson	No	TBD
4	Old	P.M.	No	No	No	Titleist	Yes	TBD
5	New	A.M.	No	Yes	No	Titleist	No	TBD
6	New	A.M.	No	No	Yes	Wilson	Yes	TBD
7	New	P.M.	Yes	Yes	No	Wilson	Yes	TBD
8	New	P.M.	Yes	No	Yes	Titleist	No	TBD

13.3.1.6 Conduct the Matrix Experiment

The robust design method can be used in any situation where there is a controllable process. The controllable process is often an actual hardware experiment. Conducting a hardware experiment can be costly. However, in most cases, systems of mathematical equations can adequately model the response of many products and processes. In such cases, these equations can be used adequately to conduct the controlled matrix experiments. The results of our golf score experiment are displayed in Table 13.5 to demonstrate the effect of using a Taguchi experimental design, orthogonal array method to minimize variability.

TABLE 13.5
L8 (2^7) Results of the Matrix Experiment

EXP #	Club Age	Time of Day	Driving Range	Golf Cart	Drinks	Ball Type	Caddy	Score
1	Old	A.M.	Yes	Yes	Yes	Titleist	Yes	84
2	Old	A.M.	Yes	No	No	Wilson	No	96
3	Old	P.M.	No	Yes	Yes	Wilson	No	89
4	Old	P.M.	No	No	No	Titleist	Yes	97
5	New	A.M.	No	Yes	No	Titleist	No	94
6	New	A.M.	No	No	Yes	Wilson	Yes	91
7	New	P.M.	Yes	Yes	No	Wilson	Yes	94
8	New	P.M.	Yes	No	Yes	Titleist	No	92

13.3.1.7 Analyze the Data to Determine the Optimum Levels of Control Factors

The traditional analysis performed with data from a designed experiment is the analysis of the mean response. The robust design method also employs an S/N ratio to include the variation of the response.

The S/N developed by Dr. Taguchi is a statistical performance measure used to choose control levels that best cope with noise. The S/N ratio takes both the mean and the variability into account. The particular S/N equation depends on the criterion for the quality characteristic to be optimized. Whatever the type of quality characteristic, the transformations are such that the S/N ratio is always interpreted in the same way: the larger the S/N ratio the better. In our simplified example, we have chosen to select our golf-playing conditions such that the signal-to-noise ratio can be considered extremely large.

There are several approaches to the data analysis. One common approach is to use statistical analysis of variance (ANOVA) to see which factors are statistically significant. Another method that involves graphing the effects and visually identifying the factors that appear to be significant can also be used. For our example, we used the ANOVA method. Table 13.6 presents the results of the pooled ANOVA, and Table 13.7 shows the totals.

TABLE 13.6
Pooled ANOVA Table

Source	df	S	V	F	S′	P%
D. Cart	1	28.125	28.125	8.46	24.80	20%
E. Drinks	1	78.125	78.125	23.50	74.80	61%
Error	5	16.625	3.325		23.275	19%
Total	7	122.875			122.875	100%

TABLE 13.7
Totals Table

	Totals	N	Means
D1 (Yes)	361	4	90.25
D2 (No)	376	4	94.00
E1 (Yes)	356	4	89.00
E2 (No)	381	4	95.25
Total	737	8	92.13

The following conclusions can be drawn from the ANOVA:

- The two most important factors were (1) drinks (61% correlation) and (2) use of a golf cart (20% correlation).
- Nineteen percent (error) of the variation was unexplained.
- Factors that were not important included age of clubs, time of day, driving range practice, type of ball, and use of a caddy.
- Drinks reduced the mean golf score significantly: yes (89); no (95.25).
- The use of a golf cart also reduced the mean golf score appreciably: yes (90.25); no (94.00).
- Drinks and the use of a golf cart reduced the mean score to 87! Average of Exp. 1 and Exp. 3, the only two that used both drinks and the golf cart, is (84+89)/2 = 86.5 or approximately 87.

14 Six Sigma Problem Solving

Jonathon L. Andell

Many consultants and references advocate Six Sigma as a means to rectify quality problems in a manufacturing environment. This application is indeed valid, yielding impressive financial results, as we shall discuss. However, there is a variety of other situations wherein Six Sigma problem-solving methodologies can help an organization, such as the following:

- Identifying and eliminating the causes of nagging problems throughout a business — the application most commonly described in articles and brochures
- Developing manufactured and service products with significant competitive edges — the realm called *Design for Six Sigma* (DFSS)
- Planning and implementing management initiatives, including Six Sigma itself — setting up Six Sigma to match the requirements of each specific business

As one might expect, achieving such divergent objectives depends on applying somewhat different tools. After all, the list starts with tactical issues dealing with things, and progresses toward strategic issues of people and organizations. In order to accommodate such diverse objectives, Six Sigma problem solving encompasses a variety of approaches.

Most organizations have individuals with excellent backgrounds in Six Sigma problem solving, even if they call it by another name. Furthermore, many managers have seen literature and attended seminars on how it works. However, it is commonplace for the state of problem solving at large to lag significantly behind what an organization's best people contribute.

The challenge, therefore, is to make excellent problem-solving teams less of an exception and more the rule. As Table 14.1 shows, quite a balancing act is involved in bringing this about.

This chapter endeavors to provide managers with some guidelines for striking such a balance. However, there are limitations inherent in such a discussion:

- No single chapter can provide enough detail to make the reader into an expert problem solver. (For that matter, nobody can become an expert simply by reading. It's like golf, sooner or later you have to put down the books and pick up the clubs.)

TABLE 14.1
The Six Sigma Balancing Act

Patience	Urgency
• Allow the process to work	• Attendance at meetings
• Accept realistic scope	• Complete assigned action items
Containment	Correction
• Protect the customer	• Identify the root cause
• Temporarily higher expenses	• Eliminate the problem for good
Executive Hands-Off	Executive Hands-On
• Analytical tools	• Infrastructure & reward system
• Challenge by implementation	• Strategic project selection
	• Resource allocations
Flexibility	Rigor
• Deal with team dynamics	• No shortcuts
• Act on findings	• Diversity on team
Autonomy	Accountability
• "Worker bees" on teams	• Participation not optional
• Trust team's intent & skill	• Zero tolerance for obstruction
• Share information	• Provide guidelines & objectives

- A detailed description of all problem-solving tools also is beyond the scope of a single chapter. Fortunately, the chapters of this handbook address the more powerful tools. This chapter serves partly as an overview for when and where each chapter's contribution might apply within the big picture.
- Emphasis remains on tactical problem solving, the first of the three broad problem-solving applications described above.

The object of this chapter is to enable managers to support Six Sigma problem solving within their organizations. The direct implication is that somebody other than managers will lead the teams, specifically the practitioners, experts, and masters described in Chapter 2, "Benefiting from Six Sigma Quality." Managers generally provide a combination of guidance and support, as we will discuss.

Numerous anecdotes are used, some to describe traditional businesses, others to illustrate how a Six Sigma organization functions. The distinction between a traditional and a Six Sigma organization is not black and white. In some cases, both kinds of anecdotes emanate from within the same firm. The reader might wish to reflect on how both kinds of examples apply to his or her business.

The chapter starts by linking problem solving to financial performance, by estimating organizational resources tied up fixing defects. Next, a few established methodologies are compared against the define–measure–analyze–improve–control (DMAIC) approach associated with Six Sigma problem solving, followed by a review of how the other chapters of this handbook fit into the overall picture of problem solving. The chapter ends with a return to the discussion of roles that was started in Chapter 2, this time considering how the roles apply to successful problem solving.

14.1 PRODUCT, PROCESS, AND MONEY

A manufactured product is a physical object, with tangible properties that enable you to test its conformance to customer requirements. When a product contains one or more defects, it is called defective. Presumably, defects are not deliberate. They ensue from flaws in the processes that create the product. A variety of process problems can lead to defects in manufactured products:

- Design errors
- Defects in the materials
- Defects in the manufacturing process
- Errors in the processes that support the factory floor

Problem-solving teams identify which process, and which aspect thereof, is responsible for the defects. They then identify and implement remedies, with the intent of preventing the defects from happening again. Later we discuss how this is done. First, however, managers will benefit from understanding the costs of fixing defective products once they occur.

14.1.1 DEFECTS PER UNIT (DPU)

Consider a product. It could be a manufactured product such as a hammer, a service product such as tax preparation, or something in between, such as automobile repair. Suppose we are able to contain every defect, meaning that the delivered product contains zero defects (though this final supposition is most unrealistic, we beg the reader's indulgence).

Over time, we produce an average of one defect per unit of deliverable product, or one DPU. Whether this is a good or a bad number depends on the complexity of the product: if a unit were one jumbo jet, one DPU would be an excellent number indeed; 1 DPU would be horrendous if a unit was a single carpet tack. Figure 14.1 shows how 100 defects might be distributed among a sample of 100 units.

This typically is modeled using the Poisson distribution:

$$Y_{TP} \cong e^{-\text{DPU}} \tag{14.1}$$

In Equation 14.1, Y_{TP} is called *throughput yield*. It is the probability that a given unit is nondefective. In Figure 14.1, DPU = 1.0, which corresponds to a value of $Y_{TP} \cong 37\%$; thus, 37 of the units contain zero defects.*

14.1.2 THROUGHPUT YIELD (Y_{TP}), K, AND R

So how does this relate to managing a business? It comes down to how much it costs the business to fix defective product. Some have called the rework process

* Over time, a process averaging 1 DPU should average approximately 37% defect-free units. However, any single sample is likely to vary somewhat from the expected value.

0	1	2	0	1	2	2	2	0	0
1	2	2	2	1	0	3	1	2	1
2	0	0	3	1	1	2	0	0	2
0	4	2	1	0	1	0	1	1	0
1	1	1	1	2	0	2	1	0	3
2	2	0	2	1	0	1	0	2	1
2	2	1	0	1	0	1	0	0	0
2	0	1	0	1	0	2	1	1	2
0	1	0	1	0	0	1	1	0	0
0	0	2	0	0	3	1	2	0	2

FIGURE 14.1 How 1 DPU might appear in 100 units.

"the hidden factory," because rework usually is mixed in with first-pass product.* Because the two product streams are mingled, computing the magnitude of the hidden factory is difficult, especially using traditional cost accounting.

Fortunately, we can use Y_{TP} to estimate this magnitude, based on the following:

$$R \cong 1 + K \cdot (1 - Y_{TP}) \tag{14.2}$$

In Equation 14.2, R represents the amount of resources required to produce and rework a product, including the 100% necessary to do everything just once. From Equation 14.1 we can tell that if DPU is low, then Y_{TP} is nearly 1. From Equation 14.2, we can see that if Y_{TP} approaches 1, then R does, too. In other words, low defect rates enable us to run our process very close to its "entitlement" level of $R = 100\%$.

However, as defect rates rise and Y_{TP} falls, we must add extra resources to handle the rework caused by the $(1 - Y_{TP})$ units that contain one or more defects. The coefficient K quantifies the extra resources.

To understand K, consider Figure 14.2, representing a ten-step process. Two defect scenarios are shown. In one, a defect is detected at step 3 and reworked at step 2. For this defect, the value of K is one step repeated out of a total of ten, or $K = 1/10 = 0.1$.

However, we also show a defect detected at step 10 and reworked at step 1. What is not shown for the rework at step 1 is whether the product can be returned immediately to step 10, or whether it must pass through the entire process all over again. The answer depends as much on the type of defect as on the type of product.

* One exception occurred on a certain automotive assembly line in Europe, where a full 1/3 of the factory floor was designated for fixing defects.

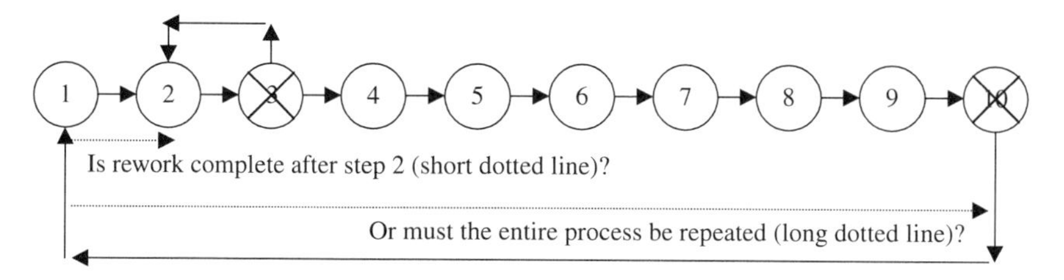

FIGURE 14.2 Various rework scenarios.

Here the value of K can be anything from 0.1 to 0.9. In fact, K can be any value greater than zero, in light of other resource requirements:

- Product disassembly
- Problem diagnosis
- Reviews, paperwork, and administrative support
- Redundant inspections
- Rework that fails to rectify the problem
- Queues
- Inventory: tracking, adjustments, expediting
- Delayed shipments
- Escaping defects

A Six Sigma problem-solving team may be able to estimate an average value of K. However, it takes a lot of work to do so. Also, process changes that reduce defect rates are likely to alter the value of K.

As a rule of thumb, consider using a value of $K \cong 0.5$. Though this tends to be on the low side of reality, the following discussion will show its impact.

14.1.3 An Example Calculation

Consider a process with DPU ≅ 2.3. Based on Equation 14.1, the resulting $Y_{TP} \cong 0.1$, meaning that only 10% of product starts completes all steps of production defect-free. Using the default value of $K = 0.5$, we can use Equation 14.2 to estimate that

$$R \cong 1 + 0.5 \cdot (1 - 0.1) = 1 + (0.5 \cdot 0.9) = 1.45$$

Thus, rework consumes an estimated 45% more resources — floor space, capital equipment, personnel, etc. — than it should take to do the job right the first time. Putting it another way, approximately 31% of the process's resources are consumed fixing defects.

Suppose this team was able to reduce defects by 75% — an accomplishment that is fairly routine in Six Sigma problem solving. Table 14.2 shows the *before* and *after* numbers. Note that the reduction in the hidden factory is 42%, which is less than the reduction in defects.

TABLE 14.2
Impact of 75% Reduction in DPU

	Before 6σ	After 6σ
Defects Detected & Reworked		
Defects per unit (DPU)	2.30	0.58
Throughput yield (Y_{TP})	0.10	0.56
R value	1.45	1.22
% Hidden factory	31%	18%
Hidden factory reduction		42%
Escaping Defects		
Total DPU (detected + estimated escaping)	2.63	0.66
Escaping DPU	0.33	0.08
Field Y_{TP}	72%	92%
Shipped units defective	28%	8%

Consider the ramifications of DPU and hidden factory.

- DPU provides ease of measurement and process information.
- Hidden factory estimates the financial impact of waste due to defects.

This indicates why Six Sigma seeks eventually to achieve even lower defect levels and how such improvements relate to financial performance.

14.1.4 Escaping Defects

Recall that we started this discussion by presuming that all defects could be detected and contained. In reality, that seldom is the case. A rule of thumb is that one stage of visual inspection detects 85 to 90% of all defects.*

Let us apply this to the process described in Table 14.2, presuming that the 2.3 DPU represent 87.5% of all defects, detected using a single visual inspection stage:

$$DPU_{\text{Actual}} \cong 2.30 \div 0.875 = 2.63$$

$$DPU_{\text{Delivered}} \cong 2.63 - 2.30 = 0.33$$

$$(Y_{TP})_{\text{Delivered}} \cong e^{-0.33} = 72\%$$

$$1 - (Y_{TP})_{\text{Delivered}} \cong 28\%$$

Thus, approximately 28% of the delivered product contains at least one defect. If customer complaint data show a lower rate, the business may have to contend

* Automated inspection systems have become popular lately. However, the reader is cautioned: though their speed is indisputable, many have fared poorly in tests of accuracy.

with customers who are silently dissatisfied. The second column shows how reducing defects by 75% cuts delivered defectives to 8%.

One can reduce defects by adding subsequent inspections, each of which should detect roughly 85 to 90% of the remaining defects. In this case, we include in our estimates the cost of inspection resources. A brief exercise in these numbers shows why quality cannot be "inspected in" as anything but a temporary containment measure.

14.1.5 Final Comments on Defects and Money

The primary mission of Six Sigma problem solving is to eliminate defects. However, the activity includes gathering defect data, which provide an estimate of the financial impact of the team's efforts. When we compare escaping defects with customer complaint data, we begin to understand how quality may be affecting more than just profits.

As a temporary measure, we can institute more inspections. However, the object is to eliminate defects. Now that we have considered the financial ramifications of defects, let us proceed to the means by which defects are prevented from recurring.

14.2 BASICS OF PROBLEM SOLVING

The literature abounds with descriptions of MAIC and DMAIC as models of Six Sigma problem solving. In truth, these are variations on themes that have been around for decades, starting with the granddaddy of them all: Shewhart's and Deming's plan–do–study–act (PDSA). The effectiveness Six Sigma problem solving is based on the same principles that make many other team-based, problem-solving approaches effective.

14.2.1 Basic Problem Solving

Consider briefly the overall activities in Six Sigma problem solving, similar perhaps to Figure 14.3. This summary does not describe any single methodology, but rather describes common aspects of the more effective approaches. Table 14.3 summarizes the activities and why they are important. Traditional problem solving is characterized by the tendency to omit or abbreviate steps. In such environments, problems tend to hide and reappear at inconvenient times.

In Figure 14.3, each row represents a community within a business, and the sequence of activities proceeds from left to right. The white box naming each activity encompasses the typical participants in that aspect of the problem-solving process. Finally, the crosshatched boxes represent groups that may be called upon periodically during a given activity.

Note the distinction between Upper Management and Middle Management. Middle management tends to be closer to immediate process supervision, so they participate more than top management. Also note that Team is separate from Operators, because one operator usually represents numerous peers in team activities.

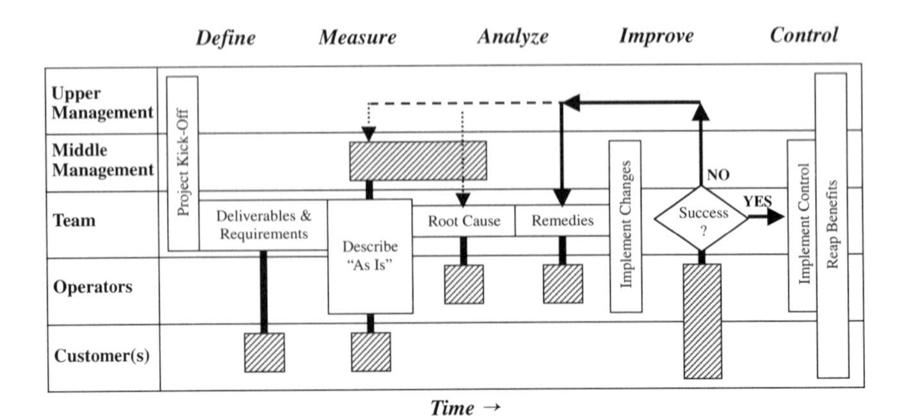

FIGURE 14.3 Effective problem solving in manufacturing.

TABLE 14.3
Steps in Effective Problem Solving

Step	Purpose(s)	Signs of Success
Project kick-off	• Common understanding ▪ Objectives ▪ Scope	• Focus on process to fix instead of "Rules of Engagement"
Deliverables & requirements	• Understand customers & needs • Fix the right problem	• Objective metrics
Describe "as is"	• Qualitative process description • Objective performance data	• Quantified measures • Cost of poor quality
Root cause	• Fix the right things	• Consensus on "Vital Few" problem causes
Remedies	• Implement the right fixes	• Consensus on "Vital Few" interventions
Implement changes	• Test drive revisions	• Process improves as hoped
Implement control	• Make improvements permanent	• Self-sustaining at improved levels
Reap benefits	• Reward contributors • Spread the message	• Wait lists to join teams • Project ideas proliferate

Finally, note that the stages of DMAIC appear across the top of Figure 14.3, but without distinct boundaries. Accomplished problem solvers recognize that hard boundaries simply don't exist.

14.2.2 Comparison of Methodologies

Between published literature, Internet sites, and consultants' offerings, the apparent variety of problem-solving methodologies can be downright intimidating. One way to classify myriad materials might be to use the following categories:

- Tools: techniques and activities used to achieve specific outcomes, such as gathering information or making decisions
- Methodologies: frameworks in which sequences of tools are selected and applied to achieve broader objectives, such as project outcomes
- Infrastructure: organizational interventions to enhance the business's abilities to benefit from methodologies and tools

The above list proceeds from the tactical to the strategic. That is, individuals can understand and apply some tools rather quickly, whereas infrastructure requires investing time and effort in both personal and organizational growth.

The above categories can be used to create a rough classification of the chapters of this handbook, shown in Table 14.4. As the table indicates, there is considerable overlap among the classifications.

At the methodology level, three approaches to problem solving are currently being used extensively: DMAIC (Six Sigma), lean manufacturing (kaizen), and Ford's eight-discipline team-oriented problem solving (also called TOPS or 8D). Ultimately, all three adhere to the precepts of Figure 14.3, along with the PDSA philosophy.

TABLE 14.4
Six Sigma Context of Handbook Chapters

<table>
<tr><th>Infrastructure</th><th>Methodologies</th><th>Tools</th></tr>
<tr><td colspan="2">Six Sigma Management (Chapt. 2)</td><td>Design of Experiments (DOE) (Chapt. 3)</td></tr>
<tr><td colspan="2">Supply Chain Management (Chapts. 16 and 17)</td><td>Measurement System Analysis (MSA) (Chapt. 9)</td></tr>
<tr><td colspan="2">Integrated Product & Process Development (Chapt. 5)</td><td>Process Analysis (Chapt. 10)</td></tr>
<tr><td>Agile Enterprise (Chapt. 1)</td><td colspan="2">Design for Six Sigma (DFSS) (Chapt. 4)</td></tr>
<tr><td>ISO 9001 (Chapt. 6)</td><td colspan="2">Design for Manufacture & Assembly (DFMA/DFSS) (Chapt. 4)</td></tr>
<tr><td>ISO 14001 (Chapt. 7)</td><td colspan="2">Theory of Inventive Problem Solving (TRIZ) (Chapt. 19)</td></tr>
<tr><td></td><td colspan="2">Theory of Constraints (TOC) (Chapt. 18)</td></tr>
<tr><td></td><td>Lean Manufacturing (Chapt. 8)</td><td>Quality Function Deployment (QFD) (Chapt. 11)</td></tr>
<tr><td></td><td>Six Sigma Problem Solving (Chapt. 14)</td><td>Robust Design (Chapt. 13)</td></tr>
<tr><td></td><td></td><td>Manufacturing Controls Integration (Chapt. 12)</td></tr>
<tr><td></td><td></td><td>Statistical Quality/Process Control (SPC) (Chapt. 15)</td></tr>
</table>

TABLE 14.5
Comparison of Problem-Solving Approaches

PDSA	8D (TOPS)	Lean (Kaizen)	6σ (DMIAC)	Purpose
Plan	Form team		Recognize	Tie quality to strategy
			Define	Prioritize projects & resources
	Describe problem	Define actual performance	Measure	Finalize project scope Understand "as is" • Requirements • Procedures • Performance
	Contain symptoms	Define desired performance		
Do	ID & verify root causes	Gather & analyze data	Analyze	Understand process behaviors • Key input variables • Sources of variation
		ID root causes		
	Choose & verify corrective actions	Remove root causes	Improve	Finalize what to change
Study	Implement permanent corrections	Change procedures to sustain gains	Control	Sustain gains
Act	Prevent recurrence	Standardize	Standardize	Become accustomed to new procedures
			Integrate	Propagate improvements
	Celebrate			Recognize & encourage success

TABLE 14.6
Applicability of Problem-Solving Approaches

Application	Ford 8D (TOPS)	Lean (KaiZen)	6σ (DMAIC)
Manufacturing quality	Strong	Strong	Strong
Lean manufacturing	Moderate	Strong	Moderate
Transactional	Moderate	Moderate	Strong
Design	Moderate	Moderate	Strong
Infrasturcture	Weak	Weak	Moderate

Table 14.5 provides a rough comparison of the steps associated with the approaches, with a brief summary of each step's purpose. Table 14.6 provides some guidelines on the strength of the tools in specific problem-solving situations. Here is a brief description of how the three methods work:

14.2.2.1 Six Sigma DMAIC

The primary topic of this chapter, DMAIC, originated as an approach to rectify quality problems on the manufacturing floor. It has also proven effective in addressing quality problems throughout an organization, including transactional and design issues. In conjunction with project management, DMAIC even supports establishing infrastructures.

As discussed in Chapter 2, the right kind of management involvement and organizational infrastructure strongly influences the degree to which problem solving affects the bottom line. Of course, this pertains to all problem-solving methodologies.

14.2.2.2 Ford 8D TOPS

Some consider this to be a variant on a method of problem solving attributed to Kepner and Tregoe. Although particularly effective at rectifying quality problems originating on the manufacturing floor, it has also had some success in design and transactional processes. Traditionally, 8D has not been a major component of management strategy; instead it is controlled closer to the teams.

14.2.2.3 Lean Manufacturing

So-called “lean” encompasses a broad range of topics, including single minute exchange of die (SMED, or quick changeover), poka-yoke (defect prevention), and kanban (“pull” system production and just-in-time inventory). The theory of constraints was developed separately from lean, but the approaches are quite compatible. The primary focus is on maximizing how efficiently the organization’s resources deliver output. Defect reduction is a means to achieve this end.

The problem-solving aspect of lean is called kaizen, in which production floor teams have extensive localized control of their process. Whereas lean often is a strategic issue for top management, kaizen tends to be controlled closer to teams. Likewise, while Lean can attack design and some transactional issues, kaizen tends to emphasize the factory floor.

14.3 SELECTING TOOLS AND TECHNIQUES

To some degree, there are two types of decisions to make when approaching selection of tools and techniques for Six Sigma.

The strategic decision occurs at the executive level: whether to favor DMAIC, 8D, lean, or some other fundamental approach to problem solving. Here, the coordinator wields considerable influence with the top staff, who must rely upon his or her judgment and impartiality.

During projects, practitioners, experts, and masters make many tactical decisions. They have “tool boxes” from which to select, along with skills to aid in the selection. Managers need enough understanding of the tools to help teams overcome obstacles against tool use. Table 14.7 shows a list of common problem-solving tools, with some ways each tool might be useful:

TABLE 14.7
Usage of Various Problem-Solving Tools

● = Highly Applicable
◑ = Moderately Applicable
○ = Slightly Applicable

	Describe Process	ID Variation Sources	Expand List	Reduce List	Predict Outcomes	Control Process	Stimulate Creativity
Affinity diagrams	◑	◑		◑		○	○
Brainstorming	◑	◑	●				●
Check sheets		●		◑	○	○	
Conditional probability analyses	◑	○		◑	●	○	
Descriptive statistics	○	◑	○	○	◑		
Design of Experiments (DoE) (Chapt. 3)		●		●	●	○	◑
Failure modes & effects analysis (FMEA)	○	◑	○	●	○	●	○
Fish-bone (cause & effect) diagram	○	◑	●	○			◑
Flow charts/S.I.P.O.C	◑	○	○	◑		◑	◑
Force field analysis	●	◑		◑	○	◑	◑
Hypothesis testing		◑		◑	●	○	
Interrelationship diagraphs				●	○		○
Measurement System Analysis (Chapt. 9)		◑	○	◑	○		
Multi-vari (nested, crossed)	●	●	○	●	◑		○
Multi-voting	○	◑		●		○	
Pairwise comparisons	●	○		●			○
Pareto charts		◑		●	○		
Poka-Yoke	◑				◑	●	◑
Qualities/Functions/Deployment (QFD) (Chapt. 11)	○	◑		●	○	◑	◑
Scatter diagrams/linear regression	○	◑		◑	●	○	
Statistical Process Control (SPC) (Chapt. 15)		●		●	◑	●	○

- *Describe the process.* In order to know how a process should be changed, we need to have a thorough understanding of how it currently operates. These tools enable us to understand the procedures used and the people involved, as well as to gather and analyze objective data regarding how well the process meets customer needs.
- *Identify potential sources of variation.* Unacceptable process behavior results in large part from excessive variation. These tools help us identify which factors cause the greatest process variation. As a result, teams focus where the payoff is greatest.
- *List expansion.* Process improvement amounts to a series of informed decisions. These tools make sure that all options are considered, so that the best options are not omitted from consideration.

- *List reduction.* Once a large list of options has been created, we employ specialized tools to select only the "vital few" for further attention.
- *Predict outcomes.* In order to know whether we have identified the key sources of variation, and whether we have implemented effective process improvements, we use tools to test our ability to control and predict process performance.
- *Process control.* Once we identify, implement, and verify changes in a process, we put in place additional procedures to make sure that the gains are sustained.
- *Stimulate creativity.* At certain junctures during problem-solving activities, we need to encourage creativity. These tools free the team from artificially restricting the options we consider to make improvements happen.

Table 14.8 classifies the tools another way: the kinds of data for which each tool is applicable. We also consider whether the tools are effective in low-volume applications, in which the process is repeated relatively few times. Here is a classification of the types of data:

- *Continuous data* include measures such as time, sizes and distances, mass, and so on. These are values that can be subdivided as necessary. Continuous data provide the greatest amount of process information per data point.
- *Rank-order data* represent relative levels of acceptability. In a foot race, this is a listing of who finished first, second, etc.
- *Attribute data* count occurrences which either happen or not. For instance, one cannot have half of a leak, or a portion of an invoice error. If we count votes for candidates, we are tallying attributes, but when we indicate who finished first, second, etc., we convert the results into rank-order data. Similarly, cycle times represent continuous data, but comparing cycle times against deadlines creates a count of delinquencies, which represents attribute data.
- *Ideas* are not data in the strictest sense, but they represent an important input to problem-solving efforts. When a team creates a brainstorming list, they are generating ideas.

14.4 MANAGING FOR EFFECTIVE PROBLEM SOLVING

What makes problem solving effective has been the subject of extensive and intensive research. The object here is to boil down the findings and add a dash of practicality. The emphasis is on how executive management balances the issues in Table 14.1, in order to derive maximum organizational benefit from problem-solving teams.

14.4.1 BALANCING PATIENCE AND URGENCY

At times we are inundated with unsolicited offers of rapid weight loss, quick college degrees, speedy prosperity, and so on. The offers prey upon people's desire for

TABLE 14.8
Matching Problem-Solving Tools with Data Types

● = **Highly Applicable**
◑ = **Moderately Applicable**
○ = **Slightly Applicable**

	Continuous	Rank Order	Attribute	Low Volume	Ideas
Affinity diagrams		◑	◑	◑	●
Brainstorming		○	○	◑	●
Check sheets	○	◑	●	◑	◑
Conditional probability analyses	●	○	◑	◑	○
Design of Experiments (DoE) (Chapt. 3)	●	○	○	●	
Descriptive statistics	●	○	◑	◑	○
Failure modes & effects analysis (FMEA)		◑	●	◑	○
Fish-bone (cause & effect) diagram		○	◑	◑	●
Flow charts/SIPOC			○	●	●
Force field anlaysis		◑	○	◑	●
Hypothesis testing	●	○	●	◑	○
Interrelationship diagraphs		◑		○	●
Measurement System Analysis (Chapt. 9)	●	○	◑	◑	
Multi-vari (nested, crossed)	●	○	○	◑	
Multi-voting		●	◑	◑	●
Pairwise comparison		●	◑	◑	○
Pareto charts	◑	◑	●	◑	◑
Poka-Yoke	○		◑	◑	○
Qualities/Functions/Deployment (QFD) (Chapt. 11)		◑	◑	◑	●
Scatter diagrams/linear regression	●	○	◑	○	
Statistical Process Control (SPC) (Chapt. 15)	●	○	●	◑	

significant outcomes to occur instantly. This common trait extends into our management of problem solving.

The preceding discussions have clarified why problem solving is, and must be, a deliberate process. For teams of five to ten people, meeting 2 hours once per week, the task typically takes 4 to 8 months.* For the most urgent of projects, management can assign a master to optimize focus, and can mandate longer and more frequent team meetings. However, these projects often have enlarged scopes, with typical time frames still stretching into 3 to 6 months.

---

* Some approaches to kaizen achieve results within a week, but (1) the team works the issue full-time for the entire week, and (2) the scope is much narrower than the typical Six Sigma project. Still, this has a valid place in context with Six Sigma, as we discussed.

At the other extreme lies the "virtual team," whose members are geographically dispersed and whose "meetings" take place by telephone, electronic mail, video conferencing, etc. It is commonplace for such teams to take 50% longer to complete comparable projects.

This alone can cause stress for executives anticipating rapid returns. Unfortunately, the issues of infrastructure compound the problem, because of what must transpire before the organization can kick off the first strategically selected project.*

- Top staff must decide what to tell the organization and must start to do so.
- Resources (people, money) must be anticipated and allocated.
- Training and other activities must be scheduled and conducted.
- Improvement priorities must be determined and disseminated.
- Projects must be assigned, launched, and completed.

Because of the above factors, the break-even time for a well-designed and well-implemented Six Sigma initiative tends to be on the order of 12 months. Virtually none break even any sooner, but poorly implemented programs have taken far longer.

The executive staff's patience will be tested in yet another way. Many organizations need more practitioners, experts, and masters than they have. Grooming new experts, practitioners, champions, and team members entails a learning curve. It's like the difference between passing classes in machining or carpentry vs. earning certification as a machinist or carpenter. People simply have to start small and work slowly at first.

Given these challenges, it may appear obvious where the "urgency" aspect comes in. Management often feels that 12 months is a long time to wait for positive cash flow, so projects are initiated almost immediately. Sometimes overlooked, however, is the need to make team support a top priority. It takes us back to the topic of resistance and the reward system: there must be zero tolerance for obstructing each team's progress.

An extreme example was a team working the logistics of ensuring that shared-ownership business jets were available when needed.** One finding was that wider access to some computer data would yield cost savings in six figures, based on DPU computations. The "owner" of the computer screens objected, and a 3-hour staff meeting ended without a decision. In an optimal Six Sigma environment, 30 minutes in the champion's office would have settled the matter in favor of the customer and the bottom line, period.

Balancing patience vs. urgency comes down to this:

- We must be patient with the process. The deliberate pace of acquiring knowledge usually is rewarded by dramatic improvements in performance — improvements that seldom come about by rushing.
- We must display urgency regarding support. Obstacles to making the process work, whether related to attendance, action items, or the empowerment to implement findings, must be overcome consistently, firmly, and promptly. Anything else will provide ammunition for those who question management's sincerity about Six Sigma.

* There is nothing wrong with initiating nonstrategic projects sooner, possibly to get the organization up and running on problem solving while the strategic work proceeds. People simply need to understand the difference between the two kinds of projects.

** Note that this is an application of Six Sigma problem solving in a business where nothing is manufactured.

14.4.2 Balancing Containment and Correction

When an organization targets a significant problem for correction, there often is a flurry of activity to "detect and contain" the problem. Some organizations respond with admirable decisiveness, directing all to drop everything until the problem is solved. This is a perfectly valid approach for certain crises.

Unfortunately, the traditional definition of *solved* may be the momentary disappearance of symptoms; attempts to rectify the underlying cause are met with "No time for that." This even happened in a corporation with quite a strong reputation among Six Sigma pundits.

This approach is reminiscent of a carnival game called "Whack a Mole." The contestant uses a plastic mallet to whack toy "moles" popping out of holes in a board. Of course, the moles keep coming back, but that's fine — if we're only playing a game. However, the cost ramifications of Table 14.2 show why Whack a Mole is no way to run a business.

Managers know that containment is tempting. It quiets noisy customers quickly. With conventional cost-accounting methods, it seems cheap. By contrast, problem solving appears slow and costly. The lessons managers must learn and live are these:

- The entire organization needs to see and understand the cost ramifications of containment, based on DPU, Y_{TP}, and escaping defects.
- Containment must be identified as no more than one aspect of true problem solving, and people must be held accountable to achieve the latter.

Nobody can do this for top management. To quote Juran, the task is "non-delegable."

14.4.3 Balancing "Hands On" vs. "Hands Off"

The most traditional of organizations condition their managers to be providers of answers. Much of this relates to Taylorism described in Chapter 2. In a Six Sigma organization, this managerial role must change.

Consider what happened in an electronics firm struggling with high particulate readings in an assembly area. The team had a strategy for determining and eliminating the source of the particles, but its manager would not hear of it. A decade before, he had solved a similar problem with a specific technical solution, and the team was directed to implement it here. After much wasted time, the team was allowed to pursue their original plan. The problem finally disappeared.

It is worth noting: this manager had had Six Sigma training and was an ardent advocate of Six Sigma. He simply reverted to a familiar pattern of behavior. Eventually he knew enough to ask the team what had been considered, to offer some ideas of his own, and to allow the team to explain its decisions. Then, he needed to allow the team to make and implement its own decisions, within the scope of its authority. Fortunately, this manager learned his lesson, and even shared the anecdote with others trying to learn Six Sigma management.

Managers in Six Sigma organizations learn to back off in the following ways:

- People and teams are treated as the experts in their own processes.
- The problem-solving methods are given a fair chance to succeed.
- Teams' conclusions are challenged primarily through implementation and ongoing monitoring.

Of course, there are ways in which managers' involvement is essential:

- Establishing the organizational infrastructure, including
  - A reward system to drive appropriate behaviors throughout the business
  - Processes and software for tracking the cost of poor quality
- Strategic guidance for project selection
- Resource allocation: making sure that departments have the personnel and funding to support Six Sigma

14.4.4 Balancing Flexibility and Rigor

People become practitioners, experts, and masters not just through training, but also by demonstrating on the job that they have the appropriate skills to add value to the business. Part of how they demonstrate aptitude is through the ability to balance rigor in applying a problem-solving methodology, with the flexibility to respond to unique characteristics of individual situations.

For example, a school used Six Sigma problem solving by assigning ten separate teams to address one topic: the high number of disciplinary interventions that were necessary. Each team had its own blend of student, faculty, and nonteaching staff, and each team selected its own sequence of problem-solving tools to employ. At the end of the exercise, each team presented a rank-order list of probable causes for the problem. What was striking is that the same three causes topped every list, though not necessarily in the exact same 1-2-3 sequence.

This outcome was a profound revelation to all. It confirmed an important lesson:

Once the framework of Six Sigma problem solving is in place, the process is robust with respect to who participates and how.

Or, putting it more simply: This stuff works!

The rigor lies in insisting that we obtain the understanding summarized in Figure 14.3 and in ensuring that the team's membership represents a diverse cross section of people living with the process and its outcomes. Of course, these aspects are primarily the responsibility of masters, experts, and practitioners, with guidance and help from champions.

Where does management come in? Managers play a vital role in ensuring that their people attend meetings and complete their action items dependably — which takes us back to prior discussions of urgency regarding support, and hands-on management of the reward system and resource allocation. Later, as executives become increasingly astute at evaluating teams, they can ask probing questions to determine how well the problem-solving process was executed.

Flexibility applies to the variety of tools that can be selected at a given time and to how an effective facilitator responds to the dynamics of her or his diverse team. Here, managers simply need to resist temptation to provide too much help, as with the particulate control team.

At an extreme, the author trained some champions in a week of Six Sigma problem solving, compared with the 5 weeks that experts received on the topic. One champion asked in effect, "How can we make sure that the experts don't mess up?" They were stunned when told that champions were not to oversee the experts, but rather to support them. The discomfort of the champions, verging on hostility, was palpable. The company had done an admirable job of preparing and rolling out technical training, but had stumbled badly in terms of cultural issues.

14.4.5 Balancing Autonomy and Accountability

In Chapter 2 we discussed empowerment as a characteristic of a Six Sigma organization. We further defined true empowerment as a state in which autonomy, accountability, and guidance are balanced effectively. The balance of autonomy and guidance is most crucial during Six Sigma problem solving.

In fledgling Six Sigma organizations, there is a tendency to assign supervisors and middle managers to problem-solving teams. Sometimes the benefit is that these people learn how Six Sigma works. At other times, though, these individuals resent interference with the process they worked so hard to put in its present form — almost as an overly protective parent might resist a child's effort to display adult independence.

Here is where practitioners, experts, and champions must have absolutely unconditional executive backing: those closest to the actual work shall be assigned to teams. Actually, this rule ensures that middle managers will have time to join teams tackling the work to which they are closest — strategic initiatives at the enterprise level. The rule also optimizes use of resources:

- Managers tackle only those problems that truly demand managerial expertise.
- Tens to hundreds of times as many resources are available for solving lower level problems.

Remember also that teams are led by trained practitioners, experts, and masters. These people have ongoing communication linkage with champions, who in turn speak to and for top management. The organization should be able to detect the situation and react when a team strays from its mission.

Here is where accountability comes in. Refer again to Figure 14.3 and Table 14.3. The Project Kick-Off is to ensure that everybody understands what is expected and why it is important (note that this balances with an autonomy issue: the team must have access to pertinent and timely information). As problem solving proceeds, team leaders periodically raise the question, "Is our present activity contributing to our achieving the initial objective?" If the answer is no, the team selects from several options:

- Adjust activities and get back on track
- Propose a revision of objectives if appropriate
- Seek assistance dealing with an obstacle if necessary

Progress reports and team presentations provide the organization with the opportunity to ensure that the team is performing as desired, and to respond to the team's issues. Accountability applies to the team, but also propagates to those who interact with the team and its process. Just as autonomy is not carte blanche for the team, neither is it stuck holding the bag for others.

14.4.6 From Distrust to Win–Win

A sad reality is that Six Sigma initiatives come to many organizations wherein distrust had been the order of the day. For management simply to declare, "It's different now," could be one aspect of Six Sigma that engenders total unanimity throughout such an organization. Unfortunately, it would be unanimous cynicism and mockery — not an auspicious structure on which to garner consensus.

A Gant chart of projects and training classes is a woefully inadequate cultural intervention. Resistance doesn't fit on a Gant chart, yet it surely must be accounted for in an organization's preparation. The plan has to include incentives for desired behaviors and outcomes, and disciplinary actions for inappropriate ones — with both "carrots" and "sticks" applicable at all organizational ranks.*

When discipline is called for, management must seek to balance fairness and consistency. It's never easy, even with the best of planning, but it's flat-out impossible without planning.

Fortunately, the problem-solving process itself creates wins for teams and for the organization. This in turn comprises a "foot in the door" of credibility for beleaguered managers. The process permits, and even demands, that people take some control over their existence. The combined messages of "Yes, you may," "Yes, you must," and "Yes, you did" carry with them the implication that management trusts its problem-solving teams, and that management intends to hold teams accountable for accepting this new mantle.

As initial inroads are made, as teams are recognized and rewarded for the improvements for which they are responsible, and as management shows that these outcomes are to become the new order of the day, more people will press to be allowed on teams. Eventually, project ideas will start to originate within the ranks, and management will have a pleasant new dilemma: how to continue empowering teams without losing control over priorities.

These claims are far from the rantings of a theorist who has never experienced dirty hands. These are tangible outcomes that have happened again and again. Dozens of

* Two points warrant mention in this regard. First: incentives can be quite powerful without being terribly expensive. Second: if top management receives substantial monetary incentives for contributions to the business, so should others. The claim that "your job is your reward" is not applied selectively in Six Sigma organizations; cost tracking based on DPU makes the task easier than ever.

team members, exposed to empowerment and the problem-solving process for the first time, have said things such as, "Finally, somebody is listening to me," and "If you think this project can make a difference, let me tell you about ...." Cynics with reputed attitude problems have blossomed into amazing resources of knowledge and commitment.

Creating a win–win culture is a challenge. Sustaining it is no less so, because it is so fragile. So why try? Because it beats the alternative any way you measure it, including the bottom line.

14.5 CONTRIBUTORS' ROLES AND TIMING

In Chapter 2 we discussed how various departments and individuals had roles in a Six Sigma organization. That discussion focused on the organizational infrastructure. For that reason, we broke the roles into "transitional" and "sustaining."

In tactical problem solving, the roles tend to be more repetitive, because at most two or three general approaches are selected across a number of projects. Some of this was discussed in conjunction with Figure 14.3, as well as Tables 14.3 and 14.5.

Table 14.9 attempts to bring together the various individuals' roles in this context. Here, the five steps of DMAIC are complimented by three more.

- *Recognize* is an outcome of establishing the organizational infrastructure from Chapter 2. It refers to the identification of high priority projects and processes.
- *Standardize* takes everything that was learned in DMAIC and makes it the accustomed way to operate and manage the process. It incorporates project management to ensure that supervisors and operators all understand and comply with the revised procedures. Training, accountability and rewards, and timing and resources all play a role.
- *Integrate* expands on *standardize* by cloning the improvements throughout the organization, beyond the original project scope. It is based on a Kepner–Tregoe concept called "extending the fix."

By now, some of this discussion of roles in problem solving sounds familiar. That would be a positive development, since it reflects on learning that has taken place. In order to add a new dimension, we will factor into the discussion how the roles impose challenges upon each community of contributors.

14.7.1 UPPER MANAGEMENT

Once a project is identified, several meetings take place. The first includes the highest executive responsible for the process in question, the champion, and the expert who will lead the team. This meeting establishes the project's parameters: time constraints, objectives, etc. It also lets the expert provide inputs on what will make the project successful: participants, obligations, and so on. If there is disagreement, the parties work to resolve issues before the team is affected.

When the team is convened, the executive, champion, and coordinator attend briefly to thank the participants and attest to the importance of the project.

TABLE 14.9
Participant Roles in DMAIC

	Participant Roles							
DMAIC Phase	**Upper Management**	**Champion & Coordinator**	**Middle Management**	**Expert**	**Team**	**Operators**	**Customer(s)**	**Purpose**
Recognize								Tie quality to strategy
Define								Prioritize projects & resources
Measure								Finalize project scope Understand "as-is" • Requirements • Procedures • Performance
Analyze								Understand process behaviors • Key input variables • Sources of variation
Improve								Finalize what to change
Control								Sustain gains
Standardize								Become accustomed to new procedures
Integrate								Propagate improvements

The executive may be called upon periodically. Often the project scope is bigger than anticipated and should be narrowed. At other times, the problem originates in an entirely different process than was thought, requiring something of a changeover of the team. Finally, there will be times when the best options require approval for various expenditures. When the team is ready to implement process changes, executive backing often helps overcome resistance.

14.7.2 Champion and Coordinator

In fledgling Six Sigma organizations, champions and the coordinator work hand in hand with top management to establish the organizational infrastructure. Once projects are underway they devote much effort to advocating on behalf of teams. They back experts' requests and advice to top management. Some teams run afoul of middle management or operators by seeking a needed but possibly unwelcome change. The champion and coordinator, and to a lesser extent the expert, serve as liaisons between the team and management. They balance technical understanding of the process in question with appreciation for change management issues.

14.7.3 Middle Management

As used here, middle management refers to people responsible for the day-to-day operation of the processes being investigated by teams. In an ideal situation, they ensure that the process is staffed sufficiently so that team members can attend meetings and complete action items. When the team recommends improvements, they use their authority to make the right changes happen. Finally, they learn what the team has found, and ensure that their people are trained and accountable to follow the new procedures.

Realistically, such managers often must balance contradictory requirements of schedules, shipment quotas, and budget constraints against what surely appears to them as a drain of vital resources. Here is where the experts, champions, and coordinator must respond effectively. These advocates must support the legitimate concerns and issues that beset middle managers, but there must be no latitude for discretionary resistance.

14.7.4 Experts

In this instance, the term experts includes practitioners and masters. Just as champions have the ear of upper management, experts become the advocates for their teams.

Experts lead teams in using methodologies and tools. Team members accustomed to traditional problem solving may challenge the approach, necessitating a balance between diplomacy and rigor.

When management must hear the team's voice, experts carry the messages — and bring back the responses. When it's time for the control, standardization, and integration phases, experts provide guidance to management on the tasks necessary to transition from old to new.

14.7.5 Team Members

Team members seldom get the recognition they deserve for their challenging roles. They learn problem-solving tools and skills. They stand up for their peers, and they also stand up *to* them and to middle management. They encounter pressure to contribute to the team, while simultaneously being pressed not to do so. On top of it all, their expertise is needed to ensure that the process is improved effectively.

14.7.6 Operators

Operators are asked to fill in for peers who get to attend team meetings. Then they are asked to add to their workloads by helping gather data whose purpose is unclear to them. Later, they are asked to change how they operate their processes. If uncertainty causes stress, this adds up to a stressful situation indeed.

On the positive side, once they benefit from some of the changes, and feel as if somebody actually cares what they think, then the organization can tap into a vast resource of knowledge and dedication.

14.6 CONCLUSION

Six Sigma problem solving defies narrow definition, because it encompasses many approaches with valid applications to a manufacturing business. It can be regarded as an umbrella under which most of this handbook can fit comfortably.

It's neither fast nor cheap. Its merit is in how much better it is for an organization's profitability than the more traditional approaches to handling problems. Not only do problems disappear, but the approach also gives management the ability to estimate before- and after-costs.

Businesses that apply Six Sigma with appropriate rigor are among the most successful in their respective fields. Truly, Six Sigma is an outstanding embodiment of the very best that capitalism can be.

15 Statistical Process Control

Paul A. Keller

15.1 DESCRIBING DATA

When it comes right down to it, data are boring, just a bunch of numbers. By themselves, data tell us little. For example: 44.373. By itself: nothing. What it lacks is context. Even knowing that it's the measurement in inches for a key characteristic, we still want more: Is this representative of the other parts? How does this compare with what we've made in the past? Context allows us to process the data into information.

Descriptive data are commonly presented as point estimates. We see point estimates in many aspects of our personal and business life: newspapers report the unemployment rate, magazines poll readers' responses, quality departments report scrap rate. Each of these examples, and countless others, provide us with an estimate of the state of a population through a sample. Yet these point estimates often lack context. Is the reported reader response a good indicator of the general population? Is the response changing from what it has been in the past?

Statistics help us to answer these questions. In this chapter, we explore some tools for providing an appropriate context for data.

15.1.1 HISTOGRAMS

A histogram is a graphical tool used to visualize data. It is a bar chart, where each bar represents the number of observations falling within a range of data values. An example is shown in Figure 15.1.

An advantage of the histogram is that the process location is clearly identifiable. In Figure 15.1, the central tendency of the data is about 0.4. The variation is also clearly distinguishable: we expect most of the data to fall between 0.1 and 1.0. We can also see if the data are bounded or have symmetry.

If your data are from a symmetrical distribution, such as the bell-shaped normal distribution, the data will be evenly distributed about a center. If the data are not evenly distributed about the center of the histogram, it is skewed. If the data appear skewed, you should understand the cause of this behavior. Some processes will naturally have a skewed distribution, and may also be bounded, such as the concentricity data in Figure 15.1. Concentricity has a natural lower bound at zero, because no measurements can be negative. The majority of the data is just above zero, so there is a sharp demarcation at the zero point representing a bound.

1-57444-300-3/02/$0.00+$1.50

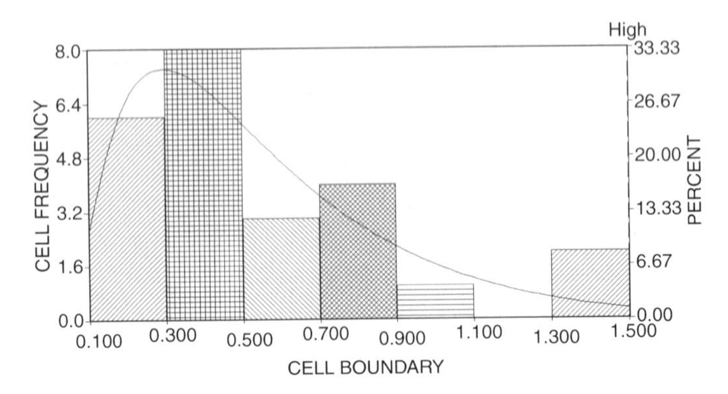

FIGURE 15.1 Example histogram for non-normal data. Concentricity. Best-fit curve: Johnson Sb; K–S test: 0.999. Kac K of fit is not significant; specified lower bound = 0.000.

If double or multiple peaks occur, look for the possibility that the data are coming from multiple sources, such as different suppliers or machine adjustments.

One problem that novice practitioners tend to overlook is that the histogram provides only part of the picture. A histogram of a given shape may be produced by many different processes, although the only difference in the data is their order. So the histogram that looks like it fits our needs could have come from data showing random variation about the average, or from data clearly trending toward an undesirable condition. Because the histogram does not consider the sequence of the points, we lack this information. Statistical process control (SPC) provides this context.

15.2 OVERVIEW OF SPC

Statistical process control is a method of detecting changes to a process. Unlike more general *enumerative* statistical tools, such as hypothesis testing, which allow conclusions to be drawn on the past behavior of static populations, SPC is an *analytical* statistical tool. As such, SPC provides predictions on future process behavior, using its past behavior as a model.

Applications of SPC in business are as varied as business itself, including manufacturing, chemical processes, banking, healthcare, and general service. SPC may be applied to any time-ordered data, when the observations are statistically independent. Methods addressing dependent data are discussed under 15.5.1, Autocorrelation.

The tool of SPC is the statistical control chart, or more simply, the control chart. The control chart was developed in the 1920s by Walter Shewhart while he was working for Bell Laboratories. Shewhart defined statistical control as follows:

> A phenomenon is said to be in statistical control when, through the use of past experience, we can predict how the phenomenon will vary in the future.

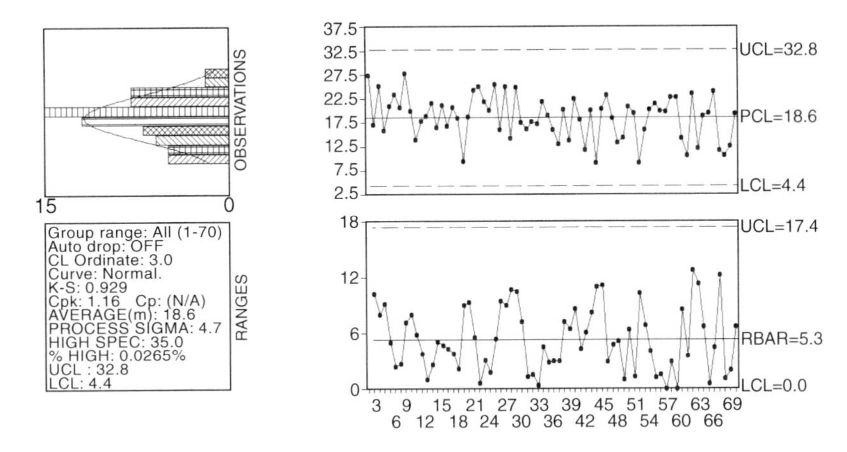

FIGURE 15.2 Example of individual-X/moving control charts (shown with histogram).

15.2.1 Control Chart Properties

Control charts take many forms, depending on the process that is being analyzed and the data available from that process. All control charts have the following properties:

- The x-axis is sequential, usually a unit denoting the evolution of time.
- The y-axis is the statistic that is being charted for each point in time. Examples of plotted statistics include an observation, an average of two or more observations, the median of two or more observations, a count of items that meet a criteria of interest, or the percentage of items meeting a criteria of interest.
- Limits are defined for the statistic that is being plotted. These *control limits* are statistically determined by observing process behavior, providing an indication of the bounds of expected behavior for the plotted statistic. They are never determined using customer specifications or goals.

An example of a control chart is shown in Figure 15.2. In this example, the cycle time for processing an order is plotted on an individual-X control chart, the top chart shown in the figure. The cycle time is observed for a randomly selected order each day and plotted on the control chart. For example, the cycle time for the third order is about 25.

In Figure 15.2, the centerline (PCL, for process center line) of the individual-X chart is the average of the observations (18.6 days). It provides an indication of the process location. Most of the observations will fall somewhere close to this average value, so it is our best guess for future observations, as long as the observations are statistically independent of one another.

We notice from Figure 15.2 that the cycle time process has variation. That is, the observations are different from one another. The third observation at 25 days is clearly different from the second observation at 17 days. Does this mean that the process is changing over time?

The individual-X chart has two other horizontal lines, known as control limits. The upper control limit (UCL) is shown in Figure 15.2 as a line at 32.8 days; the lower control limit (LCL) is drawn at 4.4 days. The control limits indicate the predicted boundary of the cycle time. In other words, we don't expect the cycle time to be longer than about 33 days or shorter than about 4 days.

For the individual-X chart shown in Figure 15.1, the control limits are calculated as follows:

$$UCL_x = \bar{x} + 3\sigma_x \tag{15.1}$$

$$LCL_x = \bar{x} - 3\sigma_x \tag{15.2}$$

The letter x with the bar over it is read "x bar." The bar notation indicates the average of the parameter, so in this case, the average of the x, where x is an observation. The parameter σ_x (read as "sigma of x") refers to the process standard deviation (or process sigma) of the observations, which in this case is calculated using the bottom control chart in Figure 15.2, the moving range chart.

The moving range chart uses the absolute value of the difference (i.e., range) between neighboring observations to estimate the short-term variation. For example, the first plotted point on the moving range chart is the absolute value of the difference between the second observation and the first observation. In this case, the first observation is 27 and the second is 17, so the first plotted value on the moving range chart is 10 (27 – 17).

The line labeled RBAR on the moving range chart represents the average moving range, calculated by simply taking the average of the plotted points on the moving range chart. The moving range chart also has control limits, indicating the expected bounds on the moving range statistic. The lower control limit on the moving range chart in this example is zero. The upper control limit is shown in Figure 15.2 as 17.4.

The moving range chart's control limits are calculated as

$$UCL = \bar{R} + 3d_3\sigma_x \tag{15.3}$$

$$LCL = MAX(0, \bar{R} - 3d_3\sigma_x) \tag{15.4}$$

Process sigma, the process standard deviation, is calculated as

$$\sigma_x = \frac{\bar{R}}{d_2} \tag{15.5}$$

For a moving range chart, the parameters d_3 and d_2 are 0.853 and 1.128, respectively.

15.2.2 General Interpretation of Control Charts

The control limits on the individual-X chart help us to answer the question posed in the section above. Since all the observations fall within the control limits, the answer is, "No, the *process* has not changed," even though the observations are clearly different.

We see variation in all processes, provided we have adequate measurement equipment to detect the variation. The control limits represent the amount of variation we expect to see in the plotted statistic, based on our observations of the process in the past. The fluctuation of the points between the control limits is due to the variation that is intrinsic (built in) to the process. We say that this variation is due to *common causes*, meaning that the sources of variation are common to all the observations in the process. Although we don't know what these causes are, their effect on the process is consistent over time.

Recall that the control limits are based on process sigma, which for the individual-X chart is calculated based on the moving range statistic. We can say that process sigma, and the resulting control limits, are determined by estimating the short-term variation in the process. If the process is stable, or in control, then we would expect what we observe *now* to be about the same as what we'll observe *in the future*. In other words, the short-term variation should be a good predictor for the longer-term variation if the process is stable.

Points outside the control limits are attributed to a *special cause*. Although we may not be able to immediately identify the special cause in process terms (for example, cycle time increased due to staff shortages), we have statistical evidence that the process has changed. This process change can occur in two ways.

- A change in process location, also known as a *process shift*. For example, the average cycle time may have changed from 19 days to 12 days. Process shifts may result in process improvement (for example, cycle time reduction) or process degradation (for example, an increased cycle time). Recognizing this as a process change, rather than just random variation of a stable process, allows us to learn about the process dynamics, and to reduce variation and maintain improvements.
- A change in process variation. The variation in the process may also increase or decrease. Generally, a reduction in variation is considered a process improvement, because the process is then easier to predict and manage.

Control charts are generally used in pairs. One chart, usually drawn as the bottom of the two charts, is used to estimate the variation in the process. In Figure 15.2, the moving range statistic was used to estimate the process variation, and because the chart has no points outside the control limits, the variation is in control.

Conversely, if the moving range chart were not in control, the implication would be that the process variation is not stable (i.e., it varies over time), so a single estimate for variation would not be meaningful. Inasmuch as the individual-X chart's control limits are based on this estimate of the variation, the control limits for the individual-X chart should be ignored if the moving range chart is out of control. We must remove the special cause that led to the instability in process variation before we can further analyze the process. Once the special causes have been identified in process terms, the control limits may be recalculated, excluding the data affected by the special causes.

15.2.3 Defining Control Limits

To define the control limits we need an ample history of the process to set the level of common-cause variation. There are two issues here.

- To distinguish between special causes and common causes, you must have enough subgroups to define the common-cause operating level of your process. This implies that all types of common causes must be included in the data. For example, if we observed the process over one shift, using one operator and a single batch of material from one supplier, we would not be observing all elements of common cause variation that are likely to be characteristic of the process. If we defined control limits under these limited conditions, then we would likely see special causes arising due to the natural variation in one or more of these factors.
- Statistically, we need to observe a sufficient number of data observations before we can calculate reliable estimates of the variation and, to a lesser degree, the average. In addition, the statistical constants used to define control chart limits (such as d_2) are actually variables, and they approach constants only when the number of subgroups is large. For a subgroup size of 5, for instance, the d_2 value approaches a constant at about 25 subgroups (Duncan, 1986). When a limited number of subgroups are available, short-run techniques may be useful. These are covered later in this chapter.

15.2.4 Benefits of Control Charts

Control charts provide benefits in a number of ways. *Control limits represent the common-cause operating level of the process.* The region between the upper and lower control limits defines the variation that is expected from the process statistic. This is the variation due to common causes: causes common to all the process observations. We don't concern ourselves with the differences between the observations themselves. If we want to reduce this level of variation, we need to redefine the process, or make fundamental changes to the design of the process. Deming demonstrated this principle with his red bead experiment, which he regularly conducted during his seminars. In this experiment, he used a bucket of beads or marbles. Most of the beads were white, but a small percentage (about 10%) of red beads were thoroughly mixed with the white beads. Students volunteered to be process workers, who would dip a sample paddle into the bucket and produce a day's "production" of 50 beads for the "White Bead Company." Another student would volunteer to be an inspector. The inspector counted the number of white beads in each operator's daily production. The white beads represented usable output that

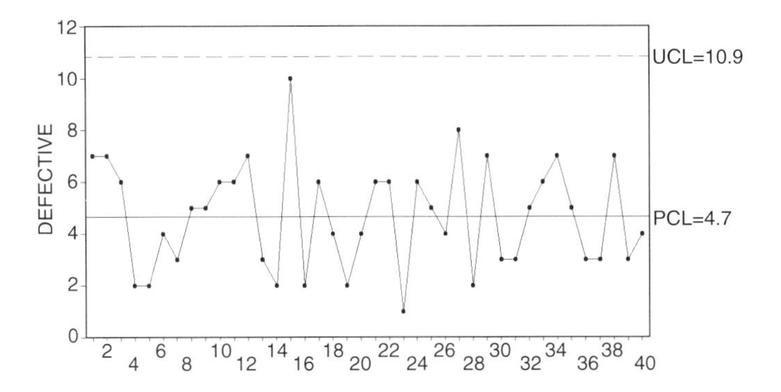

FIGURE 15.3 Example, control chart for Deming's red bead experiment. Sample size = 50.

could be sold to White Bead Company's customers, and the red beads were scrap. These results were then reported to a manager, who would invariably chastise operators for a high number of red beads. If the operator's production improved on the next sample, he or she was rewarded; if the production of white beads went down, more chastising.

A control chart of the typical white bead output is shown in Figure 15.3. It's obvious from the figure that there was variation in the process observations: each dip into the bucket yielded a different number of white beads. Has the process changed? *No! No one has changed the bucket, yet the number of white beads is different every time.* The control limits tell us that we should expect between 0 and 11 red beads in each sample of 50 beads.

Control limits provide an operational definition of a special cause. As we've seen, process variation is quite natural. Once we accept that every process exhibits some level of variation, we then wonder how much variation is natural for this process. If a particular observation seems large, is it unnaturally large, or should an observation of this magnitude be expected? The control limits remove the subjectivity from this decision, and define this level of natural process variation.

In the absence of control limits, we assume that an arbitrarily large variation is due to a shift in the process. In our zeal to reduce variation, we adjust the process to return it to its prior state. For example, we sample the circled area in the leftmost distribution in Figure 15.4 from a process that (unbeknownst to us) is in control. We feel this value is excessively large, so assume the process must have shifted. We adjust the process by the amount of deviation between the observed value and the initial process average. The process is now at the level shown in the center distribution in Figure 15.4. We sample from this distribution and observe several values near the initial average, and then sample a value such as is the circled area in the center distribution in the figure. We adjust the process upward by the deviation between the new value and the initial mean, resulting in the rightmost distribution shown in the figure. As we continue this process, we can see that we actually *increase* the total process variation, which is exactly the opposite of our desired effect.

Responding to these arbitrary observation levels as if they were special causes is known as *tampering*. This is also called "responding to a false alarm," since a

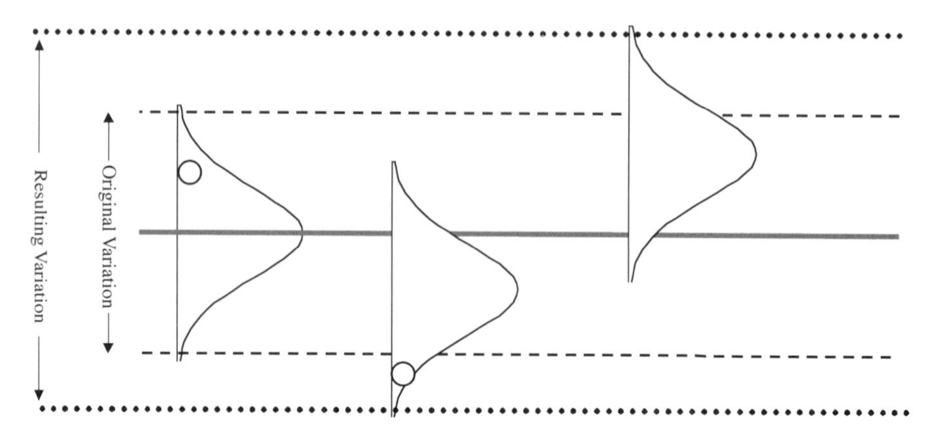

FIGURE 15.4 Tampering increases process variation.

false alarm is when we think that the process has shifted when it really hasn't. Deming's funnel experiment demonstrates this principle. In practice, tampering occurs when we attempt to control the process to limits that are narrower than the natural control limits defined by common cause variation. Some causes of this:

- We try to control the process to specifications, or goals. These limits are defined externally to the process, rather than being based on the statistics of the process.
- Rather than using the suggested control limits defined at ±3 standard deviations from the centerline, we use limits that are tighter (or narrower) than these, based on the faulty notion that this will improve the performance chart. Using limits defined at ±2 standard deviations from the centerline produces narrower control limits than the ±3 standard deviation limits, so it would appear that the ±2 sigma limits are better at detecting shifts. Assuming normality, the chance of being outside of a ±3 standard deviation control limit is 0.27% if the process has not shifted. On average, a false alarm is encountered with these limits once every 370 subgroups (= 1/0.0027). Using ±2 standard deviation control limits, the chance of being outside the limits when the process has not shifted is 4.6%, corresponding to false alarms every 22 subgroups! If we respond to these false alarms, we tamper and increase variation.

Control charts prevent searching for special causes that do not exist. As data are collected and analyzed for a process, it seems almost second nature to assume that we can understand the causes of this variation. In Deming's red bead experiment, the manager would congratulate operators when their dips in the bucket resulted in a relatively low number of red beads, and chastise them if they submitted a high number of red beads. This should seem absurd, because the operator had no control over the number of red beads in each random sample. Yet, this same experiment happens daily in real business environments. In the cycle time example shown above, suppose the order-processing supervisor, being unfamiliar with statistical process control, expected all orders to be processed at a quick pace, say 15 days. It seemed the process could deliver at this rate, because it had processed orders at or below this many times in the past. If this was the supervisor's expectation, then he or she may look for a special cause ("This order must be different from the others") that doesn't exist. Instead, he or she should be redesigning the system (i.e., changing the fundamental nature of the bucket).

Control charts result in a stable process, which is predictable. When used on a real-time basis, control charts result in process stability. In the absence of a control chart, a common reaction is to respond to process variation with process adjustments. As discussed above, this tampering results in an unstable process that has increased variation. Personnel using a control chart to monitor the process in real time (as the process produces the observations) are trained to react with process adjustments only when the control chart signals a process shift with an out-of-control point. The resulting process is stable, allowing its future capability to be estimated. In fact, the future performance of processes may be estimated only if the process is stable (see also, process capability later in this chapter).

15.3 CHOOSING A CONTROL CHART

Many control charts are available for our use. One differentiator between control charts is the type of data to be analyzed:

Attribute data: also known as "count" data. Typically, we will count the number of times we observe some condition (usually something we don't like, such as a defect or an error) in a given sample from the process.

Variables data: also known as measurement data. Variables data are continuous in nature, generally capable of being measured to enough resolution to provide at least ten unique values for the process being analyzed.

Attribute data have less resolution than variables data, because we count only if something occurs, rather than take a measurement to see how close we are to the condition. For example, attribute data for a manufacturing process might include the number of items in which the diameter exceeds the specification, whereas variables data for the same process might be the measurement of that part's diameter.

Attribute data generally provide us with less information than variables data would for the same process. Attribute data would generally not allow us to predict if the process is trending toward an undesirable state, because it is already in this condition. As a result, variables data are considered more useful for defect *prevention.*

15.3.1 Attribute Control Charts

There are several attribute control charts, each designed for slightly different uses:

- NP chart — for monitoring the number of times a condition occurs, relative to a constant sample size. NP charts are used for *binomial* data,

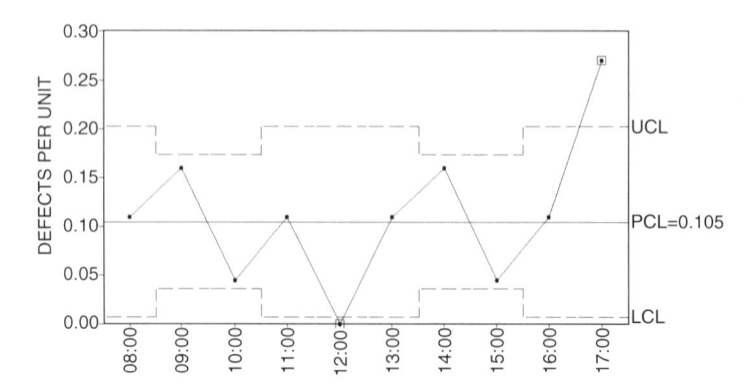

FIGURE 15.5 U control chart, number of cracks per injection molding piece.

which exist when each sample can either have this condition of interest, or not have this condition. For example, if the condition is "the product is defective," then each sample unit either is defective or not defective. In the NP chart, the value that is plotted is the observed number of units that meet the condition in the sample. For example, if we sample 50 items, and 4 are defective, we plot the value 4 for this sample. The NP chart requires a constant sample size, inasmuch as we cannot directly compare 4 observations from 50 units with 5 observations from 150 units. Figure 15.3 provided an example of an NP chart.

- P chart — for monitoring the percentage of samples having the condition, relative to either a fixed or varying sample size. Use the P chart for the same data types and examples as the NP chart. The value plotted is a percentage, so we can use it for varying sample sizes. When the samples vary by more than 20% or so, it's common to see the control limits vary as well.
- C chart — for monitoring the number of times a condition occurs, relative to a constant sample size, when each sample can have more than one instance of the condition. C charts are used for *Poisson* data. For example, if the condition is a surface scratch, then each sample unit can have 0, 1, 2, 3 … etc., defects. The value plotted is the observed number of defects in the sample. For example, if we sample 50 items and 65 scratches are detected, we plot the value 65 for this sample. The C chart requires a constant sample size.
- U chart — for monitoring the percentage of samples having the condition, relative to either a fixed or varying sample size, when each sample can have more than one instance of the condition. Use the U chart for the same data types and examples as the C chart. The value that is plotted is a percentage, so we can use it for varying sample sizes. When the samples vary by more than 20% or so, it's common to see the control limits vary as well. An example of a U chart is shown in Figure 15.5.

15.3.2 Variables Control Charts

Several variables charts are also available for use. The first selection is generally the subgroup size. The subgroup size is the number of observations, taken in close proximity of time, used to estimate the short-term variation. In the cycle-time example at the beginning of the chapter, the subgroup size was equal to one, since only one observation was used for each plotted point.

Sometimes we choose to collect data in larger subgroups because a single observation provides only limited information about the process at that time. By increasing the subgroup size, we obtain a better estimate of both the process location and the short-term variation at that time.

Control charts available for variables data include

- Individual-X/moving range chart (a.k.a. individuals chart, I chart, IMR chart). Limited to subgroup size equal to one. An example was provided in the previous sections, with the calculations used to develop the chart (Equations 15.1 through 15.5). Those calculations are valid for many applications, as long as the distribution of the observations is not severely non-normal. The chart has been shown to be fairly robust to departures from normality, but data that are severely bounded can cause irrational control limits. Figure 15.6a shows cycle-time data on an individual-X/moving range chart using the standard calculations. The lower control limits are calculated as a negative number, which clearly cannot exist for cycle-time data in the real world. Figure 15.6b provides the same data on an individual-X/moving range chart that uses a fitted curve to calculate control limits with the same detection ability as a normal distribution's ±3 sigma limits. These revised control limits allow us to detect process shifts (in this case, improvements to the process) that would go undetected using the standard calculations. Other techniques for dealing with non-normality include data transformations, such as the Box-Cox transformation.
- X-bar chart. Used for subgroup size two and larger. The plotted statistic is the average of the observations in the subgroup. The average value has been shown to be insensitive to departures from normality, even for a subgroup size as small as three or five, so the control limits need not be adjusted for non-normal process distributions.

  X-bar control limits are calculated as follows:

$$UCL = \bar{\bar{x}} + \frac{3\sigma_x}{\sqrt{n}} \tag{15.6}$$

$$LCL = \bar{\bar{x}} - \frac{3\sigma_x}{\sqrt{n}} \tag{15.7}$$

The letter x with the two bars over it is read "x double bar." Because the bar notation indicates the average of the parameter, x double bar is the

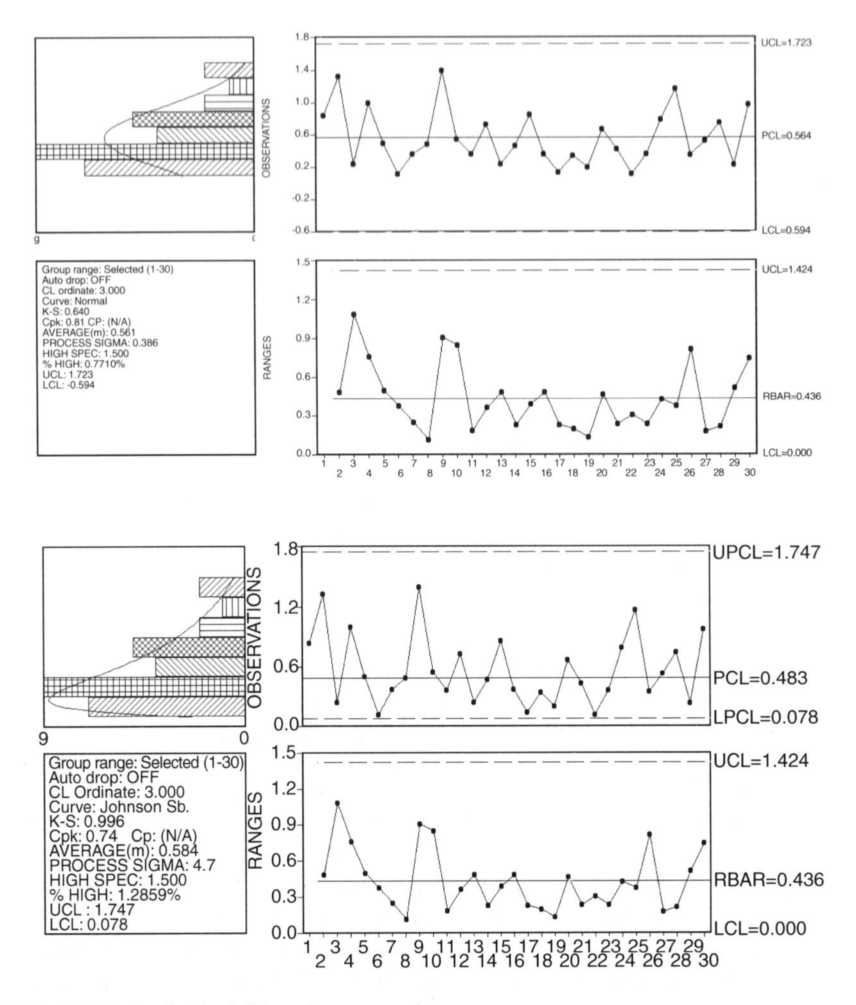

FIGURE 15.6 Individual X/moving range charts.

average of the subgroup averages. The process sigma σ_x (read as "sigma of x") is calculated using either the range chart or the sigma chart. The range and sigma charts, like the moving range chart described earlier, are used to estimate, and detect instability in, the process variation.

- Range chart. Plots the range of observations (i.e., largest minus the smallest observation) within the subgroup. Because it attempts to estimate the variation within the subgroup using only two of the observations in the subgroup (the smallest and largest), the estimate is not as precise as the sigma statistic described below. The range chart should not be used for subgroup sizes larger than ten because of its poor performance. Its popularity is due largely to its ease of use before computers. Its control limits are calculated as in Equations 15.3 through 15.5, where the parameters d_3 and d_2 are found in reference tables, such as in Montgomery and Runger.

- Sigma chart. Plots the sample standard deviation of observations within the subgroup, where x-bar$_j$ is the average of the jth subgroup, and n is the subgroup size:

$$S_j = \sqrt{\frac{\sum_{i=1}^{n}(x_i - \bar{x}_j)^2}{n-1}} \qquad (15.8)$$

The sigma chart is always more accurate than the range chart. The sigma chart's control limits are calculated as follows:

$$UCL_S = \bar{S} + 3\sigma_x\sqrt{1-c_4^2} \qquad (15.9)$$

$$LCL_S = MAX(0, \bar{S} - 3\sigma_x\sqrt{1-c_4^2}) \qquad (15.10)$$

Process sigma, the process standard deviation, is calculated as:

$$\sigma_x = \frac{\bar{S}}{c_4} \qquad (15.11)$$

- Other charts. The EWMA (exponentially weighted moving average) chart and the CuSum (cumulative sum) chart each have unique properties that make them preferable for particular situations. Both charts are robust to departures from normality, so they can be used for the bounded process of Figure 15.6. Another valuable characteristic is their increased sensitivity to small process shifts, as an alternative to increasing the sample size. Although the plotted statistics are inconvenient to calculate by hand, the use of computer software to generate the charts allows ease of use comparable to any of the other charts.

15.3.3 Selecting the Subgroup Size

Control charts rely upon rational subgroups to estimate the short-term variation in the process. This short-term variation is then used to predict the longer-term variation defined by the control limits.

A rational subgroup is simply "a sample in which all of the items are produced under conditions in which only random effects are responsible for the observed variation" (Nelson, 1988). As such, a rational subgroup has the following properties:

- The observations composing the subgroup are independent. Two observations are independent if neither observation influences, or results from, the other. When observations are dependent on one another, we say the process has *autocorrelation*, or *serial correlation* (these terms mean the same thing). Autocorrelation is covered later in this chapter.

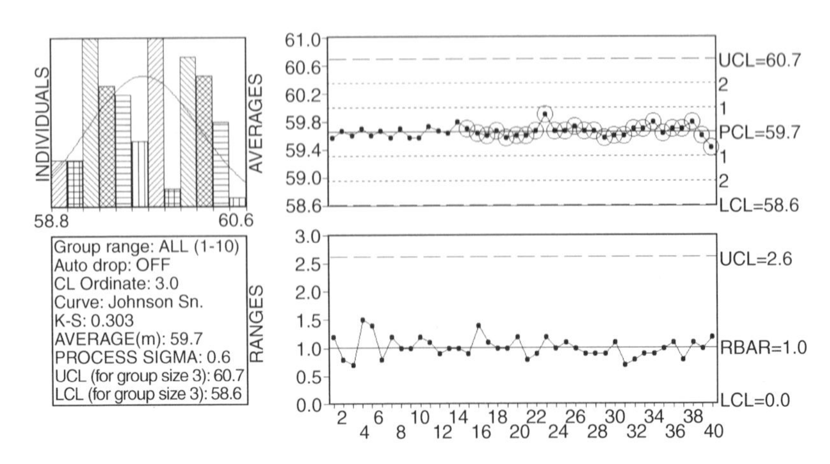

FIGURE 15.7 Irrational subgroups hug the centerline of this X-bar chart of fill weight.

- The subgroups are formed from observations taken in a time-ordered sequence. In other words, subgroups cannot be randomly formed from a set of data (or a box of parts); instead, the data composing a subgroup must be a "snapshot" of the process over a small window of time, and the order of the subgroups would show how those snapshots vary in time (like a movie). The size of the small window of time is determined on an individual process basis to minimize the chance of a special cause occurring in the subgroup (which, if persistent, would provide the situation described immediately below).
- The observations within a subgroup are from a single, stable process. If subgroups contain the elements of multiple-process streams, or if other special cause occur frequently within subgroups, then the within-subgroup variation will be large relative to the variation between subgroup averages. This large within-subgroup variation forces the control limits to be too far apart, resulting in a lack of sensitivity to process shifts. In Figure 15.7, you might suspect that the cause of the tight grouping of subgroups about the X-bar chart centerline was a reduction in process variation, but the range chart fails to confirm this theory.

These data, provided by a major cosmetic manufacturer, represent the fill weight for bottles of nail polish. The filling machine has three heads, so subgroups were conveniently formed by taking a sample from each fill head. The problem is that the heads in the filling machine apparently have significantly different average values. This variation between filling heads caused the within-subgroup variation (as plotted on the range chart) to be much larger than the variation in the subgroup averages (represented graphically by the pattern of the plotted points on the X-bar chart). The X-bar chart's control limits, calculated from the range chart, are thus much wider than the plotted subgroups.

The underlying problem then is that the premise of a rational subgroup has been violated: we tried to construct a subgroup out of apples and oranges. But all is not

TABLE 15.1
Average Number of Size n Subgroups to Detect k Sigma Shift

n/k	0.5	1	1.5	2	2.5	3
1	155	43	14	6	3	1
2	90	17	5	2	1	1
3	60	9	2	1	1	1
4	43	6	1	1	1	1
5	33	4	1	1	1	1
6	26	3	1	1	1	1
7	21	2	1	1	1	1
8	17	2	1	1	1	1
9	14	1	1	1	1	1
10	12	1	1	1	1	1

lost (fruit salad isn't so bad). We've learned something about our process. We've learned that the filler heads are different, and that we could reduce overall variation by making them more similar. Note the circles that highlight subgroups 16 and on. The software has indicated a violation of run test 7, which was developed to search for this type of pattern in the data (see Run Tests).

This type of multistream behavior is not limited to cosmetic filling operations. Consider the potential for irrational subgroups in these processes:

- A bank supervisor is trying to reduce the wait time for key services. She constructs a control chart, using subgroups based on a selection of five customers in the bank at a time. Because she wants to include all the areas, she makes sure to include loan applications as well as teller services in the subgroup.
- An operator finish-grinds 30 parts at a time in a single fixture. He measures five parts from the fixture for his subgroup, always including the two end pieces. His fixture is worn, so that the pieces on the two ends differ substantially.

Many times the process will dictate the size of the rational subgroup. For example, the rational subgroup size for service processes is often equal to one. A larger subgroup, taken over a short interval, would tend to contain dependent data; taken over a longer interval, the subgroup could contain special causes of variation.

The safest assumption for maintaining a rational subgroup is to use a subgroup size of one. Since data usually have some associated costs, smaller subgroups are generally cheaper to acquire than larger subgroups. Unfortunately, smaller subgroup sizes are less capable of detecting shifts in the process. Table 15.1 shows the average number of subgroups necessary to detect the shift of size k (in standard deviation units), based on the subgroup size n. For example, if we observe the process a large number of times, then on average a subgroup of size n = 3 will detect a 1 sigma shift in nine subgroups. As you can see from the table, small subgroups will readily detect relatively large shifts of 2 or 3 sigma, but are less capable of readily detecting smaller shifts. This demonstrates the power of the X-bar chart.

15.3.4 Run Tests

Run tests, developed by Western Electric with some improvements by statistician Lloyd Nelson (Nelson, 1984), apply statistical tests to determine if there are any patterns or trends in the plotted points. Some of the patterns are due to process shifts, while others are due to sampling errors. The run tests increase the power of the control chart (the likelihood that shifts in the process are detected in each subgroup). They are specifically designed to minimize an increase in false alarms.

Run tests 1, 2, 5, and 6 are applied to the upper and lower halves of the chart separately. Run tests 3, 4, 7, and 8 are applied to the whole chart.

- Run test 1 (Western Electric) — a subgroup beyond 3 sigma. Provides an indication that the process mean has shifted.
- Run test 2 (Nelson) — nine consecutive subgroups same side of average. (Note: Western Electric uses eight consecutive points same side of average.) Provides an indication that the process mean has shifted.
- Run test 3 (Nelson) — six consecutive points increasing or decreasing. Provides an indication that the process mean has shifted (a trend).
- Run test 4 (Nelson) — fourteen consecutive points alternating up and down. Provides an indication of sampling from a multi-stream process, as alternating subgroups are sampled from separate processes.
- Run test 5 (Western Electric) — two out of three consecutive points beyond 2 sigma. Provides an indication that the process mean has shifted.
- Run test 6 (Western Electric) — four out of five consecutive points beyond 1 sigma. Provides an indication that the process mean has shifted.
- Run test 7 (Western Electric) — fifteen consecutive points between plus 1 sigma and minus 1 sigma. Provides an indication of either decreased process variation or stratification in sampling. If each subgroup contains observations from multiple process streams, then the within-subgroup variation would be larger than the variation seen from subgroup to subgroup, causing the control limits to be much wider than the plotted subgroup averages. See Figure 15.7 in rational subgroups for an example of this condition.
- Run test 8 (Western Electric) — eight consecutive points beyond plus 1 sigma and minus 1 sigma (both sides of center). Provides an indication of sampling from a mixture. The subgroups on one side of the mean are from a different process stream than the ones on the other side of the mean.

Keep in mind that the subgroup that first violates the run test condition does not usually indicate when the process shift occurred. For example, when run test 2 is violated, the shift may have occurred nine points (more or less) prior to the point that first violated the run test. An additional example of this is evident from Figure 15.7, discussed previously.

15.3.5 Short-Run Techniques

Short-run analysis combines data from several runs into a single analysis. Short run is typically used to analyze processes with an insufficient amount of data available from a given product or service classification to adequately define the characteristics of the process. In manufacturing, for instance, you may only produce 30 units of a given part, and then reset the machine for a different part. Although the process is fundamentally the same (if it is acted upon by the same causal system), the first part may be 1 inch in diameter, plus or minus 1/8 inch, and the second part 5 inches in diameter, plus or minus 1/8 inch. This difference in nominal size prevents you from charting the raw measurements from the different parts on the same chart.

In the same way, in a service application, the amount of time to resolve a customer complaint may be influenced by the type of complaint, such as 1 day for correcting the shipping of an incorrect item vs. 5 days for correcting an incorrect billing. In either case, we are interested in statistically significant changes in our system, relative to either a nominal value (which we define) or an average value (which the system defines).

Thus, if we assume that the process is influenced by a common set of causes, regardless of the run (i.e., part number, complaint type, etc.), then we could use a single control chart to define the operating level for all runs. To do this, we must standardize each observation based on the properties of its run. Standardization can be performed a number of ways, as explained below (Pyzdek, 1992a).

- Nominal control charts. Created by simply subtracting the nominal value of the run from the observation. The nominal value is usually the midpoint of the specification limits, the target value, or the historical average observed from past studies. However, the nominal charting method must be used only if it can be safely assumed that each run has the same amount of variation. This method of standardization is useful for any subgroup size, and subgroup size may vary. The standardization equation is as follows, where x_i is the observed value, *nominal* is the nominal value for the particular run, and z_i is the standardized value:

$$z_i = x_i - nominal \tag{15.12}$$

- Stabilized control charts. As mentioned above, the nominal control chart is valid only when each run has the same amount of variation. In manufacturing, even when two parts are produced by the same process, the effects of the process may increase the process variation based on the specific run-to-run differences. For example, it may be that the machine setup is not as rigid for larger parts. In the same way, the variation in time to resolve a billing complaint may be much larger than a shipment complaint, because more departments may be involved. When the level of variation is not similar for all runs, then we must standardize relative to both the nominal value and the variance. The standardization equation is as follows, where x_i is the observed value, *nominal* and *range* are the nominal and calculated standard range values, respectively, for the particular run, and z_i is the standardized value:

$$z_i = \frac{x_i - nominal}{range} \tag{15.13}$$

In either case, inasmuch as the short-run standardization is done to the raw observations, the standardized values can be used with any control chart or other analysis tool.

15.4 PROCESS CAPABILITY AND PERFORMANCE INDICES

Process capability indices attempt to indicate, in a single number, whether a process can consistently meet the requirements imposed on the process by internal or external customers. Process capability indices are only meaningful if the data are from a controlled process. The reason is simple: process capability is a prediction, and you can predict only something that is stable. To estimate process capability, you must estimate the location, spread, and shape of the process distribution. One or more of these parameters are, by definition, changing in an out-of-control process. Therefore, use process capability indices only if the process is in control for an extended period.

Process performance, on the other hand, tells us about a specific sample of observations. Whereas process capability indices use the *process sigma* statistic (from the control chart) to estimate variation, process performance indices use the *sample standard deviation* statistic to estimate variation. Thus, the process performance index is valid only for the sample in question, telling us whether the sample meets customer requirements. As mentioned above, the process capability index indicates the long-term potential of the process to meet requirements so long as it is maintained in control. For each of the capability indices below, a corresponding performance index can be calculated by replacing the process sigma with the sample sigma in the formula. The notation for the index then also changes: C_p becomes p_p; C_{pk} becomes p_{pk}; C_{pm} becomes p_{pm}.

A number of capability indices have been developed that assume normality of the data. In the absence of normality, a data transformation can be performed to achieve normality of the transformed data. One such technique uses the family of Johnson distributions (Pyzdek, 1992b), which unfortunately require computer computation. When both are available, compare the non-normal and normal indices, and test which assumption (normal or not) fits the data better.

C_p. Compares the tolerance to the spread of the distribution, expressed as ±3 sigma. Note that the sigma value is the *process sigma*, calculated using the control charts.

Normal distribution:

$$C_p = \frac{High\,Spec - Low\,Spec}{6\sigma_x} \tag{15.14}$$

Non-normal distribution:

$$C_p = \frac{High\ Spec - Low\ Spec}{ordinate_{0.99865} - ordinate_{0.00135}} \tag{15.15}$$

C_{pk}. A measure of both process dispersion and its centering about the average.

$$C_{pk} = MIN(C_{pl}, C_{pu}) \tag{15.16}$$

where

$$Cp_l = -\frac{Z_l}{3} \tag{15.17}$$

$$Cp_u = -\frac{Z_u}{3} \tag{15.18}$$

Normal distributions:

$$Z_l = \frac{\bar{\bar{x}} - Low\ Spec}{\sigma_x} \tag{15.19}$$

$$Z_u = \frac{High\ Spec - \bar{\bar{x}}}{\sigma_x} \tag{15.20}$$

where x-double bar is the grand average and σ_x is process sigma.

Non-normal distributions:

$$Z_l = Z_{normal,p} \tag{15.21}$$

$$Z_u = Z_{normal,1-p} \tag{15.22}$$

$Z_{normal,p}$ and $Z_{normal,1-p}$ are the *z*-values of the normal cumulative distribution curve at the *p* percentage point and the $1 - p$ percentage points, respectively.

C_{pm}. A measure similar to the C_{pk} index that also takes into account variation between the process average and a target value. If the process average and the target are the same value, C_{pm} will be the same as C_{pk}. If the average drifts from the target value, C_{pm} will be less than C_{pk}.

TABLE 15.2
Parts per Million Defect Rates for C_{pk}

C_{pk}	One-Sided Spec	Two-Sided Spec
0.25	226627	453255
0.5	66807	133614
0.7	17864	35729
1.0	1350	2700
1.1	483	967
1.2	159	318
1.3	48	96
1.4	13	27
1.5	3	7
1.6	1	2
2	0.00099	0.00198

$$C_{pm} = \frac{C_p}{\sqrt{1 + \frac{(\bar{\bar{x}} - T)^2}{\sigma_x^2}}} \tag{15.23}$$

where T is the process target, x-double bar is the grand average, and σ_x is process sigma.

15.4.1 Interpretation of Capability Indices

When interpreting capability indices, remember that the process must be in control for the capability index to have any meaning. If the process is not in a state of statistical control, use the process performance index.

Most practitioners consider a capable process to be one that has a c_{pk} of 1.33 or better. A process operating between 1.0 and 1.33 is considered marginal. Many companies now suggest that their suppliers maintain even higher levels of c_{pk}. A c_{pk} exactly equal to 1.0 would imply that the ±3 sigma process variation exactly meets the specification requirements. Unfortunately, if the process shifted slightly, and the out-of-control condition was not immediately detected, then the process would produce output that did not meet the requirements. Thus, an extra 0.33 is allowed for some small process shifts to occur that could go undetected. Table 15.2 provides an indication of the level of improvement effort required in a process to meet these escalating demands, where "PPM Out of Spec" refers to the average defect level measured in parts per million.

A capability index is a statistic, subject to statistical error. A Monte Carlo simulation (Pignatiello and Ramberg, 1993) involving 1000 different trials of 30-piece samples showed that when the true capability equaled 1.33, nearly 20% of the trials indicated a capability less than 1.2. Similarly, if the true capability was 1.0, more than 10% of the trials indicated that the capability was 1.2 or greater.

Confidence limits are provided below for each of the capability indices. These calculated values can be added or subtracted from the calculated capability index to indicate the range of values expected from random samples of a stable process.

- C_p:

$$CL = \frac{3\,C_p}{\sqrt{2\,(n-1)}} \tag{15.24}$$

where n is the subgroup size.

- C_{pk}:

$$CL = 3\left(\frac{C_{pk}^2}{2\,(n-1)} + \frac{1}{9n}\right)^{\frac{1}{2}} \tag{15.25}$$

where n is the subgroup size.

- C_{pm}:

$$CL = \frac{3C_{pm}}{\sqrt{n}}\left(\frac{1+2z^2}{2(1+z^2)^2}\right)^{1/2} \tag{15.26}$$

where n is the subgroup size and

$$z = \frac{\bar{\bar{x}} - T}{\sigma_x^{\;2}} \tag{15.27}$$

15.5 AUTOCORRELATION

Standard control charts require that observations from the process are independent of one another. Independence implies that the particular value of an observation in time cannot be predicted based on prior data observations. For example, in Deming's red bead experiment shown in Figure 15.3, observing a particular value of, say 7 red beads, does not provide us with any information to predict the next observation. Our best estimate of every sample is the process mean. In contrast, we can use the current temperature of an oven that is being warmed to 350° to predict the temperature 1 minute later. We say these temperature data are dependent and autocorrelated (serially correlated).

Examples of autocorrelation in practice include

- Chemical processes. When dealing with liquids, particularly in large baths, samples taken close together in time are influenced by one another. The factors influencing the first observation are carried over in the large mass

of liquid to maintain a similar environment that carries over into subsequent temperature observations for a period of time. Subgroups formed over a small time frame from these types of processes are sometimes called homogenous subgroups, because the observations within the subgroups are often nearly identical, except for the effect of measurement variation.
- Service processes. Consider the wait time at a bank. The wait time of any person in the line is influenced by the wait time of the person in front of him or her.
- Discrete part manufacturing. Although this is the classic case of independent subgroups, when feedback control is used to change a process based on past observations, the observations become inherently dependent.

In constructing X-bar charts, recall that the subgroup is used to estimate the short-term average and variation of the process. The average short-term variation (R-bar) is then used to estimate the control limits on both the X-bar and range charts. If the process is in control, or stable, then the average short-term variation provides a good indication of long-term variation. Therefore, in forming subgroups, a convenient rule to remember is that the short-term (or within subgroup) variation must be comparable to the long-term (or between-subgroup) variation. In practical terms, the potential causes of within-subgroup variation (machines, materials, methods, manpower, measurement, and environment) should be comparable to those causes that exist between subgroups.

If we tried to construct subgroups from these autocorrelated processes, the short-term variation would typically be much smaller than the longer-term variation. This causes the control limits to be unnaturally tight, increasing the chance that the control chart will indicate a process shift when the process has NOT shifted (a false alarm). Responding to these false alarms is tampering, which increases overall process variation. If control limits on an X-bar chart are particularly tight, with many out-of-control points, autocorrelation should be suspected.

Consider now a subgroup created from a sample of each head of a six-head machining operation (or six order processors performing the same procedure). In these examples, the observations would show correlation between every sixth observation. The differences between the machine heads (or order processors) would cause the subgroup range to be large, resulting in excessively wide X-bar control limits. This was shown in Figure 15.7.

These examples point out how control limits could be either too large or too small, resulting in failure to look for special causes when they really do exist, or searching for special causes that don't exist. The important point to note here is that these errors are not caused by the methodology itself, but rather by ignoring a key requirement of the methodology: independence.

15.5.1 Detecting Autocorrelation

The scatter diagram in Figure 15.8A shows the correlation, or in this case the autocorrelation, between each observation and the one observed immediately (one

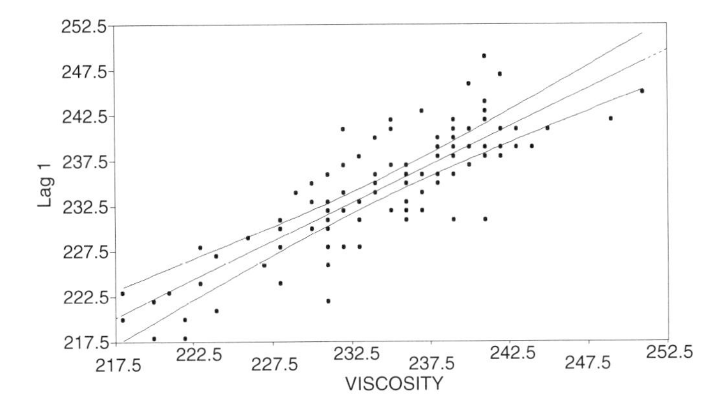

FIGURE 15.8A Viscosity vs. itself, one sample apart.

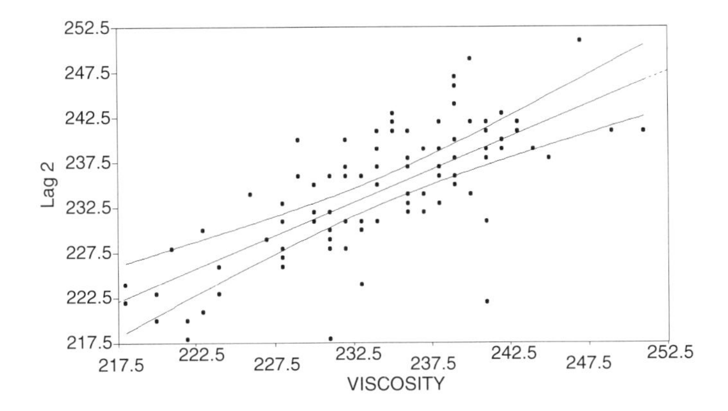

FIGURE 15.8B Viscosity vs. itself, two samples apart.

period or lag) following it. Each period is 1 minute or, in other words, one sample was taken every 60 seconds.

The scatter diagram in Figure 15.8B shows the autocorrelation using observations made two periods apart, or 2 minutes between samples. Figures 15.8C and D, respectively, show 5 and 10 minutes between samples. As seen by the plots, the influence of an observed temperature on the temperature 1 minute later is stronger than on temperature readings made 10 minutes later.

Although scatter diagrams offer a familiar approach to the problem, they are a bit cumbersome to use for this purpose, because you must have separate scatter diagrams for each lag period. A more convenient tool for this test is the autocorrelation function (ACF), which plots the autocorrelation at each lag, as shown in Figure 15.9, indicating departures from the assumption of independence.

The ACF will first test whether adjacent observations are autocorrelated; that is, whether there is correlation between observations 1 and 2, 2 and 3, 3 and 4, etc. This is known as lag one autocorrelation, because one of the pair of tested observations lags

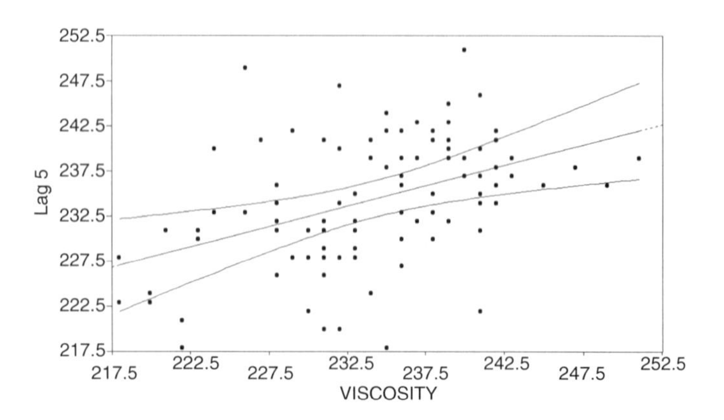

FIGURE 15.8C Viscosity vs. itself, five samples apart.

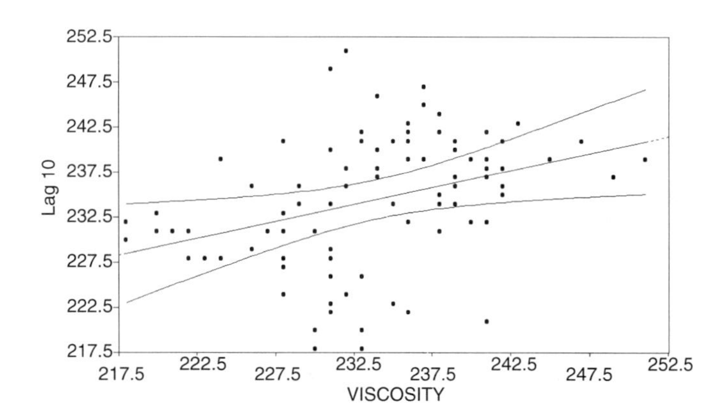

FIGURE 15.8D Viscosity vs. itself, ten samples apart.

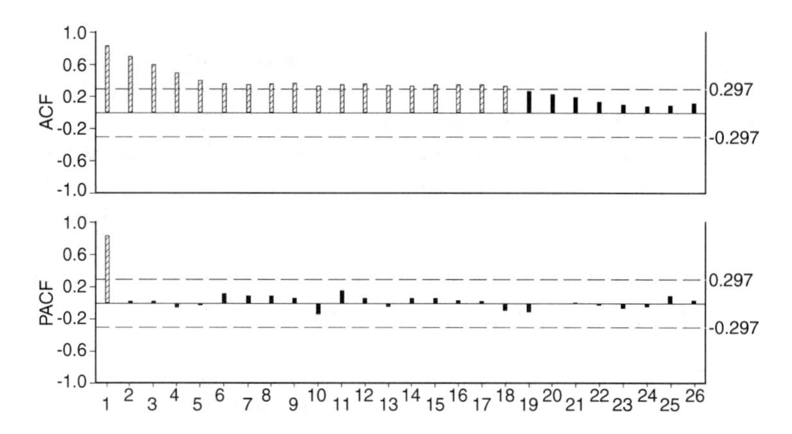

FIGURE 15.9 Autocorrelation function for viscosity data.

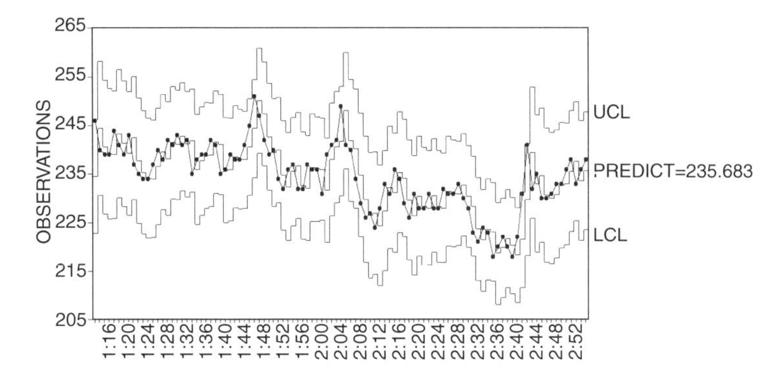

FIGURE 15.10 Moving centerline EWMA chart using viscosity data.

the other by one period or sample. Similarly, it will test at other lags. For instance, the autocorrelation at lag 4 tests whether observations 1 and 5, 2 and 6, … 19 and 23, etc. have been correlated. In general, we should test for autocorrelation at lags 1 to lag $n/4$, where n is the total number of observations in the analysis. Estimates at longer lags have been shown to be statistically unreliable (Box and Jenkins, 1970).

In some cases, the effect of autocorrelation at smaller lags will influence the estimate of autocorrelation at longer lags. For instance, a strong lag 1 autocorrelation would cause observation 5 to influence observation 6, and observation 6 to influence 7. This results in an apparent correlation between observations 5 and 7, even though no direct correlation exists. The partial autocorrelation function (PACF) removes the effect of shorter lag autocorrelation from the correlation estimate at longer lags. This estimate is valid to only one decimal place.

ACFs and PACFs each vary between plus and minus one. Values closer to plus or minus one indicate strong correlation. The confidence limits are provided to show when ACF or PACF appears to be significantly different from zero. In other words, lags having values outside these limits (shown as lined bars in Figure 15.9) should be considered to have a significant correlation.

15.5.2 Dealing with Autocorrelation

Autocorrelation can be accommodated in a number of ways. The simplest technique is to change the way we take samples, so that the effects of process autocorrelation are negligible. To do this, we have to consider the reason for the autocorrelation.

If the autocorrelation is purely time based, we can set the time between samples long enough to make the effects of autocorrelation negligible. In the example above, by increasing the sampling period to greater than 20 minutes, autocorrelation becomes insignificant. We can then apply standard X-bar or individual-X charts.

A disadvantage of this approach is that it may force the time between samples to be so long that process shifts are not detected in a reasonable (economical) time frame. Alternatively, we could model the process based on its past behavior, including the effects of autocorrelation, and use this process model as a predictor of the process. Changes in the process (relative to this model) can then be detected as special causes. Specially constructed EWMA (wandering mean) charts with moving centerlines, such as is shown in Figure 15.10, have been designed for these autocorrelated processes.

When the autocorrelation is due to homogeneous batches, as in a chemical process, we might consider taking subgroups of size one using an individual-X chart. In this case, each plotted point represents a single sample from each batch, with only one sample per batch. Now the subgroup-to-subgroup variation is calculated using the moving range statistic, which is the absolute value of the difference between consecutive samples. An enhancement to this method is to take multiple samples per batch, then average these samples and plot the average as a single data point on an individual-X chart. This is sometimes referred to as a batch means chart. Each plotted point will better reflect the characteristics of the batch, because an average is used.

REFERENCES

Box, G. E. P. and Jenkins, G. M., *Time Series Analysis: Forecasting and Control.* Holden-Day, San Francisco, 1970.

Duncan, A. J., *Quality Control and Industrial Statistics.* 5th ed., Homewood, IL, Richard D. Irwin, 1986.

Montgomery, D. C. and Runger, G. C., *Applied Statistics and Probability for Engineers.* 1st ed., John Wiley and Sons, New York, 1994.

Nelson, L. S., The Shewhart control chart: Tests for special causes, *J. Qual. Technol.*, 16(4), 237–239, 1984.

Nelson, L. S., Control charts: Rational subgroups and effective applications, *J. Qual. Technol.,* 20, 1, 1988.

Pignatiello, J. J., Jr. and Ramberg, J. S., Process capability indices: Just say "no," *47th Annu. Congr. Trans.,* ASQC Press, Milwaukee, WI, 1993.

Pyzdek, T., *Pyzdek's Guide to SPC, Volume Two: Applications and Special Topics,* ASQC Press, Milwaukee, WI, Quality Publishing, Tucson, AZ, 1992a.

Pyzdek, T., Process capability analysis using personal computers, *Qual. Eng.*, 4(3), 419–440, 1992b.

Western Electric Company, Inc., *Statistical Quality Control Handbook,* 2nd ed., Western Electric, New York, 1958.

16 Supply Chain Management

Douglas Burke

16.1 INTRODUCTION

Supply chain management has a quaint ring to it. It conjures images of an industrial economy with warehouses, transportation systems, suppliers, and assembly lines. In the world of manufacturing, this bricks-and-mortar vision is still fairly accurate despite all the click-and-order hype associated with cyberspace. Manufacturing enterprises around the world living in this traditional vision are experiencing change at a rapidly increasing pace. Some of the changes they face are fiercely competitive markets, shorter and shorter product life cycles, heightened customer expectations, and a diminished ability to raise prices even on high-demand products.

With these changes come enormous business pressures. Pressure to find more effective ways to shorten the concept-to-delivery cycle. Pressure to drive out inefficiencies in all their processes. Pressure to develop and execute a strategic plan that will anticipate and address these changes. Only by aggressively seeking process improvements and enhancements to cost, quality, productivity, and customer satisfaction can companies hope to survive these changes.

As manufacturers seek the mechanisms for survival, they turn their attention to the supply chain, seeking to capture improved efficiency. Currently, considerable activity in manufacturing is focused on eliminating inefficiencies through supply chain management. Abramson (1999) reports that inventory being held across the retail supply chain at any one time amounts to $1 trillion. Of those inventories, 15 to 20% ($150 to 200 billion worldwide; $40 to 50 billion in the United States) could be eliminated through improved supply chain management in the form of planning, forecasting, and replenishment. Anderson, Britt, and Donavon (1997) report that companies now recognize the importance of meeting customer needs. By using supply chain management, companies can tailor products and services to specific customers and win customer loyalty. This loyalty translates into profits. Xerox has found satisfied customers six times as likely to buy additional Xerox products over a period of 18 months than dissatisfied customers. Other benefits that can be gained through supply chain management are improved cash utilization (how soon after delivery do you get paid?), flexible schedules, shortened schedules, delivery of product or services at the time of need, and price advantages.

How can a business gain all those advantages through supply chain management? It is first necessary to have an extremely effective Six Sigma process to protect against losing customers due to product nonperformance. Next, a mature and effective lean

1-57444-300-3/02/$0.00+$1.50

manufacturing program must be in place to ensure the maintenance of minimum inventory levels while the manufacturing processes are still consistently delivering product to the customer on time. Finally, the company needs to integrate Six Sigma and lean manufacturing across the entire supply chain by including supply chain management in its strategic planning process. Strategic planning, lean manufacturing, and Six Sigma are covered in separate sections of this book. The remainder of this chapter focuses on contemporary issues that exist in supply chain management, the more traditional topics of inventory management and control, and the importance of synchronizing supply to demand.

16.2 DEFINING THE MANUFACTURING SUPPLY CHAIN

There are probably as many definitions of a supply chain as there are practitioners of supply chain management (SCM). Poirier and Reiter (1996) define the supply chain as a system of organizations that delivers products and services to its customers. This supply chain model can be illustrated as a network of linked organizations that has a common purpose of delivering product and services through the best possible means.

Another supply chain definition, developed by Kearney (1994), shows linked groups of enterprises that work synchronously to acquire, convert, and distribute goods and services to the customer. Kearney also captures the need to distribute new designs through the network, ensuring a rapid response to the dynamic requirements of the market.

Though Copacino (1997) never presents a concise definition of the supply chain, he alludes to it as all the players and activities necessary to convert raw materials into product and deliver them to consumers on time and at the right location in the most efficient manner. In this supply chain model, the major business processes of a manufacturing company are composed of suppliers, manufacturing, distribution retailing, and consumers. He extends this model by showing the demand-and-supply chain as integrating functions to the major business processes.

Walker and Alber (1999) define the manufacturing supply chain as the global network used to deliver products and services from raw materials to end customers through an engineered flow of information, physical distribution, and cash.

Mohrman (1999) defines the supply chain as the business, capital, material, and information associated with the flow of goods. The total supply-and-demand chain extends from natural resources through a network of value-added steps and transport links until it reaches the ultimate consumer.

Different practitioners developed these definitions for different reasons. Although it would seem that they are completely different, closer examination of these definitions reveals common key themes that can be used to develop our own definition. This definition will be generic enough to be applicable to any manufacturing supply chain.

One key concept is that the supply chain is a network of linked companies and organizations. This network has a broad span that starts with obtaining natural resources and ends when the product or service reaches the ultimate customer. Finally, the dynamics of a supply chain involve the conversion of natural resources into a product or service that is delivered to a customer. With this, we can develop our definition of a generic manufacturing supply chain:

> A supply chain is a dynamic network of interlinked organizations that converts natural resources into products or services that are delivered to the consumer at the right place and at the right time.

A simple graphical model of this supply chain is shown in Figure 16.1. From this illustration, we see that the supply chain starts when a supplier (or suppliers) converts natural resources into usable materials for the manufacturing company. Usable materials can be raw material, such as steel bar stock, if the manufacturing company is a machine shop or subassemblies if the manufacturing company is a personal computer-manufacturing firm. After all the necessary resources are supplied to the manufacturing firm, they are converted into the end product for which the customer ultimately pays. A logistics organization, not depicted in Figure 16.1, is necessary to ensure the proper delivery of the end product to the consumer.

To better illustrate this supply chain model, let's look at it in the context of the aerospace industry. In the aerospace industry, a jet engine manufacturing supply chain can be a very complicated group of companies. Suppliers would start by purchasing raw aluminum and steel stock and converting it into castings and forgings. Other suppliers may take those castings and forgings and machine them, adding gears, splines, shafts, and motors to create mechanical subassemblies. These subassemblies are then delivered to the engine-manufacturing firm where they are assembled into complete and functional jet engines. These engines are tested, packaged, and shipped to the consumer through the logistics network.

Inventory of all types can be found at all stages of the supply chain. As illustrated, raw material inventories are typically accumulated at the beginning. Work-in-process inventory in the form of subassemblies and partially assembled jet engines will accumulate at the manufacturing stage. Finished goods inventory in the form of completed jet engines can accumulate in the logistics network, at the distribution centers, and at the customer's site.

Another interesting aspect of the manufacturing supply chain is that information flows in the opposite direction of the product. Products and services typically flow from suppliers to the manufacturing firm. From there the products and services are transported to the customer through a logistics network. Conversely, information about consumption patterns, points of sales, and demand forecasts flows from the customer to the manufacturing firm. From there the manufacturing firm disseminates the information and flows it down to the appropriate suppliers.

From this we can conclude that a supply chain is a very complex group of suppliers, manufacturing firms, and logistics organizations that must work together

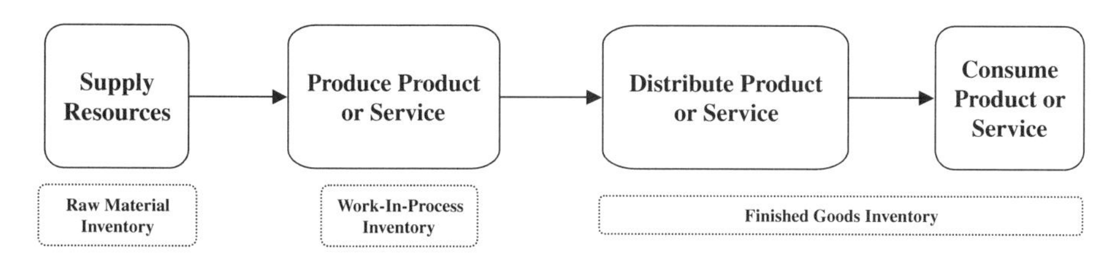

FIGURE 16.1 Generic manufacturing supply chain model.

to accomplish a common goal. The manufacturing supply chain also needs an efficient information technology organization that can quickly and accurately move information down the supply chain. Finally, it is apparent that the only way to deal with the complexity of the supply chain is to have an effective supply chain management philosophy. Without a common management philosophy among the elements of the supply chain, it is very difficult to define and accomplish the supply chain goal(s). In the next section we define supply chain management and how it should be used to synchronize all the elements of the supply chain.

16.3 DEFINING SUPPLY CHAIN MANAGEMENT

Have you ever tried to define supply chain management (SCM) to someone? About the time you compare SCM to logistics management or materials management, you notice that your audience has lost interest. It is readily apparent that SCM is not easy to define. The fact is, there are many practitioners' definitions of SCM. Let's look at some of them to see if we can come up with one of our own.

One practitioner defines SCM as the driving force that oversees the relationships across the entire supply chain. In this definition, SCM is responsible for obtaining the necessary information to run the business, to get product delivered through the business, and to get the revenue that generates profits for the business. This definition also mentions the need for SCM to consider the entire supply chain.

Another SCM practitioner provides a much broader definition. He or she sees SCM as coordinating, scheduling, and controlling procurement, production, inventories, and deliveries of products and services to customers. It includes the everyday administration, operations, logistics departments, and processing information from customers to suppliers.

Yet another definition positions SCM as the organization responsible for making, selling, and delivering products to the customer. This definition goes further by requiring collaboration among all members of the supply chain to manage sensitive strategic planning as well as the flow of information.

A more detailed definition of SCM starts by calling SCM a set of approaches that must be utilized to efficiently integrate suppliers, manufacturers, warehouses, and stores. This is necessary to ensure that product is manufactured and distributed at the right quantities, to the right location, and at the right time. The results can be measured in minimized systemwide costs and satisfied customers.

A manufacturing-specific definition goes as follows: SCM is the driving force in ensuring that the manufacturer and its suppliers work together to make a product or service available to the marketplace for which the customer will pay. This convolution of companies, functioning as one extended enterprise, makes optimum use of shared resources to achieve operating productivity. The result is a product or service that is high quality and low cost, and is delivered on time to the marketplace.

Our last definition is probably the simplest and most concise. SCM is the mechanism that links all the players and activities involved in converting raw materials into products. These players and activities are responsible for delivering those products to customers at the right time, at the right place, and in the most efficient way.

By looking at all these definitions, we can develop some common themes. First, we see that each definition emphasizes management across the entire supply chain. In other words, SCM should be pervasive from suppliers to customers. Second, the definitions use words such as *coordinate, link, oversee, collaborate,* and *integrate*. This implies that management across the supply chain must be used to synchronize each of the individual elements of the supply chain. Finally, each element of the supply chain must have a common goal. What is the goal? In every definition, the concept of manufacturing a high-quality, low-cost product or service and delivering it on time to the right customer is mentioned. Additionally, each definition has the customer central to SCM, so customer satisfaction should be a goal. Now, putting all these elements together we develop the following definition of SCM:

> SCM is the mechanism that synchronizes all the individual elements of the supply chain. SCM must ensure that the supply, production, and delivery of a product or service always meets the customer's requirements for cost, quality, and performance. This means that the product must be low cost and high quality, and be delivered to the right customer at the right time.

From this definition, we see that there are some important topics that require more discussion, such as supply chain synchronization, inventory management, logistics network configuration, strategic partnering, and information technology's role in SCM. These topics are all central to our definition of SCM, and we discuss them in the sections that follow.

16.4 CRITICAL ISSUES IN SUPPLY CHAIN MANAGEMENT

Recent developments in SCM have spawned numerous books, articles, and academic publications addressing the current issues facing SCM. One issue that appears in almost every publication on the topic of SCM is the need to integrate the entire supply chain. Many managers recognize that integrating the supply chain can improve both cost and customer satisfaction. Supply chain integration is necessary simply because it allows a firm to match the supply of a product to the product's consumption pattern. Synchronization of supply to demand has many cost benefits. We discuss the details of synchronizing supply to demand in a later section but first, the issues of supply chain integration are discussed.

Integration of every link in the supply chain has proven to be very difficult for many reasons. One reason is that the supply chain system for any firm is in a state of constant change and evolution. Another difficulty with integration is related to the complexity of the supply chain. There are so many organizations and facilities in a supply chain that there will always be conflicting objectives and a lack of communication. Two common approaches for addressing integration issues exist. The first is for a firm to take advantage of information technology, which helps to simplify the supply chain and improve communication. The second approach is for firms to form strategic alliances among all partners in the supply chain.

The use of information technology is always identified as the enabling force for accomplishing supply-to-demand synchronization. With the proper information, all the links in the supply chain can maintain minimum costs and still meet customer demand. With this information, a firm can also develop accurate forecasts, which are imperative when matching the demand for a product with the supply of materials in the overall supply chain.

Another SCM issue that is related to supply chain integration is the need for a company to develop strategic partners throughout the supply chain. If a firm has successfully established a product demand-to-supply synchronization through its supply chain, then there must be some level of coordination and partnering within each component of the supply chain. Later in this chapter we present some of the most common strategic partnering approaches used by modern manufacturing firms.

Finally, the more commonly discussed issue of SCM is configuring the logistics network. Let's assume that a typical manufacturing firm produces a product from several plants and distributes the product to a set of geographically dispersed customers through a network of warehouses. The issue here is that the firm needs to determine the optimal number and location of warehouses. Optimization in this area means determining the appropriate number, size, location, and inventory of each warehouse. This, of course, assumes the manufacturing plants and customers remain geographically fixed. An analytical approach to address network configuration and inventory management are presented later in this chapter. The remainder of this chapter is dedicated to summarizing what critical issues face SCM and what has been proposed to address those issues.

16.4.1 Supply Chain Integration

Integration of the supply chain is difficult because of its dynamic nature and conflicting objectives, but not impossible, as major companies in the semiconductor, consumer retail, and chemicals industries have demonstrated. How do companies successfully integrate their supply chains? Research on hundreds of manufacturing companies shows that the most common approach is to first establish lines of communication across the entire supply chain, then to establish strategic partnerships among all partners in the supply chain. A firm's information technology (IT) department is the key functional area for providing the ability to communicate across the supply chain. Strategic partnering has been a common practice for many years; however, it is typically only practiced in the procurement department and used in isolation. Later in this chapter we summarize some of the common types of partnerships and how partnerships should be formed across the entire supply chain.

Before a supply chain can be integrated, there must be open sharing of information for coordinated operational planning. The sheer magnitude of data and information that can be shared is enough to clog the flow of products through a supply chain. So, what are the roles of IT in SCM? One role is to provide access to information through a seamless link from the beginning to the end of the supply chain. Another is to provide a centralized hub of all available information.

16.4.1.1 Information Technology

Effective use of information to integrate the supply chain has been recognized as an important focus of SCM since the early 1990s (Copacino, 1997). Much of the current interest in information technology is motivated by the ability to apply sophisticated analytical methods to supply chain data to glean savings. Also, much interest is developing from opportunities provided by electronic commerce, especially through the Internet.

Information linkages among all partners in a supply chain must be developed and implemented. Supply chain managers also need analytic capabilities for logistics network modeling, routing and scheduling, production scheduling, and logistics simulations. Information systems must be multifunctional so they can handle the complexity of the supply chain. Speed and accuracy of transaction handling are also important. All functional areas such as manufacturing, warehousing, transportation, and logistics must use real-time systems and accurate data-capture technologies. Decision support systems are also needed to make strategic, tactical, and operational decisions. Considering these needs, information technology is the most important enabling function for developing an integrated supply chain. Because the supply chain spans the entire network from supplier to customer, our discussion of information technology will encompass systems internal to an individual company as well as external systems that transfer information between companies.

16.4.1.2 Information Access

One goal of information technology in any supply chain is to provide access to information through a seamless link from suppliers of raw materials through manufacturing and ultimately to the customer. Figure 16.2 illustrates the flow of information through the supply chain. Note that the flow of information is opposite to the flow of products through the supply chain.

This link provides access to information concerning the location or status of a product anywhere in the supply chain. With this link a firm can plan, track, and accurately estimate lead times based on actual data. Of course, this necessitates access to data that reside in systems physically located at different companies as

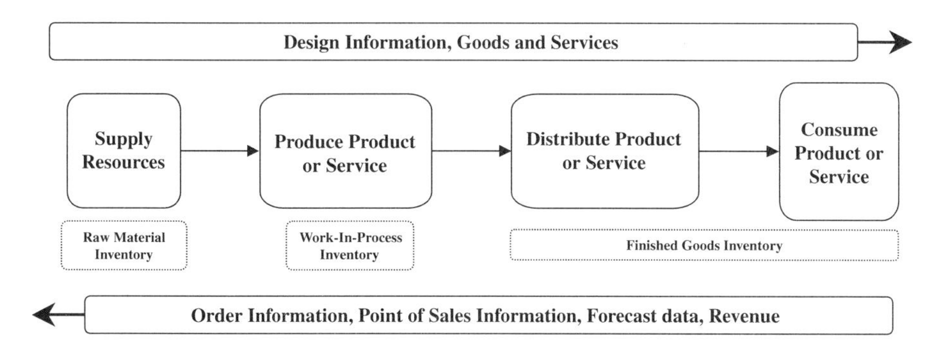

FIGURE 16.2 Information and product flow through the supply chain.

well as at geographically separated systems within the same company. Another key aspect of this link is to assure the availability of information so rational, timely decisions can be made. Information systems also need to be proactive. For example, if the delivery of an order is delayed, a mechanism must be in place that will automatically notify interested parties so they can adjust schedules or seek alternative sources of the product.

Companies in the personal computer manufacturing industry have made the most advances in developing this information access link. Take a look at the IT infrastructure within any major personal computer manufacturer today and you will find an order-tracking system that provides real-time information on the whereabouts of an order. This information is available to all internal organizations, all external suppliers, and to the customer. It is this type of information access that every manufacturing company must strive to obtain.

16.4.1.3 Centralized Information

Another information technology goal of SCM is to provide a centralized hub of all available information. In most companies, each information system is isolated from other information systems within that company. Manufacturing, logistics, and customer service work with a shop-floor control system, accounting works with another system, quality has a separate system, sales and marketing use yet another system, and customer service has their own system. Figure 16.3 illustrates a typical IT systems configuration. Occasionally, some crucial bits of information will cross the lines between systems, but it usually takes a lot of effort and it is rarely accomplished in a timely manner.

In an ideal world, all information requested by anyone in the supply chain would be accessible at one location with a robust mode of access (e.g., fax, phone, or Internet). There hasn't been a single manufacturing company researched that has achieved this goal. Some industries, such as banking, are close (Bramel and Simchi-Levi, 1997) but none of them has a centralized hub for information access.

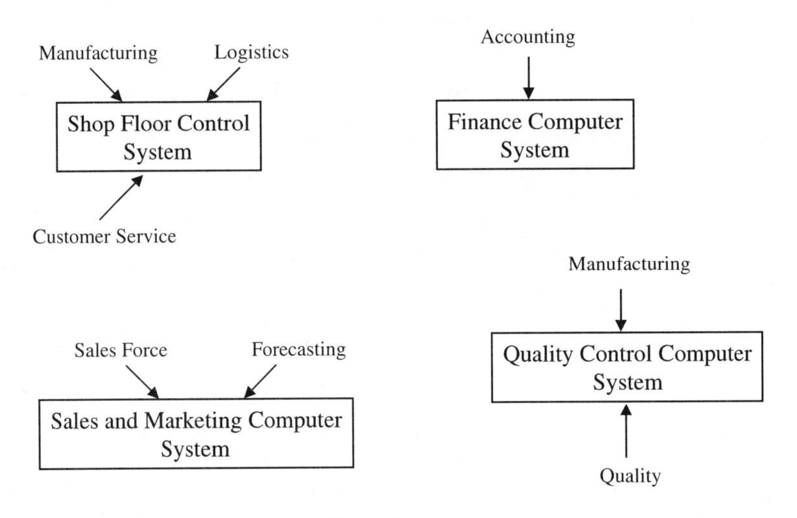

FIGURE 16.3 Typical IT systems configuration.

16.4.1.4 IT Development and Strategic Planning

Now that we know the importance of IT in integrating the supply chain, how can a company access its current stage of development and plan for the future? The very complexity of the supply chain implies that there is no simple and inexpensive answer to this question. Most companies do not introduce IT innovations because it is not obvious if there will be a return on the investment. This is truly shortsighted. Every company should take two simple steps in IT relative to SCM: assess its current level of IT development, then create a corporate-wide vision to get to the next level of development and beyond. This chapter provides a simple way to assess your company's current level of development, and you can use the strategic planning topics in this book to develop and achieve your corporate-wide vision.

16.4.2 Strategic Partnering

It may not always be effective for one firm to perform all key business functions internally. Even if a firm has the resources available to perform a particular manufacturing task, another firm in the supply chain may be better suited to perform that task. Sometimes a combination of physical location in the supply chain, resources, and core competency determines the most appropriate firm in the supply chain to perform a manufacturing function. Once the appropriate firm to perform a task has been identified, steps must be taken to ensure that the function is actually performed by that firm.

From our research, firms typically rely on one, or a combination, of three basic approaches to ensure that a manufacturing-related function is completed:

- Committing internal resources. If a company does not have the resources or core competency internally, then it must acquire a firm that does. In either case, this gives the manufacturing concern total control over all aspects of the way that particular business function is performed. On the other hand, acquisitions can be very difficult, lengthy, and expensive.
- Developing short-term external arrangements. Most business transactions are accomplished through this type of arrangement. If a firm needs a specific part, resource, or service, it will either purchase or lease it. This is typically the most effective arrangement for all parties involved. However, this kind of arrangement is only short term and it rarely, if ever, leads to long-term strategic advantages.
- Developing strategic partners. This approach, if done properly, results in long-term partnerships between two companies. In most cases, the problems of committing internal resources or acquiring those resources can be avoided by developing strategic partners. Additionally, developing strategic partners can lead to the commitment of more resources than can be freed up with short-term external arrangements. Ultimately, this approach allows risks and rewards to be shared by all partners, along with the benefits of a stronger, healthier business.

For the remainder of this section, we focus on two of the most common strategic partnering agreements used in SCM.

16.4.2.1 Supplier Partnerships

Supplier partnerships are the most common form of strategic partnerships used by today's manufacturing companies. This type of partnership is formed between the suppliers of resources and the manufacturing firm. The simplest type of supplier partnership is one in which the manufacturing firm shares customer demand information to assist the supplier in production planning. The most complicated partnership is one wherein the supplier has complete ownership and management responsibility of the inventory until it is sold to the customer.

In a basic supplier partnership, the supplier receives customer demand data from the manufacturing firm. The supplier uses these data to synchronize its production rates and inventory levels to the customer's requirements. In this partnership, the manufacturer is responsible for individual customer orders. However, the customer demand information is used by the supplier to improve forecasting and scheduling.

In a more advanced partnership, the supplier receives customer demand data to prepare shipments at previously agreed-upon intervals to maintain specific levels of inventory. As this partnership matures, suppliers gradually decrease inventory levels at the manufacturing firm, resulting in predictable inventory reductions.

Finally, in the most advanced partnership, the supplier decides on the appropriate inventory levels and inventory policies to maintain those levels. In the early stages, the manufacturer approves supplier decisions, but eventually this form of oversight should be eliminated. This type of partnership has been used successfully in retail, department store, and discount department store industries.

Clearly, a supplier–manufacturer partnership requires a certain level of trust; without this trust the affiliation will fail. In some cases, the partnering supplier must be trusted to manage a large segment of the supply chain. In other cases, the supplier must be trusted to manage the manufacturer's inventory as well as its own. Finally, in every partnership, confidential information, which could serve competing manufacturers or suppliers, must pass between all firms safely.

16.4.2.2 Logistics Partnerships

Another type of partnership used in SCM is logistics partnerships. These involve the use of a firm outside the manufacturing company to perform all or a portion of the manufacturing firm's materials management and product distribution function. These partnerships involve commitments that are generally longer term than supplier partnerships. A good provider of logistics services must be able to perform multiple functions, because it will be required to manage across many stages of the supply chain. Because of the complexity and multifunctional nature of this type of partnering, it is used mostly by large firms. A logistics partnership contract is usually a major, complicated business decision. Many considerations are critical in deciding whether a company should enter into a logistics partnership. The two most important considerations are knowledge of its own costs and ownership of assets.

The most basic issue in selecting a logistics provider is to know your own costs so they can be compared with the cost of using an external provider. In many cases it is necessary to use modern cost-accounting methods that will track both direct and indirect costs back to specific products and services.

There are advantages and disadvantages to consider when using asset-owning vs. non-asset-owning logistics providers. Asset-owning logistics providers are typically large, have access to an extensive customer base, and can provide economies of scope and scale. Some of the disadvantages to consider when using asset-owning logistics providers are that they are typically bureaucratic, they will favor their own company's organizations when awarding work, and it may take a long time for them to make a decision.

Non-asset-owning logistics providers are usually capable of being more flexible in many areas that the manufacturing firm cannot. One area that a non-asset-owning partner can be flexible is in the technology it uses to provide services. Another flexibility is in the area of geographic location. Flexibility in the services provided can also be obtained by entering a partnership with a non-asset-owning logistics provider. All these flexibilities allow freedom to mix and match providers. This type of provider will also typically have low fixed costs and specialized expertise. Loss of control is the most commonly cited disadvantage of using this type of strategic partner.

Other types of partnerships can be developed. However, for manufacturing firms, logistics and supplier partnerships are the most commonly used choices to manage the supply chain more efficiently and effectively.

16.4.3 Logistics Configuration

Issues typically discussed on the topic of logistics configuration, without exception, focus on strategic decisions concerning the warehousing and distribution aspect of SCM. Specifically, the strategic decisions every manufacturing firm must address are

- Proper customer access from each warehouse
- Proper product allocation in each warehouse
- Proper mode of transporting product from each warehouse
- Appropriate number and location of each warehouse
- Appropriate size of each warehouse

Making proper decisions about these issues is necessary to minimize costs across the entire supply chain. To address these issues, a manufacturing firm must follow these steps:

1. Gather data
2. Estimate cost
3. Develop a warehouse network model
4. Optimize the model

The data are gathered and costs are estimated and used as inputs to a network model. Once the network model is developed, it can be used to analyze and optimize the current warehouse configuration. Ultimately, the analysis will help the supply chain manager make the strategic decisions presented at the beginning of this section.

16.4.3.1 Data Gathering

A typical logistics network configuration problem involves large amounts of data, including information on customers, existing warehouses, distributors' facilities, manufacturing facilities, and transportation. Each of these categories can be further stratified as follows:

- Customers — location, product demand by customer location, shipment size, and frequency by customer, customer service expectations, and requirements
- Warehouses and distributors — location, inventory carrying costs, operating costs, labor costs
- Manufacturers — location of facilities, order processing costs, location of suppliers
- Transportation — costs for typical and special modes

This suggests that the amount of data involved in any logistics network modeling effort could be overwhelming. For example, a typical aerospace manufacturing firm has warehousing capacity at manufacturing sites in three, four, or five states from coast to coast. A medium-sized personal computer manufacturing firm has between 5000 and 150,000 customer accounts and from 50 to 10,000 different products flowing through the supply chain. For this reason, it is necessary to consolidate the data-gathering effort by using data reduction techniques. One useful data reducing technique is to develop logical families for the data and summarize the data by these families. Customer location and product type are the two most commonly used data families.

Customers located close to each other can be grouped into a single family. An effective technique that is commonly used is to group customers by zip code. In one example, a company was able to consolidate customers located at 3220 sites scattered across the Untied States into 217 more uniformly distributed groups.

Product families can also provide similar opportunities for data reduction. In many cases, products might differ only in minor characteristics such as packaging material, product model, and product style or shipment size. These products can typically be grouped into the same product family.

In most cases, using simple techniques such as developing customer and product families can reduce the time and resources required in the data-gathering phase of model development. After the appropriate data have been collected, it is time to move on to cost estimation.

16.4.3.2 Estimating Costs

The next step in developing a logistics network model is to estimate the important costs. Note that we do not recommend attempting to estimate all costs, which in some cases could result in "analysis paralysis." Important costs typically fall into two broad categories: transportation and warehousing.

Transportation costs can be divided into two primary components. Actual transportation rates are a function of distance and volume. Transportation costs also differ as to whether a company uses an internal or external fleet. Two other transportation choices are exception and commodity freight, which can be used to provide less expensive but more specialized transportation rates.

As stated earlier, transportation rates are a function of the distance between two points. Therefore, the accuracy of estimating transportation rates is only as good as the estimate of the distance between two points. One formula that can be used to estimate distances is shown in the formula for estimating transportation distances (Equation 16.1).

$$D = 69\sqrt{((long(a) - long(b))^2 + ((lat(a) - lat(b))^2} \tag{16.1}$$

The distance, D, is measured in miles. The value 69 is the approximate number of miles for every degree of latitude. This formula assumes that the distance between point a and point b is relatively short. When measuring longer distances, we need to consider the curvature of the earth. The United States Geological Survey has developed an approximation that can be used to do this. The formula shown in Equation 16.1, modified with this approximation, is presented in Equation 16.2 to estimate long transportation distances.

$$D = 2{*}69 \sin^{-1} \sqrt{\sin\left(\frac{lat(a) - lat(b)}{2}\right)^2 + \cos(lat(a)) * \cos(lat(b)) * \sin\left(\frac{long(a) - long(b)}{2}\right)^2} \tag{16.2}$$

Both of these formulas are very accurate for estimating distances. However, they tend to underestimate actual road distances. To account for this inaccuracy, we can multiply D by a correction constant, C; C can assumed to be 1.3 for metropolitan areas and 1.14 outside metropolitan areas. With these formulas we can estimate distances that, in turn, enable us to estimate transportation costs.

Another transportation cost that needs to be estimated, if applicable, is the cost of using trucks owned by the company vs. using trucks owned by a fleet company. Estimating costs when using a company-owned fleet is relatively simple. It involves annual maintenance fees on a per-truck basis, annual quantities delivered per truck, capacity of each truck, and the annual distance traveled per truck. These data are then used to calculate the cost per mile per SKU for the entire fleet.

When an external fleet is used, estimating transportation cost is more complicated. Most fleet service providers base the price for transporting goods on distance and quantity. Generally, the fleet service provider breaks the United States into zones and provides a document or database of cost per mile per truckload from one zone to another. An important aspect of this type of transportation cost is that the costs are not linear. In other words, it is typically less expensive to transport a truckload of material from Reno, NV to Los Angeles, CA than it is to transport that same truckload from Los Angeles to Reno. This type of transportation cost structure is very common in the manufacturing industry. However, other types of cost structures are employed by external fleet providers.

Another type of transportation cost structure is based on basic freight rates. The fleet service provider develops a set of freight rates based on the characteristics of the product being shipped and the distance between origin and destination. From these two items the cost per unit weight is calculated. Other types of transportation cost structures are employed by fleet service providers; however, the two methods discussed in this section cover the most commonly used methods.

The other primary cost category is warehouse and distribution costs. Warehouse costs can be incurred at the manufacturing plant, at a warehouse, or at a distributor's site. Regardless of where the cost is incurred, there are two primary cost components: handling costs and storage costs. Handling costs encompass labor and fixed costs such as utilities. Storage costs encompass all aspects of inventory, which is primarily holding costs.

16.4.3.3 Logistics Network Modeling

Once the appropriate logistics data have been collected and the appropriate costs estimated, the data can be used to develop a logistics network model. The most common type of logistics network modeling employed by companies today is an operations research model. This type of model is a static model that requires knowledge of linear programming to obtain an optimal solution. The following example describes this type of modeling.

Let's assume the following simple logistics network:

- Three manufacturing plants produce the same product.
- Each plant can produce 100,000 units per year at the same cost per plant.
- Two warehouses have the same costs.
- Two customer locations have annual demands as shown in Table 16.1.
- Manufacturing plants ship only to warehouses; no direct shipments to the customer.
- Logistics costs are defined in Table 16.2, wherein the cost to ship one unit from plant 3 to warehouse 1 is $3.

TABLE 16.1
Annual Customer Demand

Customer 1	200,000
Customer 2	100,000

Now let's define this network mathematically:

- Let P_1, P_2, and P_3 represent the three manufacturing plants.
- Let W_1 and W_2 represent the two warehouses.

TABLE 16.2
Logistics Costs

	Plant1	Plant 2	Plant 3	Customer 1	Customer 2
Warehouse 1	1	2	3	3	5
Warehouse 2	2	1	4	4	2

- Let C_1 and C_2 represent the two customer locations.
- Let $F\{P_iW_j\}$ represent the flow of product from plant i to warehouse j where i = 1,2,3 and j = 1,2.
- Let F{Wi,Cj} represent the flow of product from warehouse i to customer j where i = 1,2 and j = 1,2.

Now, to develop the linear programming model, we need to define the objective we are trying to optimize. In this case, the objective is to minimize the total logistics costs, which can be described mathematically in Equation 16.3:

$$1F\{P_1W_1\} + 2F\{P_1W_2\} + 2F\{P_2W_1\} + 1F\{P_2W_2\} + 3F\{P_3W_1\} + 4F\{P_3W_2\} + 3F\{W_1C_1\} + 5F\{W_1C_2\} + 4F\{W_2C_1\} + 2F\{W_2C_2\} \quad (16.3)$$

The objective notation shown above is subject to the following manufacturing capacity constraints:

$$F\{P_1W_1\} + F\{P_1W_2\} + F\{P_2W_1\} + F\{P_2W_2\} + F\{P_3W_1\} + F\{P_3W_2\} \leq 300{,}000$$

and the following warehouse constraints:

$$F\{P_1W_1\} + F\{P_2W_1\} + F\{P_3W_1\} = F\{W_1C_1\} + F\{W_1C_2\}$$

$$F\{P_1W_1\} + F\{P_2W_1\} + F\{P_3W_1\} = F\{W_1C_1\} + F\{W_1C_2\}$$

and the following customer-demand constraints:

$$F\{W_1C_1\} + F\{W_2C_1\} = 200{,}000$$

$$F\{W_1C_2\} + F\{W_2C_2\} = 100{,}000$$

This model is a classic example of a linear programming model. Solving this problem can be accomplished by using the well-known simplex algorithm. Many of the more popular personal computer spreadsheets have built-in utilities that will solve these types of problems. We will not get into the details of specific solutions to linear programming problems. Many textbooks on the topic of operations research are available that will guide the reader to a solution and ultimately an optimized logistics model.

16.5 INVENTORY MANAGEMENT

Inventory management has been important in manufacturing for a long time. Poor management of inventory will have a significant impact on customer service and costs throughout the supply chain. Unfortunately, the complexity of the supply chain has made managing inventory difficult.

Many reasons exist for a manufacturing firm to hold extra inventory, listed below are a few of the most common reasons:

- Buffering against changing customer demand
- Buffering against uncertainty in availability of supplied resources
- Taking advantage of lower transportation costs for large shipment quantities

Several forms of inventory are found across the supply chain. At the beginning, raw material inventory is apparent. At the manufacturing plants, work-in-process inventory can be found. Finished-goods inventory fill the end of the supply chain. Each type of inventory in the supply chain needs a control method. Efficient inventory control hinges mainly on a manufacturing firm's ability to, first, accurately forecast customer demand and then, have an effective inventory ordering process. The next two sections cover forecasting methods and reorder policies.

16.5.1 FORECASTING CUSTOMER DEMAND

One difficult aspect of inventory control is matching the inventory order quantity to the demand forecast. Because customer demand is uncertain, accurate forecasting is critical to determining the optimum order quantity. A typical forecasting system is driven by information created at the customer end of the supply chain. These forecasts are seasonally smoothed estimates based on 1 to 5 years of sales history data. Changing customer demand, use of historical data, and smoothing techniques all contribute to uncertainty in the demand estimate. This uncertainty leads to slow execution times, a need to discount products to have them consumed, dependence on inventory to obtain supplies, excess paperwork, and redundant costs. In every case study, we could not find a company that consistently relied on the data forecast beyond the first few days of each time period covered by the estimate.

If the forecasting process is inaccurate, different parts of the organization will operate using varying forecasts. Sales will create forecasts reflecting desired sales to meet goals, manufacturing will modify the forecasts to reflect what it feels the customer really wants and flow this new set of forecasts to its suppliers, then finance will operate with a third set of self-created forecasts. Unfortunately, organizations will never achieve unity and harmony if each part is working with a different set of forecast numbers. Clearly, an accurate demand-forecasting system is essential to the supply chain. It is required to drive the financial planning of the manufacturing firm. It is also a good starting point to determine which inventory has to be available to meet the estimated demands for a specific time period. Planning and scheduling also have to be based on some form of projected demand that is fairly accurate.

Many consumer products manufacturing firms operate with a surprisingly high rate of monthly forecast errors — in the range of 25 to 60%. Some of the better companies have exhibited much lower error rates — in the range of 15 to 20% (Copacino, 1997). These "best-in-class" companies typically follow one or more of the following best practices in demand forecasting:

- Long-term and short-term forecasting — the tools used, the planning, and the level of detail must be different for each type of forecast.
- Mandatory communication among sales, marketing, and manufacturing through the mechanism of regular, periodic planning meetings. Meetings should be structured and should follow a strict agenda concerning deliverables, the primary deliverable based on a team consensus of the best forecast for a specified time period.
- Organizational responsibility — companies must have a formal forecasting process and a specific forecast owner. This person is responsible for the management and performance of the forecasting process.
- Finance as the driving function of forecasting — too many companies allow the demand forecast to be influenced primarily by financial considerations. This shortsighted forecasting practice typically results in conservative demand estimates, which the sales force is confident they can exceed. What is not apparent to the business functions is how this inaccurate forecast affects inventory planning, raw materials purchasing, and customer service. The appropriate practice is to develop a true point estimate of operational forecast or a confidence interval around a point estimate.
- Sufficient analytical support — a good set of analytical forecasting tools is essential for accurate forecasting.
- Forecast error tracking — before a firm can improve its forecasting system, it must first be able to measure how well (or poorly) it is currently performing. Measuring and tracking forecasting errors will help in identifying the underlying causes of errors and ultimately will allow a company to measure the effects of any forecast-improvement initiatives.

Using the most effective forecasting techniques is crucial to forecast performance. There are numerous references in this area, and new commercial software is becoming available almost on a daily basis. These software packages include a variety of statistical models for longer-term forecasting as well as product life-cycle models for new product forecasts. This field of SCM is new and so dynamic that it is too difficult to list the "best in show." Many of the references listed at the end of this chapter have information on demand-forecasting software.

Additionally, several world-class companies have leveraged electronic linkages with their customers to improve their forecasting performances. These companies are linked electronically with their customers to obtain data on current sales rates and inventory levels. This information assists the manufacturing firms and their suppliers in understanding the real demand for their products.

A good forecasting system should also be flexible enough to recognize that forecasting is not a precise science, nor is it a cure-all that will resolve or eliminate supply chain concerns. However, it is an important element of SCM that can enhance supply chain performance by making the inventory ordering process more accurate and efficient.

16.5.2 Inventory Ordering Policy

Ultimately, a manufacturing firm must develop an inventory order policy that will effectively meet the forecasted customer demand. Numerous order policies are available to a manufacturing firm. One of the classic inventory ordering policies is the economic lot-size model.

In 1915 Ford W. Harris introduced the economic lot-size model to the manufacturing industry. This model is a simple inventory ordering policy that weighs the trade-offs between ordering and storage costs. When using the model, the goal is to find the optimal order policy that minimizes annual purchasing and carrying costs while simultaneously meeting customer demand. This model assumes that the demand is constant, order quantities are fixed, setup costs are fixed, lead time is zero, initial inventory is zero, and the planning horizon is infinite. Although the economic lot-size model allows us to understand some of the underlying difficulties of managing inventory, it does not take into account the effects of demand uncertainty, initial inventory, variable order costs, multiple order opportunities, and safety stock. Although this reorder policy has some impractical assumptions, it does provide some important insights into the dynamics of most reorder policies.

First, an optimal policy strikes a balance between inventory holding costs and setup costs. In other words, the optimal order quantity will be the point where inventory-setup cost equals inventory-holding cost. Second, inventory cost is robust relative to order quantities. In other words, when order quantities change over time, the effect on the setup costs and inventory-holding costs is relatively small. These two important insights apply to almost all reorder policies used by today's manufacturing firms.

Other, more modern reorder policies provide a more sophisticated approach to inventory management, such as the min/max policy, cross-docking, and continuous replenishment. Although these policies are an improvement, they still exhibit some of the same difficulties as the economic lot-size model. One of the fundamental difficulties with all these reorder policies is that they assume a single facility's managing inventory to minimize cost only at the facility. A better inventory management policy must consider the supply chain as a whole and the effect of uncertainty in customer demand.

Instead of this isolated objective, the main objective in a typical supply chain should be to reduce cost across the whole supply chain. Hence, it is very important to account for the interaction between each of the various facilities and the impact on the inventory policy employed by each facility. We develop a simple reorder

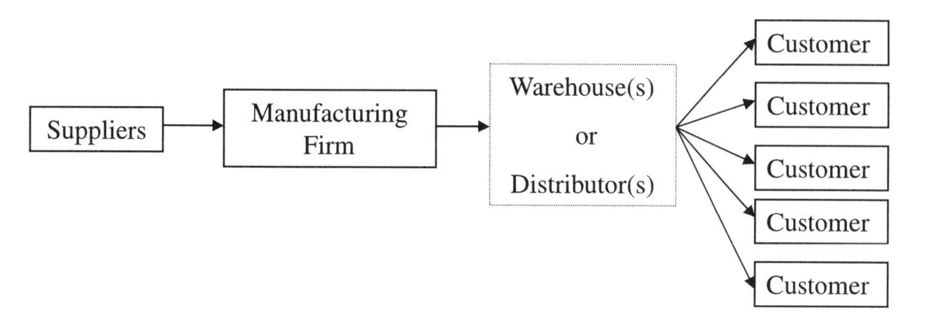

FIGURE 16.4 Single manufacturing firm, multiple customers model.

policy concept to account for customer demand uncertainty when determining the appropriate reorder quantities while accomplishing a system-wide cost reduction. This is best illustrated with an example.

For this example, let's consider a single manufacturing facility servicing multiple customers. In some cases, but not every case, there may be a warehouse or distribution center between the manufacturing facility and the customer. This portion of the supply chain is illustrated in Figure 16.4.

In this model, an inventory reorder policy for any facility in the chain is based on the cumulative inventory at each level. The *cumulative inventory* is defined as the inventory at any level of the system plus the entire inventory held at downstream levels. For example, the cumulative inventory at the manufacturing facility equals the inventory on hand plus all inventory in transit to the warehouse or distributor (if applicable), any inventory held at the warehouse or distributor, and any inventory in transit to the customers. Now, to determine the reorder policy we must define some terms.

Whenever inventory at any facility falls below a certain level, say L, an order to buy or produce enough product to bring the inventory level to U is placed. L is typically referred to as the reorder point and U is the order-up-to-level. The equations for calculating these values are shown below in the reorder point and order-up-to-level Equations 16.4 and 16.5, respectively.

$$L = (t_{cum} \times D_{avg} \times z \times D_{std.dev.}\sqrt{t_{cum}}) - I_{cum} \quad (16.4)$$

$$U = \max\left\{\sqrt{2C_{fixed} \times D_{avg}\frac{1}{C_h}}, t_{cum} \times D_{avg}\right\} + \left(z \times D_{std.dev.} \times \sqrt{t_{cum}}\right) \quad (16.5)$$

Where

t_{cum} is the cumulative lead time from the supplier to the customer in days
D_{avg} is the average daily demand of all customers
$D_{std.dev.}$ is the standard deviation of all customer daily demands
I_{cum} is the cumulative inventory as defined above

C_{fixed} is the fixed cost of one unit of product
C_h is the inventory holding cost for I_{cum} units
z is a constant based on the standard normal distribution, representing the probability that the customer demand will be met (no stockouts). Table 16.3 shows these constants.

To illustrate the use of these equations, let's say a manufacturing firm has the data on customer demand shown in Table 16.4.

From Table 16.4, we calculate D_{avg} equals 1.249 and $D_{std.dev}$ equals 0.614. Now assume the manufacturing firm knows that the cumulative lead time, t_{cum}, is 120 days from when the supplier gets the order to when the final product is delivered to the customer. Let's also assume the manufacturing firm has access to information which indicates that there are 50 units of cumulative inventory, I_{cum}, between it and the customer. The inventory holding cost, C_h, for 50 units is $1.00 per day. The fixed cost, C_{fixed}, is $10.00 per unit. Finally, the manufacturing firm has promised its customers that it will meet their delivery requirements 95% of the time, hence, z equals 1.645. Now, substitute the information into Equations 16.4 and 16.5, the equations for L and U:

TABLE 16.3
Table of Z-Values

Probability of Meeting Customer Demand (%)	z
90	1.282
91	1.341
92	1.405
93	1.476
94	1.555
95	1.645
96	1.751
97	1.881
98	2.054
99	2.326

TABLE 16.4
Data on Customer Demand

	Monthly Demand	Daily Demand (Monthly demand ÷ 30.4)
Customer 1	50	1.645
Customer 2	35	1.151
Customer 3	9	0.296
Customer 4	47	1.546

$$L = 120 \times 1.249 \times 0.614 = 92$$

$$U = \text{Max}\{176.6, 149.9\} + \{11.1\} = 176.6 + 11.1 = 187.7 = 188$$

From these equations, it can be determined that the reorder policy for this manufacturing firm is when inventory falls below 92 units, they will order enough raw materials and notify their suppliers to manufacture 96 units (188 – 92).

16.6 SYNCHRONIZING SUPPLY TO DEMAND

Most of the supply systems we have studied require improvement because of sluggish execution times, the need to drive consumption by discounting goods, dependence on inventory to obtain supplies, and excessive or unnecessary paperwork. Most of these difficulties arise because of the way product moves through the typical supply chain: forward from supplier to consumer, in a "push" manner. In the *push system,* suppliers start the process by building inventories of their products and enlisting the sales force to push inventory toward the manufacturers for consumption by the customer. This method of meeting customer demand requires a great deal of working capital and tends to build large amounts of unnecessary inventory.

In a supply chain that pushes its inventory, manufacturing decisions are typically based on long-term forecasts. Usually, the manufacturer depends on months or years of order data received from the customer, the distributor, or the warehouses to forecast customer demand. Therefore, it takes the supply chain a long time to react to the changing marketplace, which can lead to

- Unmet customer demand
- Obsolete inventory throughout the supply chain

Additionally, the variability of orders from the customer and the warehouses, if applicable, is much larger than the variability in customer demand. Increased variability can lead to

- Large safety stock quantities leading to excessive inventories
- Large, variable production batch sizes
- Obsolete product
- Inefficient resource utilization

Finally, in a push-based supply chain we often find increased transportation costs, high inventory levels, and high manufacturing costs due to the need for emergency production changeovers.

Alleviating some or all of the problems associated with a push-based supply chain requires converting to the use of true customer consumption to trigger production orders. This type of supply chain is called a *pull system* or pull-based supply chain. Many companies provide innumerable excuses for their reluctance to migrate to a pull system. The inability to pull replenishment directly from consumption seems to be the basic complication in their conversion plans. Unfortunately, most organizations are so locked into the traditional forecasting system that change to a pull system is very difficult, if not impossible, to achieve.

In a pull system, manufacturing is driven by customer demand. This means that manufacturing is matched with actual customer demand rather than being a forecast. To accomplish this, the supply chain must use rapid information technology to transfer information about customer demand throughout the supply chain. Doing this will lead to

- Decreased lead times
- Decreased inventory throughout the supply chain
- Decreased system-wide variability

Utilizing a pull-based supply chain will typically result in a significant reduction in inventory levels, improved resource management, and reduced costs throughout the entire supply chain.

A pull-based supply chain is not perfect. A pull system is difficult to implement, especially when long lead times are unavoidable, because the manufacturing firm is unable to react to demand information in a timely manner. Additionally, in a pull system, it is more difficult to take advantage of economies of scale, especially in transportation.

In some cases, a combination push-pull system may be appropriate. Using this strategy, the early stages of the supply chain are run on the traditional push system and the later stages use the pull system. One way to accomplish this is by producing in bulk at the first stages, and then segregating these products based on customer demand at the final stages of the supply chain.

Clearly, the amount of inventory and the working capital needed to support the push system can be reduced through the development of a closer linkage of the partners in the supply chain to true customer demand. Great opportunities exist to reduce inventories and cycle times when the supply chain activities from supply to consumption are clearly understood and synchronized.

REFERENCES

Abramson, G., Savings galore?, *CIO Enterprise,* September 15, 1999.

Anderson, D. L., Britt, F. E., and Donavon, J. F., The seven principles of supply chain management, *Supply Chain Manage. Rev.,* Spring 1997.

Bramel, J. and Simchi-Levi, D., *The Logic of Logistics: Theory, Algorithms and Applications for Logistics Management*, Springer-Verlag, New York, 1997.

Copacino, W. C., *Supply Chain Management, The Basics and Beyond*, 1st ed., St. Lucie Press, Boca Raton, FL, 1997.

Deutsch, C. H., New software manages supply to match demand, *New York Times,* December 16, 1996.

Fisher, M. L., Hammond, J., Obermeyer, W., and Raman, A., Making supply meet demand in an uncertain world, *Harv. Bus. Rev.*, pp. 83–89, May–June 1994.

Fisher, M. L., What is the right supply chain for your product?, *Harv. Bus. Rev.*, pp. 105–117, March–April 1997.

Compete, http://www.ascet.com/ascet/wp/wpHakanson.html
Hax, A. C. and Candea, D., *Production and Inventory Management*, Prentice-Hall, Englewood Cliffs, NJ, 1984.
Kearney, A. T., *Management Approaches to Supply Chain Integration*, Feedback Report to Research Participants, Chicago, IL, 1994.
Kuglin, F. A., *Customer-Centered Supply Chain Management*, 1st ed., AMACOM, New York, 1998.
Mohrman, M., Supply chain management puts dollars back in business, *Forbes* Small Business Tech Center, online ed., Forbes.com, October 6, 1999.
Poirier, C. C. and Reiter, S. E, *Supply Chain Optimization: Building the Strongest Total Business Network*, 1st ed., Berrett-Koehler, San Francisco, 1996.
Ross, D. F., *Competing through Supply Chain Management*, Chapman & Hall, New York, 1998.
Walker, W. and Alber, K., Understanding supply chain management, Vol. 99, No. 1, online ed., *APICS-TPA*, January, 1999.
Weber, J., Just get it to the stores on time, *Bus. Week*, pp. 66–67, March 6, 1995.

17 Supply Chain Management — Applications

Douglas Burke

In the previous chapter we presented many of the details necessary for effective SCM. In this chapter we present four pointed case studies that give the reader a view of what is currently being done to improve SCM. All the case studies are based on documented research; however, the names of the actual businesses have been changed. In some instances, the case studies presented are a compilation of numerous examples from a specific field or industry.

The first case study is centered in the retail industry, which was the first to recognize the importance of improving SCM to gain market share and business advantages. This case highlights one company's innovative approach to inventory management. This well-known retail chain took advantage of today's information technology to establish a "pull-through" supply chain, resulting in dramatic reductions in inventory and improved customer satisfaction.

The second case study focuses on how a truck-manufacturing firm used local partnering to improve its SCM and gain business advantages. This case follows the evolution of what started as a simple partnering agreement but ended as a synergistic coupling of two good companies, leading to results far in excess of what any one company could obtain alone. You will see the importance of trust and shared benefits in this type of simple partnering agreement.

The third case study focuses on the grocery industry. This case shows how advanced partnering agreements can span the entire supply chain and benefit more that just a few firms in the network. This case points out some different aspects of SCM because of the short shelf life of products, the need for short lead times, and close promotional management to smooth variations in demand.

The final case demonstrates the supply chain improvement effort of a computer-manufacturing firm, moving from the initial data analysis for determining areas of improvement through the implementation of the improvements. What is important about this case is the use of interdisciplinary teams that had to be cross functional to accomplish the established goals. You will also see how SCM improvement efforts must fit into a company's strategic plan.

17.1 OPTIMUM REORDER CASE STUDY

Working with a number of customers from the consumer retail industry, Company A developed an effective and industry-recognized three-step supply chain improvement

1-57444-300-3/02/$0.00+$1.50

system. This technique, appropriately named the *constant refill program* (CRP), was adopted as the company's vision for responding to the growing need for faster, more accurate stock replenishment while maintaining a high level of customer satisfaction. Company A continues to claim that CRP has changed the long-established relationships between customer and supplier. By implementing CRP, Company A has simplified and streamlined the reorder process, reaping improvements in efficiency and effectiveness. Additionally, the improved order response process eliminated steps that were not adding value to the customer, reducing costs and cycle times.

The CRP process begins with orders received from the customer distribution centers via EDI (electronic data interchange) along with Company A's inventory and receipts. To obtain optimum reorder quantities, these orders are accumulated and transmitted to the customer headquarters site, which represents the pull-through demand. Then the demand is compared with the inventory to calculate the optimum reorder quantities. After making summary analyses and adjustments due to promotions and other pricing activities, the headquarters group routes the actual orders back to the distribution centers and Company A's headquarters.

Orders specific to individual plants are then sent from Company A to the individual manufacturing sites to start the production process. After the needed products have been manufactured, dedicated carriers with dedicated delivery schedules move the newly manufactured goods to the customer distribution centers. The newly manufactured goods and the on-hand inventory are then sent to stores for specific customers. In this system, inventories are controlled by keeping the stocks in the distribution centers at minimal levels and shifting dependence to the flexibility of the manufacturing systems at Company A plants to meet most of the store needs.

Clearly, the actual process is more detailed than this. However, all the functions that make the system work are represented. The total effect is a system in which both the customer and Company A benefit from a lower cost, speedier process, and the ultimate consumer gets a more diverse product mix at a lower price. Company A-documented customer benefits include

- Reduced customer-owned inventories and better utilization of distribution center space
- Greater than 65% reduction in customer warehouse inventory
- Elimination of paperwork and reduction of administrative costs by using electronic interchanges
- Improvement of store service levels to over 99% on specific products by providing the correct inventory quantity and mix to the customer
- A tripling of inventory turns, with a one-time cash-flow increase of almost $0.25 million resulting from lower working capital tied up in warehouse inventory.

Company A recorded these benefits:

- A greater than four-point increase in market share
- An almost 30% increase in orders
- An average of 8% vehicle utilization improvement

- A decrease of over 50% on returns and refusals
- An average 30% reduction in damaged goods
- Improved customer service and satisfaction

By implementing this program, Company A now takes the incoming point of sales (POS) transaction data and determines what needs to be shipped even before an actual order is created by the customer. Substantial advantages are gained in planning and scheduling by this capability. For example, products such as diapers typically suffer from wide variations in scheduling. Customer needs depended on the particular type of diaper being consumed: regulars, absorbent, or super-absorbent grade. Under CRP, manufacturing fluctuations have been significantly reduced because the plant knows which types are being pulled out of the system. By having Company A and its customers involved in the CRP process, they both can offer the ultimate customer lower retail prices by passing through system savings. Additional advantages are improved product freshness, reduced out-of-stock situations, and decreased package damage.

Recently a large computer-manufacturing firm purchased Company A's system. This confirms that the CRP system is one of the leading-edge practices that will help to drive supply chain improvement to reality. We expect that other firms will develop this type of system and that it will be expanded to the upstream side, including the suppliers of the materials needed to make the products. Clearly, the opportunity for system improvements has started to be practical. System improvement requires that the companies involved are able to recognize the types of enhancements that can be worked out by cooperative efforts between suppliers and producers. The same methods can be used between suppliers and manufacturers as are used between manufacturers and stores.

Company A is now working to introduce the next version of its supply chain improvement process — termed *smooth logistics* — and developing tomorrow's solutions for today's SCM problems.

17.2 BASIC PARTNERING CASE STUDY

The next time you are on the road, take note of the number of large tractor-trailers pulling their loads across the United States. A large manufacturer of the tough and durable machines that pull these trailers formed an alliance with a large supplier of tires. The supplier (let's call it the Tire Company) provides an excellent example of how to make a true partnering effort successful. This is a true case study and presents a successful business endeavor for both parties.

Following a corporate spinoff, the truck manufacturer made a strategic decision that developing a partnering concept could offer special business advantages when evaluated against its traditional supplier relationships. Management first performed a serious internal review of existing procedures and supplier relationships. The results were eye opening but not surprising, as you can see below:

- The dominant purchasing strategy was price buying.
- The procurement base was fragmented.

- Profits for both buyer and seller had limited opportunities for growth.
- Relationships with all their suppliers were mainly adversarial, neutral at best.
- Customer satisfaction was neglected.

A three-pronged approach was developed to find specific opportunities for improvements in these traditional circumstances. The three areas selected for improvement were tires, engines, and drivetrains. Multifunctional teams were selected to start the improvement process. Team members included people from quality, marketing, manufacturing, production control, planning, finance, engineering, and purchasing. The first team, formed to investigate drivetrains, selected the Joni Corporation as a partnering candidate. Another team was formed to investigate tires; they came up with the focus of this case study.

The first investigation of the original tire improvement effort had the following results:

- The supply base was fragmented.
- The Tire Company was the largest supplier.
- Tires represented the third-largest cost item.
- Tires had a high pull-through percentage.
- The return of product was low.
- Purchasing of tires was centrally controlled.

From these findings it was clear that there were ample opportunities to improve this segment of the trucking business. At this point, the objectives of the improvement effort needed to be established. One objective, which was determined early in the project, was to develop a partnering arrangement with a key supplier or suppliers. The team recognized that it must make certain that additional profits for the truck firm would be part of the results. The team quickly moved toward the use of common resources for mutual benefit to achieve that goal. The team developed the following project objectives:

- Develop a better understanding of the tire market
- Identify potential business opportunities
- Motivate suppliers to offer better and more comprehensive proposals
- Have a positive and significant impact on profits

The team started communicating its goals and expectations to various tire companies. To convey the importance of the improvement effort, the teams chose to communicate this through site visits and formal presentations, as opposed to written or verbal media. Time was also spent interviewing dealers, customers, and truck firm managers to make certain that they were not missing input from any of the important stakeholders. These visits took the team to training facilities, the truck firm's assembly operation, tire plants, research and development centers, headquarters locations, test tracks, and trucking firms.

After the visits were completed, the field of potential partnering suppliers was reduced to four. Potential partnering candidates were selected by using a formal objective evaluation procedure. Large quantities of data went into the solicitation for proposals. Using a complex and focused table of deciding factors, the truck firm developed a scoring system that led to selection of the finalist: Tire Company.

Tire Company's director of sales and the truck firm's manager of supplier relationships were intimately involved in this segment of the process. They openly reported that many factors were key in building the original partnering relations. An early consideration was whether the decision-making processes at both firms were compatible. Data-gathering ability and the possibility of building a trusting relationship were other crucial considerations. Ultimately, what the truck firm wanted was to make sure that it moved in the right direction and that both firms were comfortable with the new alliance they were about to form. This cautious initial planning was necessary because, if successful, the arrangement would be used as a model for other alliances.

The partnering proposal that was eventually implemented satisfied all the truck firm's "must-have" criteria and the most important of the "wants" criteria. From the partner's perspective, the proposal presented Tire Company's expectations in terms of obtaining a growing share of the truck firm's business. Also stated in the agreement was the fact that the partnership was to be open-ended and could be terminated by either of the two partnering firms. Basic staffing and office commitments were outlined, providing resources to the core implementation group and a full-time partnering team that would direct the development of the alliance.

At the initial meeting of the joint tire group, a mission statement and goals were developed. The mission statement clearly established the groundwork for a successful partnering situation. It begins, "A business partnership is defined as a joint business alliance wherein two companies agree to favor each other's business activities." The mission statement adds, "Each partner must dedicate resources in capital, people, and facilities in order to support future business and growth in profit." Finally, the mission statement elaborates, "Progress is not measured by the success of a single firm but [is] measured by the success of both firms which identify, prioritize[,] develop, and implement the cooperative efforts of both companies." This progressive and eloquent mission statement was endorsed and signed by top-level executives in both firms.

Early in the process, team members suggested hundreds of improvement projects without restraint or comment. These projects were grouped by their relationship to either strategic goals or a functional work group. They were then evaluated and prioritized by members of full-time business teams. Evaluation and prioritization were accomplished by using a simple point system based on risk, timing, required resources, and potential benefits.

When the business management team was formed for the joint activities described in the previous paragraph, care was taken to get a true cross section of disciplines. The thought was that a multidisciplined team was necessary to avoid compromising one area within the company for the benefit of another. The original team included full-time participation from the disciplines shown in Table 17.1.

TABLE 17.1
Disciplines Represented in the Original Team

Truck Firm	Tire Company
Truck marketing	Engineering
Tire and wheel purchases	General product sales
General product purchases	Replacement tire sales
Partnership management	Truck tire marketing
Parts marketing	

Team roles were established as follows:

- Communicate results and promote the value of the alliance
- Provide a forum to address strategic issues
- Develop and implement partnering business plans
- Manage all aspects of the team process
- Provide leadership and support to working groups

It is interesting to note that the degree of empowerment given to the team was much higher than you would typically see in a traditional improvement effort. This supports the fact that the level of senior management endorsement was very high from both companies.

Next, working groups were set up and staffed with core members and ad hoc members participated when needed. These teams were responsible for performing myriad tasks, including reporting progress, forming task groups, developing action plans, generating innovative solutions, and acquiring necessary resources.

One working team generated results that demonstrate the success that can be generated from a true partnering association. The on-time assembly (OTA) team was established to design a better system for mounting final tire assemblies. The OTA team was easily up to the task and eventually developed an innovative solution.

The team first focused on an analysis of the current systems and procedures in the procurement and assembly areas. Tires and rims were typically ordered by the truck firm and stored with relatively low inventory levels. The rims were then painted or surface treated and sent to the mounting area that had dedicated factory floor space. The team discovered that many errors occurred in this area when the tires were mounted, such as improperly mounted, low-pressure, or out-of-balance tires. Although the percentage of defects seemed small, there were also ample opportunities for savings through reduction in floor space and increased throughput in the assembly area, along with the elimination of the previously mentioned errors.

As we will see, the solution can be showcased as a model of partnering principles. Before the partnering arrangement, the truck firm did most of the work. They received all tires and rims, did the painting and mounting, and produced the final assemblies. Because of the significant rejection rate of the final assemblies, 10 days' inventory of tires and rims had to be maintained as a form of safety stock. Under the new conditions, the truck firm suggested that Tire Company assume responsibility for the tires and rims, the painting, the mounting, and the final balancing processes. The truck firm's major wheel supplier was contacted and brought into the picture as part of the partnering arrangement. This firm had the core competency in wheel and painting expertise, so an expansion of the alliance was quickly executed. Tire Company established that the wheel supplier would be the manufacturer of choice for the rims and quickly formed a second partnering arrangement with the firm.

In order to get this newly formed alliance under one roof, a new facility was built near the truck firm's plant where tires and wheels could be sequenced for easier, error-free assembly. Robotic arms were set up to flawlessly apply paint to the wheels. Other robots lubricated the tires for proper mounting. Finally, tires were automatically inflated to a particular vehicle specification using a computer-controlled program.

Next, a fully computerized balancing station was installed, ensuring that customers would receive perfectly balanced tires on every wheel assembly. After the assembly process, the completed assemblies were stacked so that installation could be done sequentially on designated trucks. Final assemblies were loaded automatically via a computer-controlled conveyor into trailers, which were then continuously transported to the assembly plant. There they were off-loaded and put on a conveyor belt that fed the assembly lines. At the truck plant, technicians removed the finished units at the point of need and installed them on the appropriate trucks.

Implementation of this activity required the combined strengths of both Tire Company and the wheel supplier. The results were impressive: a high-quality, tire-and-wheel assembly process with a clear competitive advantage. Benefits to the truck firm include improved finished tire-and-wheel assemblies, increased shop floor space, and reduced inventories. The OTA team, which was spun off into an individual corporation, has expanded its business base and now ships units to Canada on a just-in-time basis. What once was a 10-day supply of finished inventories has been reduced to a supply of hours, which represents a savings for all three parties. Also, producing detailed business plans generated mutual savings. One unexpected benefit has been that revenue from the savings due to partnering has been used to buy a test truck, which is now used for experimenting with other new products that will ultimately lead to more customers and higher revenues for all firms involved.

Without the trust demonstrated by the open communications that developed when the partnership was put together, this alliance would have never worked out. Subsequently, the truck firm has offered to colocate Tire Company's personnel to further facilitate partnering interchanges. Another necessary aspect of this case was the training of the joint team members, which immediately improved communications between all levels of the firms. This is obviously a case with a win–win ending. Both companies knew how to apply partnering the way it was meant to be applied, and reaped more benefits together than either could have gained by itself.

17.3 ADVANCED PARTNERING CASE STUDY

Jerry's, a Midwest-based grocery firm, has more than 100 stores within a 35-mile radius of a major metropolitan area. The firm is noted for its leading-edge position among its peers in the industry and its willingness to look at innovative changes that will improve its systems.

During Jerry's usual annual strategic planning, the topics on the agenda centered on how new initiatives could be generated without substantial cost increases to the firm. They wanted to build on the current effort to develop efficient consumer response techniques and develop a model of efficiency specifically designed for the retail grocery business. Because of recent successful initiatives Jerry's has been implementing, the vice president was interested only in building onto existing initiatives rather than developing totally new systems or processes.

A decision was made to pilot an effort based on an advanced partnering solution. With that in mind, supplier ABC and a bakery, Sonja Corporation, were invited to participate. Jerry's would provide distribution resources from its own center and let its grocery stores be the focal point of retail sales. Jerry's sent out letters of invitation to participate in a pilot effort and each firm gladly accepted the opportunity.

Jerry's also developed a proof-of-concept paper, which was sent to each participant. This paper generated additional topics of discussion from all involved parties. This discussion format was used in conjunction with a questionnaire asking for objectives and expected deliverables. This solicited preliminary ideas from the group with regard to the validity of the pilot project and helped to identify those areas that needed further exploration.

When participants from each company met to discuss the pilot, consensus was quickly achieved on the validity of the exercise. Furthermore, the group developed a process map of the interconnecting relationships among the firms. Brainstorming led to the creation of more than 50 potential improvement areas. These possibilities were refined into roughly half that number of critical issues, with action teams developed to start working on them. The action teams were formed to accomplish the following:

- Develop electronic data interchanges (EDIs) that would benefit the pilot members
- Develop and analyze a flowchart for the order-handling process
- Develop and analyze a flowchart for the forecasting and planning process

Next, team assignments were made, realistic timetables were established, and action teams went out to find savings across the full supply-chain network. Initially, a list of benefits was developed that included

- Reduced transportation costs
- Improved cash flows
- Reduced administration costs
- Improved customer-service levels
- Reduced inventory

A list of available process data was developed that included promotional impact, price, cost, packaging, quantities, product, dates, and customer or consumer requirements.

Next, the teams met to develop a list of actions to meet these goals. Each team developed a high-level map of the process it was considering. Some of these maps were lengthy, but in most cases, for the first time, the team members could plainly see the interaction of activities necessary to supply product to the stores. Product cycle length was the first area of clarification because the actual estimate of cycle length far exceeded the perceptions brought to the exercise. Some of the key areas proposed for improvement were

- Products being handled an excessive number of times
- Fill rates of less than 100%, in spite of having more than 3 months' inventory
- A lack of consistency in measuring fill rates
- Accuracy of distribution-center forecasting
- Self-imposed redundant or unnecessary inspections
- Excessive paperwork
- Excessive items out of stock
- Handling of promotional items
- More effective and efficient EDI transactions
- Excessive scrap in the form of damaged and spoiled goods
- Excessive shrinkage, overshipments, and material-system waste
- An ineffective system for handling reconciliations
- Elimination or better utilization of infrastructures in the distribution network
- Excessive flaws at point of sales (POS)
- Need to use POS data to estimate stock replenishment levels

The teams developed the list of action deliverables from the opportunities listed in the previous paragraph. One product, cakes, was selected for the actual study. The reasons for selecting one product type were to keep the stock keeping units (SKUs) at a manageable level, and to develop the system around a product with seasonal variations and high inventory costs. Review steps were established to monitor the progress of each team. This also allowed the use of good program management tools and ensured that resources were allocated to each team. These review meetings spawned many needed items for each action team, such as

- A cost-benefit analysis — including payback for the actions
- A list of objectives
- A defined scope for each action team
- Recommended improvements
- A means of measuring progress
- A timeline for completion

One team had the task of improving the forecasting and planning flow. From the mapping exercise the team discovered that the lead time from the start of baking the cakes to when the packaged cakes were stocked on Jerry's shelves was more than 5 months. With this type of important information, the team was able to prepare an action item list intended to redesign the process for beneficial business results. Another key finding from this joint project was the existence of many weeks of safety inventory, necessary to cover inefficiencies in the existing supply chain network. Safety inventory was also needed so Jerry's could make changes in Sonja Corporation's manufacturing schedules due to promotional activities.

The promotional activity situation was particularly interesting. Essentially, the response of the manufacturing facility needed for the promotions to work caused variances in the production schedules. These schedules, established from earlier forecasts, were overridden by promotions that resulted in significant additional costs. The team discovered that by developing a closer liaison among the parties feeding back information on the promotions, they could mitigate the need to make so many adjustments to the manufacturing schedules. Hence, the variations could be lessened (or even eliminated) by using data to coordinate the timing of the promotions and feedback on the progress of the promotions.

From following the teams' analyses, recommendations, and implementations, these preliminary results were identified:

- Decreased manufacturing variability due to better promotions management.
- Average cycle-time improvements of 50%.
- More successful promotions due to better management.
- Almost 1 month of inventory was eliminated across the entire supply chain network.

More important, the ABC Company, the Sonja Corporation, and Jerry's have established a synergistic working relationship, built on trust, that can be expanded as they seek other areas of potential improvement. This is the essence of any advanced partnering initiative. If the cost is kept to a minimum and the potential savings are shared, future work together will be self-funding and self-perpetuating.

17.4 SCM IMPROVEMENT CASE STUDY

This final case spans most of the elements of supply chain management. It illustrates how a large organization took the necessary steps and effort to discover and implement significant improvement across a global network of supply chain activities.

The Computer Company (CC) is a worldwide producer of computer products, with annual sales of nearly $20 billion. CC has a business presence on all continents through a network of more than 50 companies and approximately 100,000 employees. This story involves the North American segment of CC, which has nearly $5 billion in sales and is organized into five operating divisions. One of the divisions, the service division, controls the accounting, logistics, and purchasing functions, which are core areas for finding supply chain enhancements.

This specific improvement effort started with CC and other leading computer product companies redefining how to gain a competitive advantage. The search for a competitive advantage is a common initiating function for companies seeking improvement through innovative management of their supply chains. The first discovery by CC was that their existing performance measures were nonresponsive and inadequate for managing future market conditions. To illustrate this point, we can look at three measures — order lead times, order completeness, and on-time delivery.

Order lead times are always considered an important measure. CC's lead time is defined as the time from the receipt of the order to its shipment to the distribution center. The industry benchmark was researched and estimated to be a 4-day-maximum cycle. Furthermore, the benchmarking study revealed that the future lead-time requirement would soon shrink to a 24-hour-maximum cycle. Clearly, a quantum improvement would be necessary to meet this new target.

Another metric, order completeness, was important because it indicated the need to eliminate back orders. Traditionally, order completeness is defined as 100% completion of any order. However, CC's current measure of order completeness showed it to average just over 80%. Once again, CC knew that in the future there would be a need for a much higher order-completeness percentage. In fact, the new average had to be closer to 98% in the near future.

Most companies typically monitor a third metric, on-time delivery. CC's on-time delivery was defined as time (in hours or days) late after the customer due date. However, it is not uncommon to see companies monitor on-time shipments as opposed to on-time delivery. Future requirements indicated that this metric was moving to a just-in-time requirement as defined by the customer.

In order to pursue the necessary changes to meet these and future requirements, the service group of CC benchmarked other successful firms. The one fact that was common among all the companies that CC looked at was an unambiguous focus on the customer. This customer focus results in a customer-driven strategy that makes them preferred suppliers. This was an eye-opening conclusion for CC, and ultimately it created a three-tiered customer-focused strategy for CC.

1. Supply chain resources and technology would be adjusted to respond more effectively to consumer needs.
2. Because consumer needs change quickly, CC would link its internal and external resources to create faster operations.
3. Optimization of services and reduction of costs would have to drive the creation of cost-effective and innovative solutions for worldclass distribution.

Because CC was operating in a global environment, additional constraints were identified and managed by all parties involved in the improvement effort. Some of the important constraints included

- Procurement and transportation would have to be conducted more effectively and across many boundaries, cultures, and languages.
- Differences in currency, language, documentation, and conversions to metric, and multiple customer requirements would need to be accommodated.
- Compliance and quality, as controlled by regulations, were different and would further complicate solutions.

All these factors were worked into the improvement effort, along with an additional requirement to help disseminate how the improvement effort would focus on quality. The final requirement, that all information would be communicated across the entire supply chain, was necessary so that all partners had access to customer requirements. This meant that any partner in the supply chain would have accurate information on any customer's order, specifically, exactly when an order arrived, when it shipped, where in the process it was at any given time, and causes for being late. Clearly, shipment and cycle-time metrics were to be integrated into this new system.

Part of the supporting infrastructure of this supply chain included an information system with data for replenishment, procurement, manufacturing, inventory management, distribution, order fulfillment, and logistics. Although these functions crossed different companies, members of each function looked at planning, scheduling, and execution to identify where the areas of improvement existed. Representatives from each function came together in a team effort to analyze their respective portions of the supply chain. They looked at all activities to identify the areas having the best opportunities for implementing effective change. After numerous meetings, reviews, and the development of alternative solutions, specific improvement areas were selected and action items were developed.

Next, customers and suppliers were brought into the effort. Their roles were to help create the improvements being targeted and to make absolutely sure that changes would not adversely affect critical customer requirements. At that point, supply chain improvement teams were formed around customers and suppliers to link participants in the supply chain through processes and systems. This would be the best way to manage supply and demand among customers, CC, and all suppliers. The key team ingredients were

- Horizontal process improvements rather than typical vertical management, which is more concerned with local turf issues
- Sales and operations-planning teams taking recommendations from the joint teams to redesign existing processes
- Training and education in supply chain–management methodologies for all participants

The strategic thought process was to shift from a short-term, local-focus orientation to a long-term, partnering, information and savings sharing, global, on-time supply chain. Everybody knew that this kind of shift would require the redesign and reengineering of processes as well as training on the new systems that would be developed. The preliminary meetings and discussions in this three-part improvement effort led to the creation of a supply chain flowchart. The original data developed by the internal team were now matched with input from the external sources as the three-part effort went in search of significant improvements across the total supply network.

Many areas were studied and redesigned for enhanced values. A few examples will illustrate the depth of the work. A map of the order-fulfillment process was developed, which allowed the team to find significant improvement opportunities, including inventory reduction, improved routing, and improved transportation and replenishment. Another effort to document the replenishment-planning workflow was undertaken, leading to a positive impact on master production scheduling, materials planning, global procurement, sales, and operations planning. Close scrutiny was given to the conventional materials-purchasing process, which contained a large number of non-value-added steps. It was necessary to change the buyer's job description to buyer and planner. The order entry function was used to verify product availability then to check pricing and credit before sending a confirming order to procurement. Next, purchase orders were issued to cover the parts called out on the bill of materials. Under the existing system, this sequence had met actual consumption demands only some of the time.

The new replenishment process focused on a model of usage that was developed by one of the teams. This model created daily inventory replenishment needs from historical data that were compared with daily movement data. All the order entry information now comes from the usage model. Some of the information the model provides includes pricing, credit, and production needs. Next, the buyer or planner creates the flow of material with the dual objectives of keeping inventories at a minimum and production at efficient levels. Some of the features of this model include global EDI from suppliers and customers, electronic order status updates, processing without invoices, and electronic booking.

Another example of the improvement can be seen in the area of customs clearance, which is the process of getting international goods through national borders and on their way to customers. The former process was characterized by

- Different processing techniques based on differences and local customs
- The use of different people in the same port of entry to handle the paperwork, entry fees, duties, and transportation
- Excessive paper chasing to track products from many countries through many ports of entry
- Excessive back-office activities for data entry and for processing redundant data
- Countless telephone calls to check the status of shipments
- Excessive process handoffs, resulting in tracking problems and introducing possible errors

With the help of the supply chain partners, the team improved this process. Its objective was to reduce handling, use outsourcing when it made sense, and automate the process where appropriate. The redesigned process had these features:

- Order status was determined electronically.
- All organizations used one standard process for all transitions.
- One person would be responsible for all processing in the United States.
- Minimal paperwork was needed to clear customs.
- All data entry was performed in a central location.

Obviously, CC's redesign efforts resulted in a simpler and more effective customs clearance process. To control the process, CC put the total process ownership into the hands of the customs broker to manage this function. The annual savings in this one area was in excess of $200,000. CC went on to obtain notable gains in many other areas.

From a review of the team results, the service division cited specific factors responsible for success in the new SCM system. It became apparent to everybody that global partnering was a critical ingredient in achieving improvement that was meaningful across the entire supply chain network. Another key ingredient for success was the allocation of resources. Wherever and whenever the need for full-time resources was identified, the participants allocated them, and they tackled the major redesign tasks. Clearly, empowered, cross-functional teams that crossed traditional company boundaries were the key ingredient for success. An additional enabling function for success was the state-of-the-art communications and information sharing which made most changes possible and practical. The effort stimulated a cultural change that required people to shed the narrow view of their jobs and to think "outside the box." This culture change allowed CC to make the kinds of improvements that benefited the entire supply chain network. Three types of benefits were documented: strategic, measurable, and economic. Finally, the goal that horizontal integration be established was achieved.

Included in the strategic benefits was the development of a flawless procurement, manufacturing, and distribution system. Diverse groups within CC came together, pooled their resources, and collectively focused on critical solutions that had a direct impact on achieving the goals laid out in the strategic plan. Measurable benefits included the following documented savings: average customer service levels of 97%, on-time delivery of at least 97%, invested inventory reduced by 30%, and administrative tasks reduced 50%. Economic benefits included the following cost reductions: manpower requirements, hardware and software expenditures, inventory-carrying costs, freight costs, and reduced warehousing costs.

CC leveraged its supply chain to take advantage of an opportunity to combine the synchronized thinking that existed from the supply base to customer consumption. It reengineered the supply chain to better meet the future needs of its customers and markets. This is yet another example of the opportunity awaiting any firm interested in obtaining a competitive advantage for future business success.

18 The Theory of Constraints

Lisa J. Scheinkopt

The whole history of science has been the gradual realization that events do not happen in an arbitrary manner, but that they reflect a certain underlying order, which may or may not be divinely inspired.

— Stephen W. Hawking

The *theory of constraints* (TOC) is a popular business philosophy that first emerged with Dr. Eliyahu Goldratt's landmark book, *The Goal*. One of the strengths of the TOC approach is that it provides focus in a world of information overload. It guides its practitioners to improve their organizations by focusing on a very few issues — the constraints of ongoing profitability. TOC is based on some fundamental assumptions. This introduction to TOC will provide you with a foundational paradigm that can enable a more effective analysis of manufacturing challenges.

18.1 FROM FUNCTIONAL TO FLOW

Imagine that I am a new employee in your organization, and it's your job to take me on a tour to familiarize me with the company's operations. What would you show me? Perhaps the scenario would look something like this.

First, we enter the lobby and meet the receptionist. Next, we walk through the sales department, followed by customer service, accounting, R&D engineering, and human resources. Then, you lead me through purchasing and production control, followed by safety, quality, legal, and don't forget, the executive offices. You save the best for last, so we go on a lengthy tour of manufacturing. You point out the press area, the machine shop, the lathes, the robots, the plating line and assembly area, the rework area, and the shipping and receiving docks.

Did you notice the *functional* orientation of the tour? I've been led on well over 1000 imaginary and real tours, and almost all of them have had this functional focus. Imagine now that we have an opportunity to converse with the people who work in each of these areas as we visit them. Let's ask them about the problems the organization is facing. Let's ask them about the "constraints." All will talk about the difficulties they face in their own functions, and will extrapolate the problems of the company from that perspective. For instance, we might hear:

- Receptionist: *"People don't answer their phones or return their calls in a timely manner."*

1-57444-300-3/02/$0.00+$1.50

- Sales: *"Our products are priced too high, and our lead times are too long!"*
- Customer service: *"This company can't get an order out on time without a lot of interference on my part. I'm not customer service, I'm chief expediter!"*
- Human resources: *"Not enough training!"*
- Purchasing: *"I never get enough lead time. Engineering is always changing the design, and manufacturing is always changing its schedules."*
- Manufacturing: "*We are asked to do the impossible, and when we do perform, it's still not good enough! Never enough time, and never enough resources.*"
- *And so on.*

What's wrong with this picture? Nothing and everything. Nothing, in that I'm certain that these good people are truly experiencing what they say they're experiencing. Everything, in that it's difficult to see the forest when you're stuck out on a limb of one of its trees.

My dear friend and colleague John Covington was once asked how he approached complex problems. His reply was, "*Make the box bigger*!" This is exactly what the TOC paradigm asks us to do. There is a time for looking at the system from the functional perspective, and there is a time for looking at a bigger box — the whole system perspective. When we want to understand what is constraining an organization from achieving its purpose, we should enlarge our perspective of the box from the function box to the value chain box.

18.1.1 The Value Chain

Let's now look at the value chain box. Pretend that we have removed the roof from your organization, and over 6 months, we hover above the organization at an altitude of 40,000 feet. As we observe, our perspective of the organization is forced to change. We are viewing a pattern. The pattern is *flow*. You may even describe this flow as *process flow*. Whether your organization produces a single product or thousands, the flow looks the same over space and time, as shown in Figure 18.1. The inside of the box represents your organization. The inputs to your organization's process are the raw materials, or whatever your organization acquires from outside itself to ultimately convert into its outputs. Your organization takes these inputs and transforms them into the products or services that it provides to its customers. These products or services are the outputs of the process. Whatever the output of your organization's process might be, it is the means by which your organization accomplishes its purpose. The rate at which that output is generated is the rate at which your organization is accomplishing its purpose. Every organization, including yours, wants to improve. The key to improving is that rate of output, in terms of purpose (*the goal*).

Actually, we can use this box to describe any system that we choose. For instance, look again at Figure 18.1. Now, let's say that the inside of the box represents your

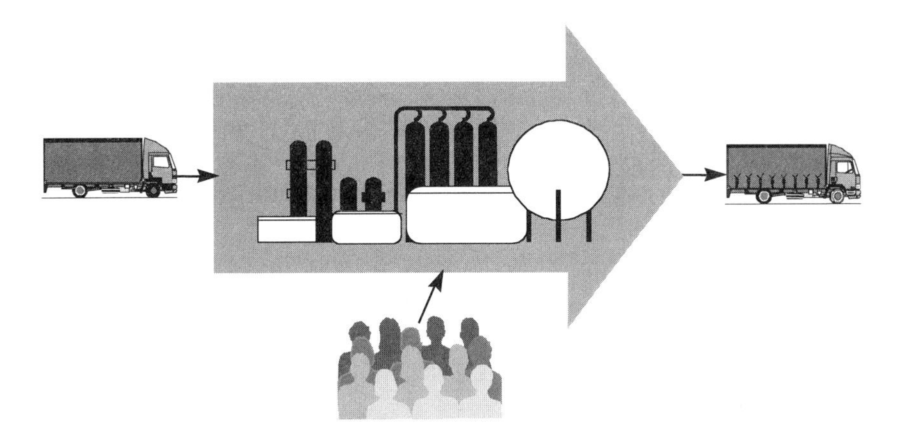

FIGURE 18.1 The 40,000 ft perspective. (Courtesy of Chesapeake, Inc., Alexandria, VA.)

department. Your department receives inputs from something outside it, and it transforms those inputs into its outputs. We can also say that the box is you, and identify your inputs and outputs. By the same token, try placing your customers and your vendors inside the box. Now try your industry, your community, your country.

18.1.2 The Constraint Approach to Analyzing Performance

In his book, *The Goal*, Dr. Goldratt emphasizes that we need to look at what the organization is trying to accomplish and to make sure that we measure this process and all our activities in a way that connects to that goal. TOC views an organization as a system consisting of resources that are linked by the processes they perform. The goal of the organization serves as the primary measurement of success. Within that system, a *constraint* is defined as anything that limits the system from achieving higher performance relative to its purpose. The pervasiveness of interdependencies within the organization makes the analogy of a chain, or network of chains, very descriptive of a system's processes. Just as the strength of a chain is governed by its single weakest link, the TOC perspective is that the ability of any organization to achieve its goal is governed by a single constraint, or at most, very few.

Although the concept of constraints limiting system performance is simple, it is far from simplistic. To a large degree, the constraint/nonconstraint distinction is almost totally ignored by most managerial techniques and practices. Ignoring this distinction inevitably leads to mistakes in the decision process. The implications of viewing organizations from the perspective of constraints and nonconstraints are significant. Most organizations simultaneously have limited resources *and* many things that need to be accomplished. If, due to misplaced focus, the constraint is not positively affected by an action, then it is highly unlikely that real progress will be made toward the goal.

A *constraint* is defined as *anything that limits a system's higher performance relative to its purpose*. When looking for its constraints, an organization must ask the question, "What is limiting our ability to increase our rate of goal generation?"

When we're viewing an organization from the functional perspective, our list of constraints is usually long. When we're viewing the organization from the 40,000-foot perspective, we begin to consider it as an interdependent group of resources, linked by the processes they perform to turn inventory into throughput. Just as the strength of a chain is governed by its weakest link, so is the strength of an organization of interdependent resources.

18.1.3 Two Important Prerequisites

TOC prescribes articulated a five-step improvement process that focuses on managing physical constraints. However, after many years of teaching, coaching, and implementing, we have identified two prerequisites that must be satisfied to gain perspective for the five focusing steps — or *any* improvement effort — that are not readily obvious: (1) define the **system** and its **purpose (goal)**, and (2) determine how to **measure** the system's purpose. Sometimes these prerequisites are just intuitive. Sometimes they're ignored because they're difficult to come to grips with. When ignored, you run the risk of suboptimization or improving the wrong things. In other words, you run the risk of system *non*improvement.

Consider the case of a multibillion-dollar, multisite, chemical company. One of our projects was to help it improve one of its distribution systems. Before we began to talk about the constraints of the system, we asked the team to develop a common understanding of the role of the distribution system as it relates to the larger system of which it is a part. They considered the 40,000-foot view of the corporation as a whole and engaged in a dialogue about the purpose of the distribution system within that bigger box. As a result, the team was able to focus on improving the distribution system not as an entity in and of itself, but as an enabler of throughput generation for the corporation.

But what are the fundamental system measures of the distribution system mentioned above? How does it know that it's doing well? Sure, we can say that ultimately they are the standard measures of net profit and return on assets. But these measures don't tell the distribution system whether or not it's fulfilling *its* role. The team identified some basic measures that looked at its impact on the company's constraint, as well as the financial measures over which the system has direct control. When this process is applied to manufacturing, the following usually unfolds.

18.1.3.1 Define the System and Its Purpose (Goal)

Given that the roots of TOC are deeply embedded in manufacturing, often the system is initially defined as the manufacturing operation, or plant. The purpose of the manufacturing operation is to enable the entire organization to achieve its goal, and it is important to have a clear definition of that goal. One goal shared by most manufacturing companies is to "make more money now as well as in the future." Although this goal may be arguable in special circumstances, making money certainly provides the funds to fuel ongoing operations and growth regardless of other stated goals. As such, making money is at least a very tight necessary condition in almost every organization. As a result, it is appropriate to continue this example using making more money now as well as in the future as the goal of the manufacturing organization. The next question to be answered is, "How do we measure making money?"

18.1.3.2 Determine How to Measure the System's Purpose

Manufacturing organizations purchase materials from vendors and add value by transforming those materials into products their customers purchase. Simply stated, companies are making money when they are creating value added at a rate faster than they are spending. To calculate making money, TOC starts by categorizing what a firm does with its money in three ways:

Throughput (T) is defined as *the rate at which an organization generates money through sales.* The manufacturing process adds value when customers are willing to pay the manufacturer more money for the products than the manufacturer paid its vendors for the materials and services that went into those products. In TOC terminology, this value added is the throughput.

Operating expense (OE) is defined as *all of the money the organization spends in order to turn inventory into throughput.* Operating expense includes all of the expenses that we typically think of as fixed. It *also* includes many that are considered to be variable, such as direct labor wages. To be profitable, the company must generate enough throughput to more than pay all the operating expenses. As such, profit is calculated simply as T – OE.

Rate of return is also an important measure of profitability. Any profit is unacceptable when it's bringing a poor rate of return on investment — and this return is greatly affected by the amount of money that is *sunk in the system.* In TOC terminology, this is *inventory.* Formally, *inventory* (I) is defined as *the money that the system spends on things it intends to turn into throughput.* Return on investment, then, is net profit (T – OE) divided by inventory (I). Inventory, as used in this equation, includes what is known as "passive" inventory such as plant and equipment. However, in improving manufacturing operations, the focus is much more on reduction of "active" inventory — the raw material, work-in-process, and finished goods needed to keep the system running.

Often, it is easy to lose sight of the goal in the process of making day-to-day decisions. Determining the impact of local decisions is complicated by the fact that measuring the net profit of a manufacturing plant in isolation from the larger system is impossible (though many organizations fool themselves into thinking they can). In practice, productivity and inventory turns may be more appropriate measures than profit at the plant level. The TOC approach to measuring productivity and turns uses the same three fundamental measures — T, I, and OE. Productivity is measured as T/OE — in essence, the ratio between money generated and money spent. Meanwhile, inventory turns are measured as T/I — the ratio between money generated and level of investment required to generate it.

The concept of allocating all the money in a system into one of three mutually exclusive and collectively exhaustive categories of throughput, inventory, or operating expense may appear unconventional at first. Why would one do such a thing? The real power lies in using T, I, and OE to evaluate how decisions affect the goal of making money. When we want to have a positive effect on net profit or return on investment, on productivity or turns, we must make the decisions that will increase throughput, decrease inventory, and/or decrease operating expense. The cause–effect connection between local decisions and their impact on the basic measures of T, OE, and I is usually much more clearly defined. These basic measures can then serve as direct links to the more traditional global financial measures.

Given three measures, one naturally takes priority over the others. One of the distinguishing characteristics of managers in TOC companies is that they view throughput as the measure with the greatest degree of leverage in both the short and long term. This is largely due to the fact that, of the three measures, opportunities to increase throughput are virtually limitless. In contrast, inventory and operating expense cannot be reduced to less than zero, and in many cases, reducing one or both may have a significant negative impact on throughput.

An overriding principle that guides TOC companies is that ongoing improvement means growth. They believe that growth doesn't happen by concentrating on what to shrink, but rather by concentrating on what to grow. That means concentrating on the means by which they choose to increase throughput. This emphasis on throughput first (inventory second and operating expenses third) is referred to as "throughput world thinking," and is often held in contrast with the common managerial obsession with cost reduction, hence the term "cost world thinking."

18.2 UNDERSTANDING CONSTRAINTS

There are three major categories of constraints: physical, policy, and paradigm. Because all three exist in any given system at any given time, they are related. Paradigm constraints cause policy constraints, and policy constraints result in physical constraints.

18.2.1 Physical Constraints

Physical constraints are those resources that are physically limiting the system from meeting its goals. Locating physical constraints involves asking the question, "What, if we only had more of it, would enable us to generate more throughput?" A physical constraint can be internal or external to the organization.

At the input boundary of the system, external physical constraints would include raw materials. For instance, if you are unable to produce all that your customers are asking of you because you cannot get enough raw materials, the physical constraint of your organization may be located at your vendor.

An external physical constraint might also be at the output boundary of the system — the market. If you have plenty of capacity, access to plenty of materials, but not enough sales to consume them, a physical constraint of your organization is located in your market.

Internal physical constraints occur when the limiting resource is a shortage of capacity or capability inside the boundaries of the organization. Although it is easy for us to relate to machines as constraints, today's internal physical constraints are most often not machines, but rather the availability of people or specific sets of skills needed by the organization to turn inventory into throughput.

Every organization is a system of interdependent resources that together perform the processes needed to accomplish the organization's purpose. Every organization has one or very few physical constraints. The key to continuous improvement, then, lies in what the organization is doing with those few constraints.

With the prerequisites of defining the system and its measures fulfilled, let's move on to the five focusing steps. These five steps can now be found in an abundance of TOC literature and are the process by which many organizations have achieved dramatic improvements in their bottom line.

18.2.1.1 The Five Focusing Steps

The five focusing steps provide a process for ongoing improvement, based on the reality — not just theory — of physical constraints.

1. **Identify** the system's constraint. For the manufacturer, the question to be answered here is, "What is physically limiting our ability to generate more throughput?" The constraint will be located in one of three places: (1) the market (not enough sales), (2) the vendors (not enough materials), or (3) an internal resource (not enough capacity of a resource or skill set). From a long-term perspective, an additional question must be answered — if not immediately, then as soon as the operation is under control by implementing focusing steps 2 and 3. That question is, "Where does our organization want its constraint to be?" From a strategic perspective, where *should* the constraint be?

2. Decide how to **exploit** the system's constraint. When we accept that the rate of throughput is a function of the constraint, then the question to be answered at this step is, "What do we want the constraint *to do*?" so that the rate of throughput generated by it is maximized (now and in the future). The following activities and processes are typically implemented in association with this step:

When the constraint is internal:

- The resource is considered "the most precious and valuable resource."
- Wasted activity performed by the constraint is eliminated, often using lean manufacturing techniques.
- People focus on enabling the resource to work on the value-added activities that it alone is capable of doing. This often means that the constraint resource off-loads other activities to nonconstraints.
- Attention is paid to setup, and efforts are made to minimize setup time on the constraint resource.
- Utilization and output of the constraint are measured. Causes for downtime on the constraint are analyzed and attacked. Care of the constraint resource becomes priority number 1 for maintenance, process engineering, and manufacturing engineering.
- Inspection steps can be added in front of the constraint to ensure that only good material is processed by it. Care is taken at the constraint (and at every step after) to ensure that what the constraint produces is not wasted.

- Often, extra help is provided to aid in faster processing of constraint tasks, such as setup, cleanup, paperwork, etc.
- Steps are taken in sales and marketing to influence sales of products that generate more money per hour of constraint time.

When the constraint is raw materials:

- The raw material is treated like gold.
- Reducing scrap becomes crucial.
- Work-in-process and finished-goods inventory that is not sold are eliminated.
- Steps are taken in purchasing to enhance relationships with the suppliers of the constraint material.
- Steps are taken in sales and marketing to influence sales of product that generate more money per unit of raw material.

When the constraint is in the market:

- The customers are treated like precious gems.
- The company gains an understanding of critical competitive factors, and takes the steps to excel at those factors.
- Steps are taken in sales and marketing to carefully segment markets and sell at prices that will increase total company throughput.

From the manufacturing perspective, this usually means

- 100% due-date performance
- Ever faster lead times
- Superior quality (as defined by customer need)
- Adding features (as defined by customer need)

Although a discussion of strategic constraint placement is a topic beyond the scope of this book, suffice it to say that there are advantages to strategic selection of an internal material flow control point. When the constraint is internal, the constraint resource is almost always selected as the control point.

To exploit the constraint or the control point, it is finitely scheduled to maximize output without overloading it. Overloads serve only to increase lead times as work queues backup in front of the constraint. The schedule defines precisely the order in which that resource will process products. It serves as the "drum" for the rest of the manufacturing organization. The drum is based on real market demand (in other words, the market demand is what pulls the schedule). This schedule serves as the backbone of an operations plan that meets due-date performance while simultaneously maximizing throughput and minimizing inventory. It is the first element of the "drum–buffer–rope" process for synchronizing the flow of product (Figure 18.2). The buffer and rope aspects are discussed in the next paragraph.

3. **Subordinate** everything else to the above decisions. Step 1 identifies *the* key resource determining the rate of throughput the organization can generate. In step

2, decisions are made relative to how the organization intends to maximize that rate of throughput: how to make the most with what it has. In this step, the organization makes and implements the decisions to ensure that its own rules, behaviors, and measures enable, rather than impede, its ability to exploit the identified constraint. *Subordinate* is the step in which the majority of behavior changes occur. It is also in this step that we define *buffer* and *rope*.

The ability of the company to maximize throughput and meet its promised delivery dates hinges first on the ability of the constraint or control point to meet its schedule — to march according to the drum. TOC also recognizes that variability — in the form of statistical fluctuations everywhere — exists in every system. It is crucial that the drum be protected from the inevitable variability that occurs. The means by which it attempts to ensure this is the *buffer*. A TOC company does not want to see its drum schedule unmet because materials are unavailable. Therefore, work is planned to arrive at the constraint or control point sometime prior to its scheduled start time. The buffer is the amount of time between the material's planned arrival time at the control point and its scheduled start time on the control point. The same concept is put to work in what is called the *shipping buffer*. In companies wherein it is important to meet the due dates quoted to their customers (can you think of any companies where it's not important?), work is planned to be ready to ship a predetermined amount of time prior to the quoted ship date. The difference between this planned ready-to-ship time and the quoted ship date is the shipping buffer.

In a TOC company, work is released into production at the rate dictated by the drum and is timed according to the predetermined length of the buffer. This mechanism is called the *rope*, as it ties the release of work directly to the constraint or control point. This third element ensures that the TOC plant is operating on a pull system. The actual market demand pulls work from the constraint or control point, which in turn pulls work into the manufacturing process.

It is important to note that at all places other than those few requiring buffer protection, inventory is expected to be moving and work center queues are minimized. There is no planned inventory anywhere else. The end result is very low total inventory in the manufacturing operation. Low total inventory in turn translates into shorter lead times, which may be used as a competitive advantage.

Several additional activities and behaviors that are required to support the *subordinate* rule include

Roadrunner mentality takes over. The analogy of the roadrunner cartoon character is used to portray the approach to work. The roadrunner operates at two speeds — full speed ahead or dead stop. In a TOC plant, if there is work to be worked on, work on it at full speed ahead (of course, the work is to be of high quality as well). If there is no work to work on, stop. Congratulations for emptying your queue. Take the time you have with no queue and use it for learning, for cleaning your work area, for helping another team member, or for working on another activity that will ultimately help the organization. It's even OK to take a break. The workers' purpose is to turn inventory into throughput, not simply to produce more inventory. Workers are responsible for ensuring that the drum of the organization doesn't miss a beat.

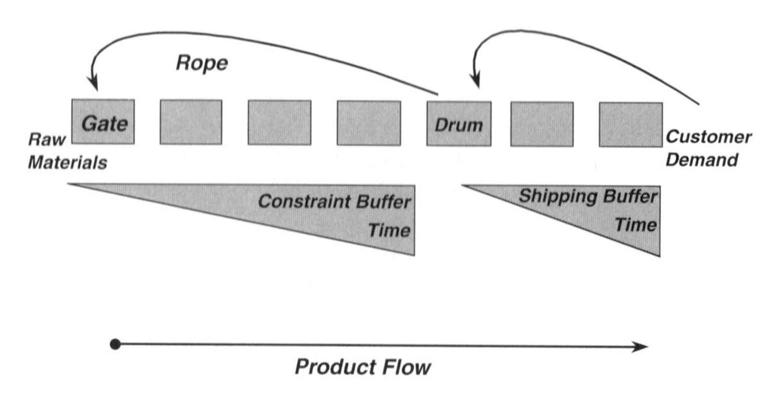

FIGURE 18.2 Synchronized flow. (Courtesy of Chesapeake, Inc., Alexandria, VA.)

Performance measures change. For instance, in many TOC companies *everybody* is measured on constraint performance to schedule. Maintenance is measured on constraint downtime. Gain-sharing programs are modified to include constraint and throughput-based measures. The old measures of efficiency and utilization are abandoned at nonconstraints.

Protective capacity is maintained on nonconstraint resources. We have already established that manufacturing organizations have both dependency and variability. Buffers are strategically placed to protect the few things that limit the system's ability to generate throughput and meet its due dates. If we have a system in which the capacity of every resource is theoretically the same, then every instance of variability (e.g., breakdowns, slow processing times, defective raw material) will result in some degree of buffer depletion. After some period of time, the buffer will be depleted enough that the constraint shuts down — because the constraint determines the rate of throughput, this is the equivalent of shutting down the whole system. If the constraint isn't working, the organization isn't generating money. Unless, of course, heroic (and expensive) efforts such as overtime, outsourcing, or customer cancellations readjust the system. In a TOC environment, additional capacity is intentionally maintained on nonconstraint resources for the purpose of overcoming the inevitable variations (instances of Murphy's Law) before the system's constraint notices. The combination of a few strategically placed buffers and protective capacity results in a predictable, stable overall system that has immunized itself from the impact of the inevitable variations that occur.

Buffer management is used as a method to ensure that constraint and shipping schedules are met, *and* to focus improvement efforts. In a TOC plant, a short 10- to 15-minute meeting occurs every shift and replaces the typical production meeting. Called a *buffer management meeting,* its participants

- Check the release schedule and keep a record of early, on-time, and late releases.
- Identify any work that is part of the planned buffer that is not yet at the buffered resource.
- Identify the current location of the missing work.

- Assign appropriate personnel (usually, someone from the current meeting) who will make sure the work moves quickly from its current location to the buffered resource. This action becomes their first priority on leaving the meeting.

The current location of the missing work and the amount of drum-time that the work represents is recorded. *This step is key to continuous improvement.* Periodically (weekly or monthly), these data are analyzed to determine where work meant for the drum is stuck most often. This becomes the focus for the improvement effort. Causes are identified and removed. Some of the "exploit" techniques are employed to ensure that wasteful activity is removed from the processes performed by that resource. If these activities don't create sufficient protective capacity (enough capacity that this resource is no longer the major cause for "holes" in the buffer), additional capacity can be acquired. The intent is to increase the velocity of the flow of material (the transformation of inventory into throughput). Once the obstruction to flow is resolved, the size of the buffer may be decreased.

4. **Elevate** the system's constraint. The foregoing three steps represent the TOC approach to maximizing the performance of a given system. In the "elevate" step, the constraint itself is enlarged. If the constraint is capacity of an internal resource, more of that capacity is acquired (additional shifts, process improvements, setup reductions, purchasing equipment, outsourcing, hiring people, etc.). If the constraint is materials, new sources for material are acquired. If the constraint is in the market, then sales and marketing bring in more business. At some stage during the elevate step, the constraint may very well move to another location in the system.

5. Don't allow **inertia** to become the system's constraint. When a constraint is broken, go back to step 1. This step reminds us to make it an ongoing improvement process. It also reminds us that once the constraint is elevated, we must ensure that sufficient protective capacity surrounds it. If the constraint changes, so must the rules, policies, and behaviors of the people in the organization.

18.2.2 Policy Constraints

Policies are the rules and measures that govern the way organizations go about their business. Policies determine the location of the physical constraints and the way in which they are or aren't managed. Policies define the markets your organization serves, they govern how you purchase products from vendors, and they are the work rules in your factory. *Policy constraints** are those rules and measures that inhibit the system's ability to continue to improve, such as through the five focusing steps.

Policies (both written and unwritten) are developed and followed because people, through their belief systems, develop and follow them. In spite of the fact that our organizations are riddled with stupid policies, I don't think that any manager ever woke up in the morning and said, "I think I'll design and enforce a stupid policy in my organization today." We institute rules and measures because we *believe* that

* Also called managerial constraints.

with them, the people in our organizations will make decisions and take actions that will yield good results for the organization.

18.2.3 Paradigm Constraints

*Paradigm constraints** are those beliefs or assumptions that cause us to develop, embrace, or follow policy constraints. In the 1980s, the people who populated many California companies believed that their companies were defense contractors. This belief enforced their policies to market and sell only to the U.S. government and its defense contractors and subcontractors. Clearly, they had the capacity as well as a wealth of capabilities that could have been productive and profitable serving non-defense-related industries. Nevertheless, the physical constraint for these companies was clearly located in the market. The result, as this industry shrank, was that many of these businesses went out of business. Their paradigm constraints prevented them from seeing this until it was too late to change the policies that would have enabled them to expand their markets and grow.

Another classic paradigm in many organizations is the goal of keeping costs and staff — particularly expensive staff — to a minimum. TOC advocates view cost from a different perspective, asking the question, *"What is the impact on throughput of adding this cost?"* In many cases — especially those where money or manpower is *added* to a constraint — the resulting analysis makes the decision extremely simple. Case in point. There once was a company whose engineering department had a backlog of more than 2 years of projects in support of the plant's production lines. Manning restrictions of corporate cost-reduction programs prevented hiring even one more engineer. This is, by the way, a perfectly defensible cost-reduction strategy; after all, engineers are expensive. However, at the same time, the queue of engineering projects contained relatively quick but lower priority projects, which would significantly improve constraint output — which in turn would increase line output. The market wanted more products, and the throughput associated with any additional output was nothing short of phenomenal. One project, designed to increase the calibration speed (the constraint on the line), would have allowed the line to produce two additional units per hour — production that could be easily sold to an eager market. Approximately $500 per unit in throughput is associated with each unit. Say that, for example, you must pay as much as $100,000 per year to hire an electrical engineer (EE) with the needed skills. Should the company hire the engineer?

The TOC-based decision would compare the $100,000 expense with the throughput that can be reasonably associated with the hiring. If the money for an additional EE was spent, what would be the impact on throughput and inventory? Completing this one project would allow the line to produce two additional units per hour. At $500 throughput each, that's $1000 per hour that won't be there until the project is completed. This project alone would pay back the engineer's annual salary in 100 hours. Four days — that's not a bad payback period for a line that runs 24 hours per day.

The reality: The *expenditure* of $100,000 was not allowed.

---

* Also called behavioral constraints.

Here is another example of physical, policy, and paradigm constraints in action, from the lens of the five focusing steps.

18.2.4 A Hi-Tech Tale

In the southwestern United States, there lives a company that manufactures high-technology electronic products for the communications industry. In this industry, speed is the name of the game. Not only must they offer very short lead times for their customers, they also must launch more and more new products at a faster and faster pace. This manufacturing organization does a very good job of meeting the challenge by blending the logistical methods of TOC with cellular manufacturing. However, though manufacturing continues to tweak its well-oiled system, the constraint of the company resides elsewhere.

1. *Identify the system's constraint(s).* When I asked the questions, "What is it that limits the company's ability to make more money? What don't you have enough of? Is there anyplace in the organization that work has to sit and wait?" — It didn't matter who I asked, from senior executives to people on the shop floor — the answer was almost unanimous: "*Engineering!*" After further checking, we learned that the specific constraint was the capacity of the software design engineers. Determining software design engineering's capacity was the key to this company's ability to increase its new product speed-to-market, and also for its ability to make improvements in existing products (in terms of manufacturability and marketability). Here was the key to this company making more money now as well as in the future. Exacerbating the issue was the fact that these types of engineers were very hard to come by, at least in this company's part of the country. Companies were stealing engineers from each other and offering large rewards for referrals. It was not difficult for software design engineers to go from company to company and raise their salaries and benefits by 25% over a year's time.
2. *Decide how to exploit the system's constraint(s).* The company obviously wanted the software design engineers to be doing software design engineering. After a little observation, the company learned some astonishing news. Would you believe that the software design engineers spent only about 50 to 60% of their time doing software design engineering? No, they were not lazy, goofing off, or playing hooky. They were working, and they were working very hard. In fact, engineering was the highest stressed, most overworked area of the company. At this point we asked, "What do the software design engineers do that only they can do, and what do they do that others are capable of doing?" Some of the tasks involved in the software design engineering function included data entry, making copies, sending faxes, attending lots of long meetings, and tracking down files, supplies, paperwork, and more. This work, though necessary work for the company, could be offloaded to other people. It meant shifting some people around, and yes, wrestling with one or two policy and paradigm constraints.

Policy: *The software design engineer does all of the tasks involved in the work that is designated "software design engineering work."*

Paradigm: *The most efficient way to accomplish a series of tasks is for one (resource) person to do those tasks. Person (or resource) efficiency is the equivalent of system efficiency.*

3. *Subordinate everything else to the above decisions.* According to the policy and paradigm constraints identified above, subordination meant that anyone feeding work to or pulling work from a software design engineer was to give that work the highest priority. Software design engineering work was no longer allowed to wait for anything or anybody, with the exception of the software design engineers. This meant that if you were a nonconstraint and you were working on something not connected to software design engineering, when that type of work came your way, you put down what you were doing and worked on the software design engineering work. Then, you went back to the task you were working on before.
4. *Elevate the system's constraint(s).* The company chose two routes to increase their software design engineering capacity. The first was to have cross-functional teams responsible for the development and launch of new products. As a result, the company reduced the necessity for much of the tweaking, because the designers are better at considering manufacturing, materials, and market criteria from the onset of the new product project. New, manufacturable and marketable products are being launched faster than ever. The policy constraint that they had to break was: *Each functional group does their part in the process and then passes the work to the next group.* Of course, this policy stems from the same efficiency paradigm that was pointed out in the preceding steps. The company has also been attacking an additional set of policy and paradigm constraints.

   Policy: *Hire only degreed engineers.*

   Paradigm: *The only way to acquire the skills of a software design engineer is by getting the formal degree.*

   Given the general shortage of software design engineers in the region, the company is putting an apprenticeship program in place. In this program, an interested nonengineer will be partnered with an engineer. Over the course of a couple of years, the apprentice will be able to acquire the software-design engineering skills that the company needs through a combination of mentoring by the engineer and some courses. This will enable engineers to offload some of their work early on, increasing their capacity to do the more difficult and specialized work. It also helps the company develop the capacity it needs in spite of the external constraints (availability of degreed engineers). At the same time, the program will help the company's people grow, leaving a very positive impact on the company's culture and on the loyalty of its employees. People feel good when they are helping and being helped by their peers.
5. *Don't allow inertia to become the system's constraint. If, in the above steps, a constraint is broken, go back to step 1.* The constraint has not yet

shifted out of software design engineering. The current challenge this company faces is to determine where, strategically, its constraint should be and plan accordingly. In other words, part of its strategic planning process should be to simulate steps 1, 2, and 3, and implement a plan based on decisions resulting from those simulations.

18.3 CONCLUSION

As you can see from the examples, the TOC approach has the initial difficulty of determining a workable goal and measures, combined with the triple challenge of addressing the physical, policy, and paradigm constraints to meeting that goal. In my work with nonprofit organizations, I have come to the conclusion that their goals and measures are extremely unclear, and this fact is the root of most of their problems. This results in goals that focus on managing the numbers, often at the expense of moving forward relative to their purpose.

For those of you who are employed by for-profit organizations, guess what? The same problem exists. Unless you're the top management, or your pay is tied directly to the profitability of the company, it's difficult to rally around the Money-is-THE-goal banner. Most people want to spend their time in meaningful ways. When companies encourage their people to enter into a dialogue aimed at discovering and clarifying their common purpose as co-members of an organization, the process of improving the bottom line becomes much easier and more fun.

I am *not* advocating that you spend an inordinate amount of time and effort doing process flow and other such diagrams to articulate these things ever so precisely before you start on the task of improving the system. I *am* suggesting that when you begin an improvement effort, you begin it with a dialogue on these important issues. (And, assuming that you want ongoing improvement, I suggest that you encourage the dialogue to be an open, ongoing dialogue.) Questions such as, "What is the system that we are trying to improve?" "What's the purpose of the system?" and "What are its global measures?" will help you take a focused and whole-system approach to your improvement efforts.

The complexity of modern organizations and systems leaves managers with an almost unlimited number of things to improve. The magnitude of the task is sufficient to paralyze even the most conscientious manager. Meanwhile, in reality, only a handful of those hundreds of potential improvements will make a real difference in achieving an organization's goal. TOC's constraint-focused approach is both logical and pragmatic. Identifying and addressing the constraints provide the fastest and lowest-cost means for increasing the throughput of any organization.

REFERENCES

At Colortree, High Performance and Low Stress Go Hand in Hand, Chesapeake Consulting, Severna, MD, 2001.

Covington, J., Help Wanted: How Can Your Business Grow When You Can't Count on Headcount?, Chesapeake Consulting, Severna, MD, 2000.

Goldratt, E. and Cox, J., *The Goal,* 2nd ed., North River Press, Croton-on-Hudson, NY, 1992.
Making the Most of Existing Resources: Productivity Takes a Giant Step at Chemical Company, Chesapeake Consulting, Severna, MD, 2001.
Moore, R. and Scheinkopf, L., *Theory of Constraints and Lean Manufacturing: Friends or Foes?,* Chesapeake Consulting, Inc., Severna, MD, 1998.
Scheinkopf, L., *Thinking for a Change: Putting the TOC Thinking Processes to Use,* St. Lucie Press, Boca Raton, 1999.

19 TRIZ

Steven F. Ungvari

19.1 WHAT IS TRIZ?

Nominally, TRIZ is a Russian language acronym for the Russian words *teoriya resheniya izobretatelskikh zadatch,* which can be translated into *the theory of the solution of inventive problems.* This title is somewhat of a misnomer, because TRIZ has moved out of the realm of theory and into a bona fide, scientifically based methodology. The development, evolution, and refinement of TRIZ have consumed some 50 years of rigorous, empirically based analysis by some of the brightest scientific minds of the 20th century.

Nevertheless, the whole notion of creativity and innovation mentioned in the context of science makes for an unusual pairing. Innovation and creativity are typically thought of as spontaneous phenomena that happen in a capricious and unpredictable way in the vast majority of people. Historically, only a precious few individuals, such as Michelangelo, Leonardo da Vinci, Henry Ford, and Thomas Edison, seem to have possessed an innate natural ability for creativity and inventiveness.

The name, the theory of the solution of inventive problems, implies that innovation and creative thought in the context of problem solving are supported by an underlying construct and an architecture that can be deployed on an as-needed basis. The implications of such a theory, if true, are enormous because it suggests that lay individuals can elevate their creative thinking capabilities by orders-of-magnitude.

19.2 THE ORIGINS OF TRIZ

The inventor of TRIZ was Genrich Altshuller, a Russian (1926–1998). Altshuller became interested in the process of invention and innovative thinking at an early age. He patented a device for generating oxygen from hydrogen peroxide at the age of 14. Altshuller's fascination with inventions and innovation continued through Stalin's regime and World War II. After the war, Altshuller was assigned as a patent examiner in the Department of the Navy. As such, Altshuller often found himself helping would-be inventors solve various problems with their inventions. In due course, Altshuller become fascinated with the study of inventions. In particular, Altshuller was interested in understanding how the minds of inventors work. His initial attempts were psychologically based, but these probes provided little if any insight on how creativity could be engineered.

Altshuller then turned his attention to studying actual inventions and in a sense reverse-engineering them to understand the essential engineering problem being

1-57444-300-3/02/$0.00+$1.50

solved and the elegance of the solution as described in the patent application. It should be noted that in the former Soviet Union patent applications (called authors certificates [ACs]) were concise documents no more that three or four pages in length. The author certificate consisted of a descriptive title of the invention, a schematic of the new invention, a rendering of the current design, the purpose of the invention, and a description of the solution.

19.2.1 Altshuller's First Discovery

The brevity of the certificates facilitated analysis, cataloguing, and mapping solutions to the problems. As the number of inventions he scrutinized grew, Altshuller uncovered similar patterns of solutions for similar problems. This was a remarkable discovery because it essentially paved the way for a scientific, standardized way to approach a problem and to incorporate a latent knowledge base as an integral element of the solution process. In other words, Altshuller discovered that similar technological problems gave rise to similar patents. This phenomenon was repeated in widely disparate engineering disciplines at different periods of time and in geographically dispersed areas.

The logical conclusion reached by Altshuller was that the possibility existed of creating a mechanism for describing types of problems and subsequently mapping them with types of solutions. This discovery led to just such a mechanism, which consisted of the 39 typical engineering parameters, the contradiction matrix, and the 40 inventive principles. These tools are covered in more detail later in the chapter.

19.2.2 Altshuller's Second Discovery

Altshuller's second enlightening discovery was made as he assembled chronological technology maps. Altshuller uncovered an unmistakable, explicit regularity in the evolution of engineered systems. Altshuller described these time-based phenomena in his lectures and writings as *The Eight Laws of Engineered Systems Evolution.* The term *laws* does not imply that Altshuller defined them as conforming to a strict scientific construction, as in the fields of physics or chemistry. The laws, though general in nature, are nevertheless recognizable and predictable; more importantly, they provide a road map to future derivatives. Today, these eight laws have been refined and expanded into more than 400 sublines of evolution and are useful in technology development, product planning, and the establishment of defensible patent fences.

19.2.3 Altshuller's Third Discovery

The third truism that emerged from Altshuller's analytical work was the realization that inventions are vastly different in their degrees of inventiveness. Indeed, many of the patents that Altshuller studied were filed simply to describe a system and provide some degree of protection. These patents were useless in Altshuller's determination to discover the secret of how to become an inventor of the highest order. To differentiate inventiveness, Altshuller devised a scale of 1 to 5 for categorizing the elegance of the solution (see Figure 19.1).

Note that only level 3 and 4 solutions are deemed to be inventive. Within the body of TRIZ knowledge, *inventive* means that the solution was one that did not

Level	Nature of Solution	Number of Trials to Find the Solution	Origin of The Solution	% of Patents at This Level
1	Parametric	None to Few	The Designer's Field of Specialty	32%
2	Significant Improvement in Paradigm	Ten to Fifty	Within a Branch of Technology	45%
3	Inventive Solution in Paradigm	Hundreds	Several Branches of Technology	18%
4	Inventive Solution Out of Paradigm	Thousands to Tens of Thousands	From Science Physical/Chemical Effects	4%
5	True Discovery	Millions	Beyond Contemporary Science	1%

FIGURE 19.1 Levels of inventiveness.

compromise conflicting requirements. For example, strength vs. weight is an example of conflicting parameters. To increase strength, the engineer will typically make something thicker or heavier. An inventive solution would increase strength with no additional weight or even a reduction in weight.

19.2.4 Altshuller's Levels of Inventiveness

19.2.4.1 Level 1: Parametric Solution

A parametric solution uses well-known methods and parameters within an engineering field or specialty. This is the lowest level solution and is not an inventive solution.

For example, the problem of roads and bridges icing over can be solved by using salt or sand, or by plowing. Calculating stress on a cantilevered structure is accomplished by using well-known mathematical formulas.

19.2.4.2 Level 2: Significant Improvement in the Technology Paradigm

Level 2 is a significant improvement in the system, utilizing known methods possible from several engineering disciplines. Although a level 2 solution is a significant improvement over the previous system, it is not inventive.

A level 2 solution of the icing problem would be required if conventional means were prohibited. This type of solution demands a choice between several variants which leaves the original system essentially intact. The roadways or bridges, for example, could be formulated or coated with an exothermic substance that would be triggered at a certain temperature.

19.2.4.3 Level 3: Invention within the Paradigm

Level 3 eliminates conflicting requirements within a system, utilizing technologies and methods within the current paradigm. A level 3 solution is deemed to be inventive because it eliminates the conflicting parameters in such a way that both requirements are satisfied simultaneously.

A level 3 solution to the conflicting requirements of strength vs. weight has been solved in aircraft by the use of honeycomb structures and composites.

19.2.4.4 Level 4: Invention outside the Paradigm

Level 4 is the creation of a new generation of a system with a solution derived — not in technology — but in science.

A level 4 solution integrates several branches of science. The radio, the integrated circuit, and the transistor are examples of level 4 solutions.

19.2.4.5 Level 5: True Discovery

Level 5 is a discovery that is beyond the bounds of contemporary science. A level 5 discovery will oftentimes spawn entire new industries or allow for the accomplishment of tasks in radically new ways. The laser and the Internet are examples of level 5 inventions.

19.3 BASIC FOUNDATIONAL PRINCIPLES

The three discoveries made by Altshuller provided the construct for the formation of the foundational underpinnings upon which all TRIZ theory, practices, and tools are built. The three building blocks of TRIZ are *ideality, contradictions,* and the maximal use of *resources.*

19.3.1 IDEALITY

The notion of ideality is a simple concept. Essentially, ideality postulates that in the course of time, systems move toward a state of increased ideality. Ideality is defined as the ratio of useful functions F_U divided by harmful functions F_H.

$$\text{Ideality} = I = \frac{\Sigma F_U}{\Sigma F_H}$$

Useful functions embody all the desired attributes, functions, and outputs of the system. In other words, from an engineering point of view, it is termed *design intent.*

Harmful functions, on the other hand, include the expenses or fees associated with the system, the space it occupies, the resources it consumes, the cost to manufacture, the cost to transport, the cost to maintain, etc.

Extrapolating the concept to its theoretical limit, one arrives at a situation where a system's output consists solely of useful functions with the complete absence of any harmful consequences. Altshuller called this state the ideal final result (IFR). The IFR is not actually calculated; rather it is a tool to define the ideal end-state. Once the end-state is defined, the question as to why it's difficult to attain flushes out the real (contradictory) problems that must be overcome.

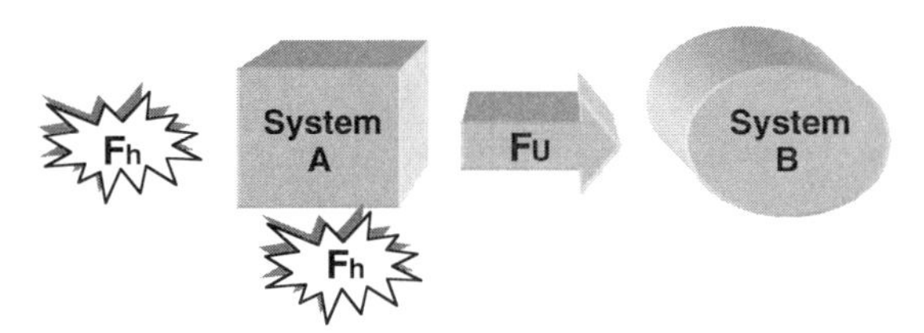

FIGURE 19.2 Typical system function. System A interacting with system B and producing a useful output but also creating harmful consequences.

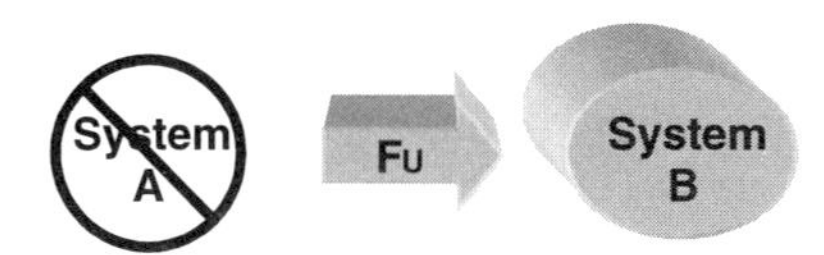

FIGURE 19.3 Ideal system function. System A does not exist, its function, nevertheless, is carried out.

One might argue that it is absurd to think of solving problems from the theoretical notion of the IFR instead of explicitly defining the current dimensions of the problem. It is, however, precisely this point of view that opens up innovative vistas by reducing prejudice, bias, and, most of all, psychological inertia (PI).

Psychological inertia is analogous to what Thomas S. Kuhn in his book, *The Structure of Scientific Revolutions,* defines as one's paradigms. Kuhn defines a paradigm as "the entire constellation of beliefs, values, techniques and so on shared by the members of a given community." The danger of paradigms is that they confine the solution space to the area inside the paradigm. An engineer competent in mechanics, for example, is unlikely to search for a solution in chemistry; it's outside his paradigm.

Dr. Stephen Covey in his best-selling book, *The 7 Habits of Highly Effective People,* offers a similar concept in habit 2, "Begin with the End in Mind." Dr. Covey stated, "To begin with the end in mind means to start with a clear understanding of your destination. It means to know where you're going so that you better understand where you are now and so that the steps you take are always in the right direction,"

The notion of ideality also postulates that a system, any system, is not a goal in itself. The only real goal or design intent of any system is the useful function(s) that it provides. Taken to its extreme, the most ideal system, therefore, is one that does not exist but nevertheless produces its intended useful function(s) (see Figures 19.2 and 19.3).

In the illustration above (Figure 19.2), the supersystem has not reached a state of ideality because the useful interaction between A and B is accompanied by some type of unwanted (harmful) functions.

An ideal system A, on the other hand, is one that does not exist; yet its design intent is fully accomplished.

In the abstract, this notion might at first blush seem fantastical, impossible, and even absurd. There is, however, a subtle yet powerful heuristic embodied in ideality. First, ideality creates a mind-set for finding a noncompromising solution. Second,

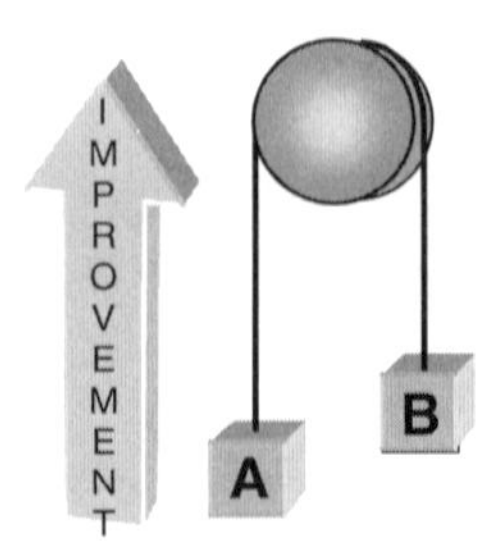

FIGURE 19.4 Technical contradiction. As parameter A improves, B is worse and vice versa.

it is effective in delineating all the technological hurdles that need to be overcome to invent the best solution possible. Third, it forces the problem solver to find alternative means or resources to provide the intended useful function. The latter outcome is similar to an organization reassigning key functions to the individuals who have been retained after a reduction in force.

19.3.2 Contradictions

The second foundation principle is the full recognition that systems are inherently rife with conflicts. Within TRIZ these conflicts are called *contradictions*. In TRIZ, an inventive problem is one that contains one or more contradictions. Typically, when one is faced with a contradictory set of requirements, the easy way out is to find a compromising solution. This type of solution, while it may be expedient, is not an inventive solution. If we return to the example of weight vs. strength, an inventive solution satisfies both requirements. Another example would be speed vs. precision. A TRIZ level 3 solution would satisfy both requirements utilizing available "in paradigm" methods, whereas a level 4 solution would incorporate technologies outside the current paradigm. In both cases, however, speed and precision would be achieved at a quality level demanded by the contextual parameters of the situation. In TRIZ, two distinct types of contradictions are delineated, technical contradictions and physical contradictions. Methods for solving technical contradictions are discussed later in the chapter.

19.3.2.1 Technical Contradictions

A *technical contradiction* is a situation where two identifiable parameters are in conflict. When one parameter is improved, the other is made worse. The two previously mentioned, weight vs. strength, and speed vs. precision, are examples (see Figure 19.4).

19.3.2.2 Physical Contradictions

A *physical contradiction* is a situation where a single parameter needs to be in opposite physical states, e.g., it needs to be thin and thick, hot and cold at the same time. This type of contradiction has, at least to the author's knowledge, never been articulated prior to the arrival of TRIZ in North America.

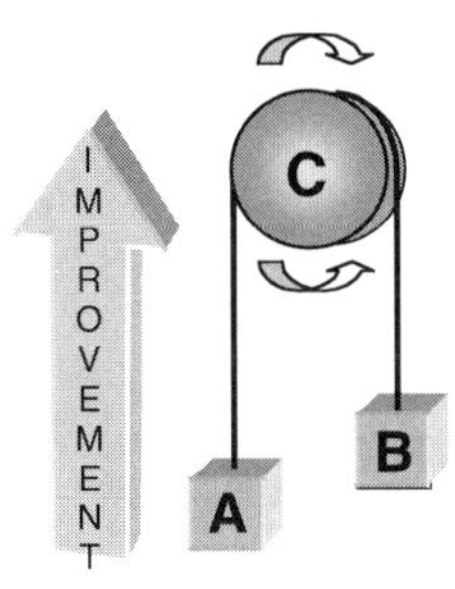

FIGURE 19.5 Physical contradiction. For A and B to improve, C must rotate clockwise and counterclockwise simultaneously.

A physical contradiction is the controlling element or parameter linking the parameters of the technical contradiction. Figure 19.5 shows the pulley (C) upon which parameters A and B rotate as the physical contradiction.

The physical contradiction lies at the heart of an inventive problem; it is the ultimate contradiction. When the physical contradiction has been found, the process of generating an inventive solution has been greatly simplified. It stands to reason that when a physical contradiction is made to behave in two opposite states simultaneously, the technical contradiction is eliminated. For example, if by some means, pulley C could rotate in opposite directions at the same time, both A and B would increase, hence eliminating the technical contradiction.

19.3.3 Resources

The third foundation principle of TRIZ is the maximal utilization of any available resources before introducing a new component or complication into the system. Resources are defined as any substance, space, or energy that is present in the system, its surroundings, or in the environment. The identification and utilization of resources increase the operating efficiency of the system, thereby improving its ideality. It is understandable that in the former Soviet Union where money was scarce necessity did in fact prove to be the mother of invention. In the West, on the other hand, system problems were often engineered out by the proverbial means of throwing money (and complexity) at the system. The utilization of resources as an "X" agent to solve the problem was and still is not widely practiced.

A practiced TRIZ problem solver will marshal any in-system or environmental resource to assist in solving the problem. It is only when all resources have been exhausted or it is impractical to use one that the consideration of additional design elements comes into play. The mantra of a TRIZ problem solver is never to solve a problem by making the system more complex. More on this when the algorithm for problem solving (ARIZ — Russian language acronym) is discussed. Table 19.1 lists the types of resources used in TRIZ.

19.4 A SCIENTIFIC APPROACH

TRIZ is composed of a comprehensive set of analytical and knowledge-based tools that was heretofore buried at a subconscious level in the minds of creative inventors.

TABLE 19.1
Types of Resources

SUBSTANCE — any material contained in the system or its environment, manufactured products, or wastes
ENERGY — any kind of energy existing in the system, any space available in the system, and its environment time intervals before start, after finish, and between technology cycles, unused or partially used
FUNCTIONAL — possibilities of the system or its environment to carry out additional functions, unused specific features and properties, characteristics of a particular system, such as special physical, chemical, or geometrical properties. For example: resonance frequencies, magneto susceptibility, radioactivity, and transparency at certain frequencies
SYSTEM — new useful functions or properties of the system that can be achieved from modification of connections between the subsystems, or a new way of combining systems
ORGANIZATIONAL — existing, but incompletely used structures, or structures that can be easily built in the system, arrangement or orientation of elements or communication between them
DIFFERENTIAL — differences in magnitude of parameters that can be used to create flux, that carry out useful functions. For example: speed difference for steam next to a pipe wall vs. in the middle, temperature variances, voltage drop across resistance, height variance
CHANGES — new properties or features of the system (often unexpected), appearing after changes have been introduced
HARMFUL — wastes of the system (or other systems) which become harmless after use

Asked to explain specifically how they invent, most are unable to provide a repeatable formula. Through his work, Altshuller has codified the amorphous process of invention. Altshuller's great contribution to society is that he made the process of inventive thinking explicit, thus making it possible for anyone with a reasonable amount of intelligence to become an inventor.

What Altshuller did for inventive thinking is not unlike what happened in mathematics with the invention of place values and the zero. Prior the modern (Hindu–Arabic) form of mathematics, the civilized Western world used Roman numerals. This system of numbers was written from left to right and used letters to designate numerical values. The number 2763, for example, is written MMDCCLXIII. The system, although somewhat awkward, was sufficient for doing simple addition and subtraction. It was nearly impossible, however, to perform calculations requiring multiplication and division. These mathematical functions were understood by only a few highly capable math wizards.

The Hindu–Arabic numbering system that used symbols and incorporated place values based on 10 was far superior and easier for the average person to learn and understand. Furthermore, the flexibility and robustness of the system allowed for the invention of algebra, statistics, calculus, differential equations, and scores of other advancements. TRIZ is the inventive analog of the Hindu–Arabic numbering system. TRIZ makes it possible for people of average intelligence to access a large body of inventive knowledge and, through analogic analysis, formulate inventive "out-of-the-box" solutions.

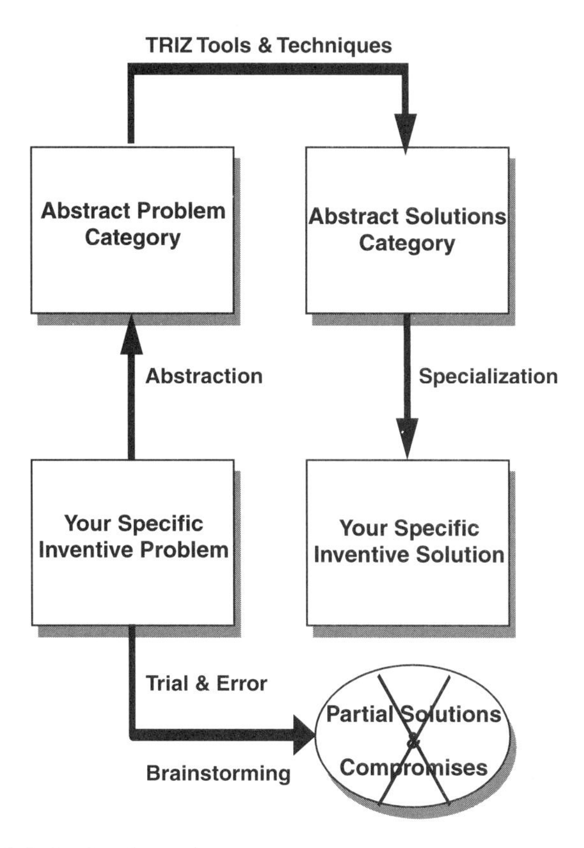

FIGURE 19.6 Solution by abstraction process.

19.4.1 How TRIZ Works

The general scheme in TRIZ is solution by abstraction. In other words, a specific problem is described in a more abstract form. The abstracted form of the problem has a counterpart solution at the level of abstraction. The connection between the problem and the solution is found through the use of various TRIZ tools. Once the solution analog is arrived at, the process is reversed, producing a specific solution. Figure 19.6 illustrates the process of solution by abstraction, and Figure 19.7 applies the process to an algebraic problem.

Assume that we were given the task of solving the problem found in the Equation, $3x^2 + 5x + 2 = 0$. Without a specific process, we would be reduced to the inefficient process of trial and error. An even more absurd method would be to try to arrive at the answer by brainstorming. Yet, brainstorming is often applied to problems that are much more complex than that shown above. This is what makes TRIZ so compelling — it provides a roadmap to highly creative and innovative solutions to seemingly impossible problems. Figure 19.7 shows the principle of solution by abstraction applied to the algebraic equation.

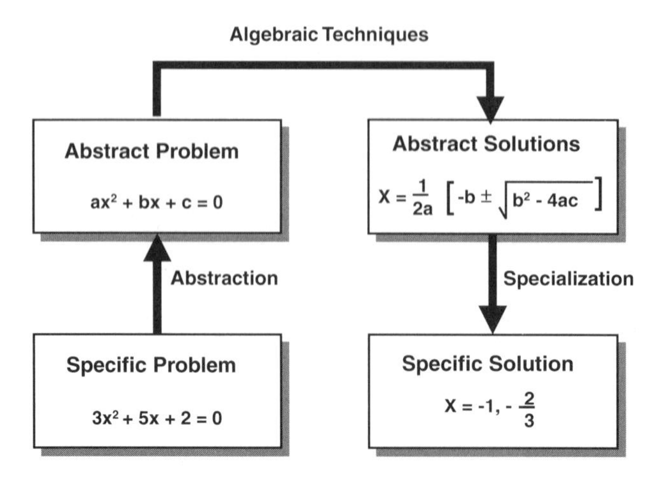

FIGURE 19.7 Solution by abstraction example.

Figure 19.6 provides the general schema for how TRIZ works. The fundamental idea in TRIZ is to reformulate the problem into a more general (abstract) problem and then find an equivalent "solved" problem. These analogs, in theory, define the solution space that is occupied by one or several noncompromising alternative solutions.

The advantage of increasing the level of abstraction is that the solution space is expanded. Solving the equation in Figure 19.8 is relatively simple, assuming knowledge of algebra. The correctness of the solution is also easier to verify because the solution space is very small, i.e., there is only one right answer! Inventive problems pose a much greater challenge than the one shown because the solution space is very large.

Figure 19.8 shows what happens when solving inventive vs. noninventive problems. An inventive problem is often confused with problems of design or engineering, or of a technological nature. For example, in constructing a bridge, the type of bridge to be built is largely an issue related to design. A cantilever bridge provides known design advantages over a suspension bridge in specific contexts, and vice versa. This is an example of a noninventive design problem. Calculating the load and stress the bridge will have to withstand is an engineering problem. Coordinating the construction and assuring that materials meet specifications and the job is on time and on budget is a technical problem. Although these problems are not insignificant by themselves, they are not inventive within the context of TRIZ because they are solvable by using known methods, formulas, schedules, etc. Furthermore, the path to the correct solution is defined and direct and, because the solution space is very small, verification of the answer is straightforward. This is not the case with inventive problems.

An inventive problem in the context of building a bridge would to be to make the bridge lighter and stronger, larger and less expensive, longer and more stable. These problems are inventive because they often have to overcome many contradictions. To reiterate, a problem is an inventive one if one or several contradictions must be overcome in its solution, and a compromise solution is not acceptable.

Several distinguishing characteristics of an inventive vs. typical problem are shown in Figure 19.8. First, the entire solution space can be quite large, containing

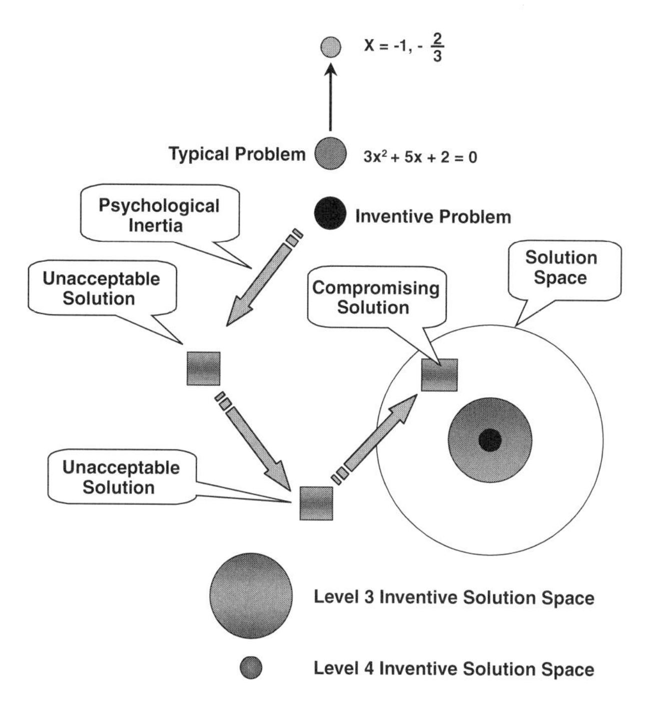

FIGURE 19.8 Solution space for inventive vs. other problems.

both noninventive and inventive solutions. The two inner concentric circles represent level 3 and level 4 inventive solutions, while the larger outer circle represents an area of noninventive solutions. Just as it is harder to hit the bullseye when shooting an arrow, so it is with hitting on an inventive solution. Why is this so?

The initial factor often driving one off the mark is psychological inertia. PI, as defined previously, presupposes a solution path as defined by a person's individual paradigms. The route to a solution is often one of trial and error and is strewn with several unacceptable solutions arrived at along the vector of one's psychological inertia. In a sense, the process of defining the current problem and then driving to a solution can be considered a "push" method for finding a solution. TRIZ is very different because one of the initial steps of the TRIZ process is to define the ideal state, i.e., the solution space found in level 3 or level 4 solutions. The articulation of the ideal solution acts to orient the problem solver and "pulls" him or her in that direction. Furthermore, TRIZ guides a person to the ideal solution through the process of abstraction and finding analogs, as discussed previously. These two fundamental elements of TRIZ serve as a powerful magnet to draw or pull one to an inventive solution, in part by providing an example of how this has been accomplished by a previous inventor.

19.4.2 Five Requirements for a Solution to be Inventive

Within the context of TRIZ, before a proposed solution is labeled as inventive, it must meet all of the stringent requirements outlined in Table 19.2.

TABLE 19.2
Requirements of Inventive Solutions

- Solution fully resolves the contradictory requirements
- Solution preserves all advantages of the current system
- Solution eliminates the disadvantages of the current system
- Solution does not introduce any new disadvantages
- Solution does not make system more complex

19.5 CLASSICAL AND MODERN TRIZ TOOLS

In the course of his analytical work, Altshuller amassed a vast body of knowledge and invented analytical methods on how to access it. The subsequent evolution of TRIZ followed logical parallel paths. The creation of a body of "inventive" knowledge gave rise to various analytical tools making it easier to catalogue and create more inventive knowledge that, in turn, spawned more sophisticated tools and so on. The end result after more than 50 years of work is a complete set of sophisticated tools and an immense knowledge base of inventive ideas, methods, and solutions that can be mobilized to attack any inventive problem. To date, to name just a few applications, these tools have been used to solve problems related to product design and development, quality, manufacturing, cost reduction, production, warranty, and prevention of product failures.

The tools of TRIZ are subdivided into two major categories. The first division is by the nature of the tool, e.g., analytical vs. knowledge base. The second differentiation is chronological, e.g., classical TRIZ vs. I-TRIZ. The classical TRIZ tools span those derived from 1946 to 1985, with Altshuller as the primary inventive force. Altshuller, for reasons of health, stopped his work in 1985. Thereafter, a protégé of Altshuller, Boris Zlotin of The Kishnev School (of TRIZ) continued developing the methodology, which for purposes of differentiation is called I-TRIZ. I-TRIZ is software based and is therefore able to automate some of the analytical work and provide graphical representations of solutions. I-TRIZ adds two additional new tools, anticipatory failure determination (AFD) and directed evolution (DE). Given length limitations, I-TRIZ is beyond the scope of this chapter. I-TRIZ is the service mark of Ideation International.

19.5.1 CLASSICAL TRIZ – KNOWLEDGE-BASED TOOLS

19.5.1.1 The Contradiction Matrix

The first of the classical TRIZ tools invented by Altshuller is the contradiction matrix. The objective of the matrix is to direct the problem-solving process to incorporate an idea that has been utilized before to solve an analogous "inventive" problem. The contradiction matrix accomplishes this by asking two simple questions: "Which element of the system is in need of improvement?" and "If improved, which element of the system is deteriorated?" This is, as has been pointed out, a technical contradiction. A portion of the 39 × 39 matrix is shown below (Figure 19.9).

Feature to Improve \ Deteriorated Feature		1 Weight of a moving object	2 Weight of non-moving object	3 Length of a moving object	●	22 Waste of Energy
1	Weight of a moving object			15,8 29,34	●	6, 2 34,19
2	Weight of a non-moving object				●	18, 19 28, 15
3	Length of a moving object	8,15 29,34			●	7, 2
4	Length of a non-moving object		35,28 40,29		●	6, 28
5	Area of a moving object	2,17 29,4		14,15 18,4	●	15, 17 30,26
6	Area of a non-moving object		30,2 14,18		●	17, 7 30
7	Volume of moving object	2,26 29,40		1,7 35,4	●	7,15 13,16
●	● ● ● ● ● ●	●	●	●	●	●
33	Convenience of use	25, 2 13,15	6,13, 1, 25	1,17 13,12	●	2,19 13
34	Repairability	2,27 35,11	2,27 35,11	1,28 10,25	●	15, 1 32,19
35	Adaptability	1,6 15,8	19,15 29,16	35,1 29,2	●	18, 15 1
36	Complexity of device	26,30 34,36	2,36 35,39	1,19 26,24	●	10,35 13,2
37	Complexity of control	27,26 28,13	6,13 28,1	16,17 26,24	●	35,3 15,19
38	Level of automation	28,26 18,35	28,26 35,10	14,13 17,28	●	23,28
39	Productivity	35,26 24,37	28,27 15,3	18,4 28,38	●	28,10 29,35

FIGURE 19.9 The contradiction matrix.

The matrix is constructed by juxtaposing 39 engineering parameters along the vertical and horizontal axies. At the intersections Altshuller filled in from one to four numerical values hinting at ways to solve the problem. The numerical values identified one of the 40 inventive principles that were culled from the knowledge base as ways in

TABLE 19.3
40 Inventive Principles (Partial List)

3. Local quality

a. Change an object's structure from uniform (homogeneous) to non-uniform (heterogeneous) or, change the external environment (or external influence) from uniform to non-uniform.
b. Have different parts of the object carry out different functions.
c. Place each part of the object under conditions most favorable for its operation.

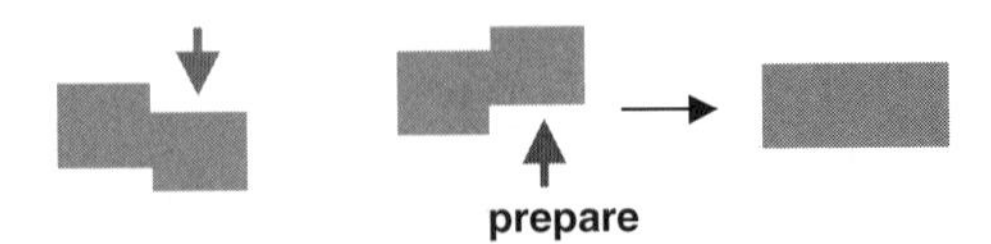

9. Preliminary anti-action

a. If it is necessary to perform some action with both harmful and useful effects, consider a counteraction in advance that will negate the harmful effects.
b. Create stresses in an object that will counter known undesirable forces later on.

13. The other way around

a. Instead of an action dictated by the specifications of the problem, implement an opposite action.
b. Make a moving part of the object or the outside environment immovable and the non-moving part movable.
c. Turn the object upside down, inside out, freeze it instead of boiling it.

which an analog to the specific problem had been solved previously. The 39 engineering parameters are general in nature and act as surrogates for the specific real parameters in conflict. The inventive principles are broad and nonspecific as the exact way in which they should be applied. In Figure 19.9 the problem is trying to improve "convenience of use" but when this is attempted, it results in waste of energy. The matrix suggests that when this type of problem is encountered, principles 2, 9, and 13 have been utilized to resolve the contradiction. Table 19.3 provides details on these three principles.

The process for using the contradiction matrix follows the general schema outlined in Figure 19.6. The steps are (1) describe the problem, (2) select the parameter most closely aligned with one of the 39 engineering parameters from the feature to improve column, (3) state your proposed solution, (4) select which feature will be deteriorated, (5) note the inventive principle(s) at the intersection, and (6) apply the inventive principle(s).

19.5.1.2 Physical Contradictions

A physical contradiction (PC) is the controlling element in the system that links the two conflicting parameters in the technical contradiction (see Figure 19.5). The PC expresses the most extreme form of contradictory requirements because the conflict must be resolved solely within a single entity. As Figure 19.5 shows, the PC (pulley C) is at the very root of the inventive problem. If it were possible to make the pulley

TABLE 19.4
Separation Principles

1.	Separation in time
2.	Separation in space
3.	Separation between the system and its components
4.	Separation upon condition
5.	Co-existence of contradictory properties

turn in opposite directions simultaneously, the technical contradiction would disappear. From a TRIZ standpoint, solving an inventive problem by satisfying the conflicting requirements of the PC results in elegant solutions with a greater degree of inventiveness.

19.5.1.2.1 Formulating and Solving Physical Contradictions

A Physical Contradiction is formulated according to the logic: "To perform function $\mathbf{F}_1$, the object must exhibit property **P**, but to perform function $\mathbf{F}_2$, it must exhibit property –**P**. The solution to physical contradictions is accomplished by incorporating principles of separation. There are five separation principles that can be used to resolve a PC (see Table 19.4).

19.5.1.2.2 An Example

The principle of separation in time can be explained by a well-known illustration used by Altshuller. Assume that one is driving concrete piles for buildings into very hard ground. To facilitate ease of driving the piles, the tip profile should be sharp. Once in place, the pile should be stable, which means the profile should be blunt. In other words, the pile should be sharp and blunt — a physical contradiction. How can this be? The problem is solved by imbedding an explosive into the sharp end of the pile and when it is in place, destroying the sharp profile by setting off the explosive. The tip profile is sharp (**P**) during time T_1 (driving into the ground) and it is blunt (**–P**) during time T_2 (in place).

19.5.1.3 The Laws of Systems Evolution

The notion of predicting future technological patterns and derivatives has been recognized as a means of creating competitive leverage. Techniques such as technology forecasting, morphological analysis, trend extrapolation, and the Delphi process have been utilized since the Second World War. All of these techniques are based on statistical probability modeling. In TRIZ, future derivatives are based on predetermined patterns of evolution that have been around since the invention of the wheel. Past evolutionary trends provide a evolutionary crystal ball for understanding how current technologies will morph over time. Altshuller termed these phenomena *laws of evolution.*

These laws represent a stable and repeatable pattern of interactions between the system and its environment. These patterns occur because systems are subject to various cycles of improvement. When a new technological system emerges, it

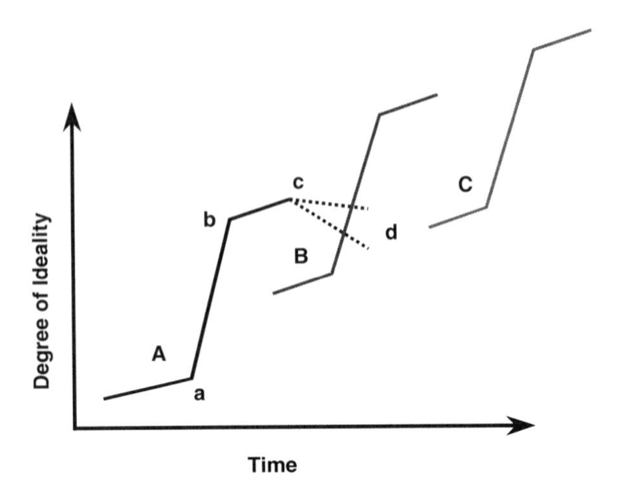

FIGURE 19.10 Life-cycle curves.

typically provides the minimum degree of functionality required to satisfy the inventor's intent. For example, the first powered flight by the Wright brothers occurred on December 17, 1903. The *Flyer,* with Orville Wright as the pilot, flew to a height of 10 feet and landed heavily after 12 seconds in the air. Today, jets are capable of flying at heights over 60,000 feet over thousands of miles at several times the speed of sound. What has happened to airplanes has been repeated in all types of engineered systems.

The way in which systems evolve can be shown on life cycle or "S" curves. Figure 19.10 shows the evolutionary picture.

From the time a system emerges to point **a,** its development is slow as it is unproven. At point **a,** the dominant design paradigm appears and the system is poised for commercialization. From points **a** to **b** the system experiences rapid improvement as commercialization and market pressures force cycles of continuous improvements. From points **b** to **c** the rate of improvement slows as the technology matures. As the system passes point **b**, the next system (**B**) is itself emerging. The abandonment of the original system in favor of the new one is governed by how much greater potential it possesses in comparison to the unrealized improvements remaining in system **A**.

Being a keen observer of inventive phenomena, Altshuller through his analysis uncovered eight describable chronologically sequenced events. He called these events the *laws of systems evolution* (see Table 19.5).

Within these eight major laws, Altshuller and his students have found numerous "sub-lines" of evolution. Given the detail that is now captured in the evolutionary knowledge base it is possible through the analysis of patents to fix where the technological system is positioned on its life-cycle curve.

Figure 19.11 shows a few of the sublines of law 4, *increased dynamism.*

One can draw an analogy between use of the laws of evolution and laws of motion. If the position of a moving object is known at a certain moment of time,

TABLE 19.5
Patterns of Technological Systems Evolution

1. Stages of evolution
2. Evolution toward increased ideality
3. Non-uniform development of systems elements
4. Evolution toward increased dynamism and controllability
5. Increased complexity then simplification
6. Evolution with matching and mismatching components
7. Evolution toward micro-level and increased use of fields
8. Evolution toward decreased human involvement

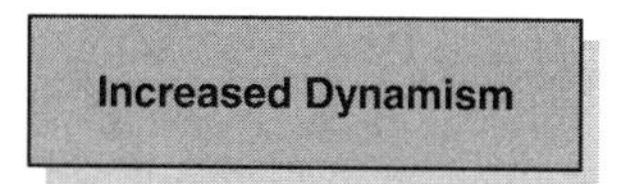

In the course of time, technological systems transition from rigid systems to flexible and adaptive ones

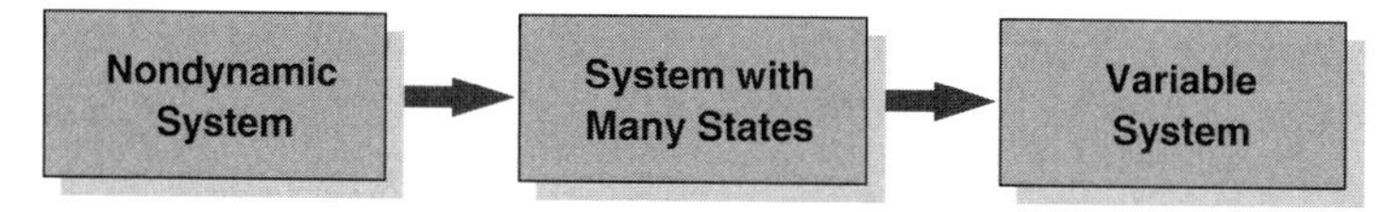

Evolution of Automotive Transmission

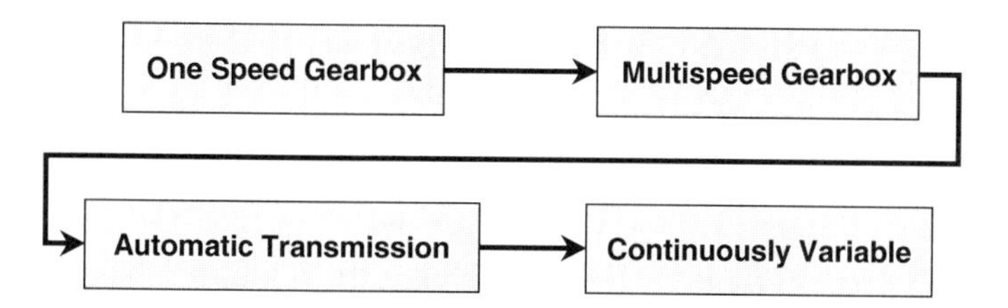

FIGURE 19.11 Increased dynamism.

any future position can be determined by solving equations containing velocity and direction. The laws of evolution serve as equations describing how the system will change as it travels through time. If the current position of the system is known, future derivatives can be calculated using the laws to indicate future positions. The implications to research and development initiatives, protection of intellectual assets, technology development strategy, patent strategies, and product development scenarios are profound.

19.5.2 Analytical Tools

In addition to the knowledge-based tools, Altshuller also developed several analytical tools. The two most widely used are substance field modeling and the algorithm for

TABLE 19.6
System of Standard Solutions

Class 1. Increasing performance
- 1.1 Synthesis of the substance field models
  - 1.1.1 Constructing the sufield field models
  - 1.1.2 Internal combined sufield model
  - 1.1.3 External combined sufield model
  - 1.1.4 Sufield model with the environment
  - 1.1.5 Sufield model with environment and additives
  - 1.1.6 Minimum regime
  - 1.1.7 Maximum regime
  - 1.1.8 Selective maximum regime
- 1.2 Destroying the sufield model

Class 2. Eliminating harmful actions
Class 3. Transition to the super-system and to the microlevel
Class 4. Eliminating problems in measurement
Class 5. Eliminating problems caused by applying standard solutions

inventive problem solving. The former is referred to as sufield and the latter according to its Russian language acronym — ARIZ.

19.5.2.1 Sufield

The object of sufield is to provide a mechanism for creating a model of a problem and a corresponding solution, as has been illustrated in the general schema of solution by abstraction (Figure 19.6). We may recall that in TRIZ a specific problem is classified and for problems in that class, analogs exist illustrating inventive solutions. It is up to the problem solver to forge a link between the real problem and the solution analog. One may wonder how this tool was invented. As is true with most of TRIZ, sufield emerged from a painstaking process of classifying problems and their corresponding solutions. Technological problems were placed into one of five classes or types of problems. These classes were further subdivided hierarchically into 76 inventive solutions. The process is not unlike classifications in biology or zoology. Table 19.6 illustrates the five classes and an exploded view of one class.

Altshuller realized the power of psychological inertia as an obstacle to objective thinking. He neutralized this by utilizing jargon-free terminology to describe the problem and illustrate the solution. The sufield model consists of three primary components: substance$_1$ (S_1, the article that is passive in nature), substance$_2$ (S_2, the tool that is active) and a field (F_i, the energy source). These three elements constitute the minimum requirements for a complete system and are shown as a triangle. The most frequently used fields in TRIZ are

- Mechanical (Me)
- Thermal (Th)

- Chemical (Ch)
- Electrical (E)
- Magnetic (M)
- Gravitational (G)

The minimum sufield model consisting of all three elements is illustrated in Figure 19.12:

The ways in which the components of a system interact with each other are shown by the use of various connecting symbols as shown in Figure 19.13.

There are four basic types of sufield models:

- A complete and effective system — a tool, article, and field
- An incomplete system — one that requires one or more elements be added to make it a complete system, e.g., a tool, an article, a field, or some combination
- A complete but ineffective system — one that requires improvement
- A complete but harmful system — one that requires a harmful effect be eliminated

Sufield illustration problem: How is it possible to measure the volume of water contained in ponds and small lakes? The volumetric characteristics of size, shoreline profile and depth vary widely from lake to lake.

When this problem was posed to a widely disparate audience, the answers ranged from guessing based on averages, to precise measurements utilizing sophisticated

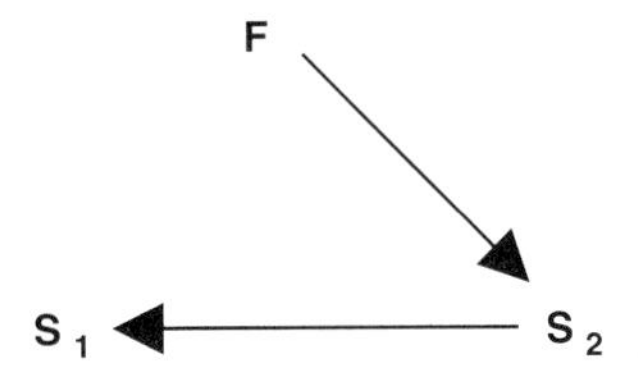

FIGURE 19.12 Minimum sufield model.

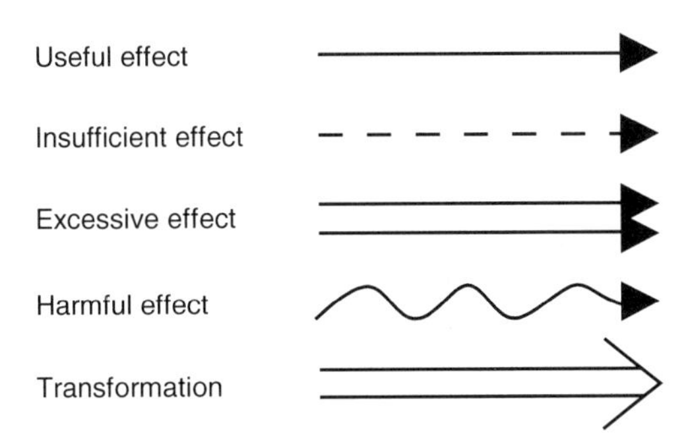

FIGURE 19.13 Sufield interactions.

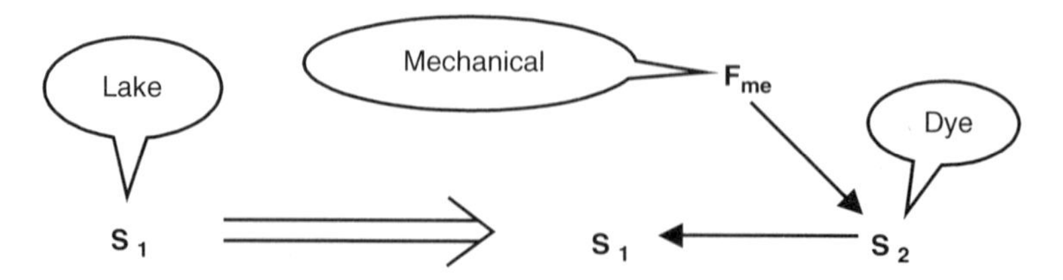

FIGURE 19.14 Sufield solution measuring volume problem.

global positioning systems (GPS) integrated with sonar mapping. None of these answers was as elegant as the one posed by a ten-year-old child.

Using sufield methodology, the solution is as follows. From a complete system point of view, the initial problem contained only one element of the three required, namely, an article (S_1) — the lake. To solve the problem, finding a tool (S_2) and a field (F_i) as shown in Figure 19.14 is required.

The proposed solution was to pour a known quantity of a highly concentrated biodegradable dye into the water, agitate it to mix evenly, and then measure the quantity of the dye in a vessel of known volume and extrapolate to determine the total volume in the lake.

The sufield transformation outlined in the measurement problem is generic and serves as an archetype for thousands of similar problems; the trick lies in recognizing this to be the case.

19.5.2.2 Algorithm for Inventive Problem Solving (ARIZ)

ARIZ (Russian language acronym) is the primary problem-solving tool in TRIZ. ARIZ was published in 1959 and revised: ARIZ-61, ARIZ-64, ARIZ-65, ARIZ-71 and ARIZ-85. Each revision improved the structure, language, and length of the algorithm. In its current state, we have a carefully crafted set of logical statements that transform a vaguely defined problem into an articulation of one with a clearly defined number of contradictions.

The assumptions designed into ARIZ are that the true nature of the problem is unknown and the process of finding a solution will follow the problem solver's vector of psychological inertia. It is precisely for these reasons that many of the steps in ARIZ are reformulations of the problem. With each reformulation, the problem is viewed from a different vantage point yielding the possibility of new and novel ideas.

In mathematics, an algorithm is a precise set of steps designed to arrive at a single outcome. There is only one right answer. No consideration is given to the personality of the problem solver nor to any changeable external conditions. The process is rote. In a broader context, an algorithm is a process following a set of sequential steps. ARIZ falls into the broader definition. ARIZ is a structured set of logic statements that guide the process of invention through a series of formulations and reformulations of the problem. It can be safely said that if a chronic technological problem persists even after many attempts to solve it, the reason is oftentimes because the wrong problem is being solved. Charles F. Kettering stated, "A problem well stated is a problem half solved." The selection of which problem to solve in an inventive situation is the starting point. It is critical that this selection is correct if there is any hope of arriving at an inventive solution in a timely manner.

A Respiratory Problem

In a CNN scientific broadcast, the narrator stated that astronauts aboard the shuttle were experiencing respiratory problems due to residual dust and other minute particulates that passed through the shuttle's filtration system. A typical (Western) response to this problem would revolve around reengineering the system to make it more efficient. If the cost of this solution was too high, another approach that might work equally as well is figuring out how to transform small particles into large particles. This is a totally different problem. The advantage of the latter is that the current system would not have to undergo a costly major redesign. Is this possible? An inventory of the resources available yields moisture in the form of water vapor and very cold temperatures outside of the shuttle. Given these resources, it is conceivable that small particles can be encapsulated in water vapor and frozen with the result that small particles are transformed into large ones, thereby allowing the filtration system to capture and retain them.

As with any systematized process, ARIZ is dependent on the innate intelligence and knowledge of the subject matter expert and the skill with which he/she utilizes the tool. The strength of ARIZ, however, is that the process of thinking inventively is stripped of psychological inertia and regulated in a stepwise fashion toward the ideal solution, or in TRIZ terms, the *ideal final result* (IFR). The result is that the innate knowledge of the inventor is leveraged so that he/she is forced into thinking inventively, e.g., into the solution space containing the most inventive ideas. Once the person is in the solution space, a number of inventive principles, analogs or substance field models promote thinking outside of the box (see Figure 19.15).

19.5.2.2.1 The Steps in ARIZ

The architecture of ARIZ is composed of three major processes that are subdivided into nine high-level steps, each with their own sub-steps. The macro- and high-level steps in ARIZ are shown in Figure 19.16.

ARIZ is designed to utilize all of the tools in TRIZ including

- Ideality
- The ideal final result
- Elimination of physical and technical contradictions
- Maximal utilization of the resources of the system
- Substance field models and standard solutions
- The 40 inventive principles

ARIZ is designed to manage the inventive process on two types of problems, micro and macro. A micro problem focuses on solving a contradiction contained within the system, while a macro problem is a redesign of the entire system. ARIZ is iterative in that the inventor is provided several alternative paths to solving a problem. If all the solutions generated at the micro level are unsatisfactory, the problem must be solved at the macro level.

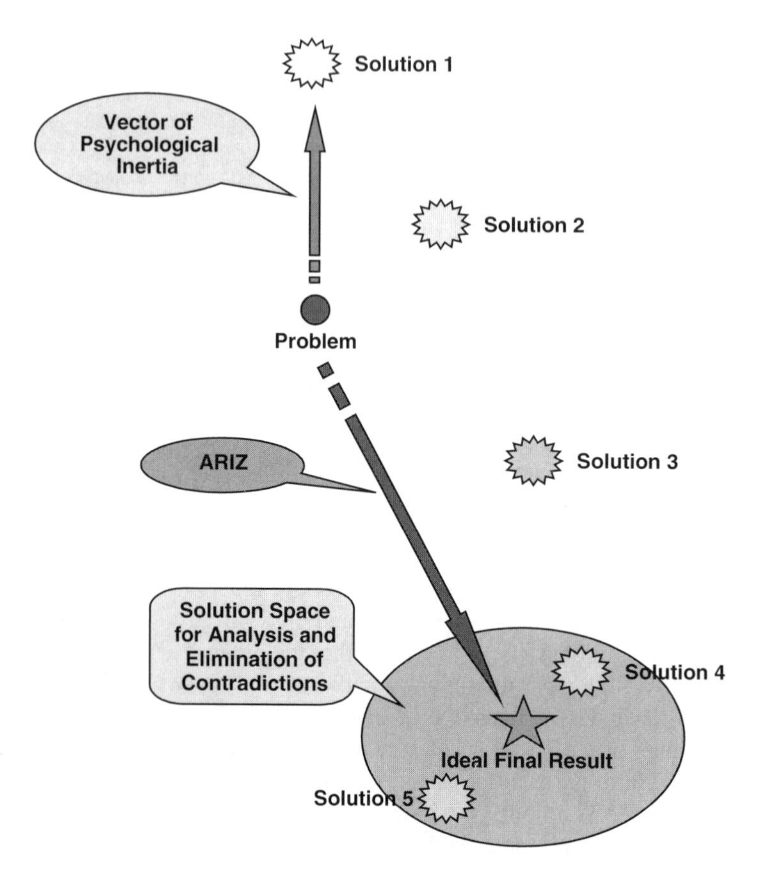

FIGURE 19.15 A respiratory problem: two perspectives.

A portion of the algorithm (Stage 1: Formulation of the problem) is detailed below.

19.5.2.2.2 Problem Analysis

19.5.2.2.2.1 Micro-Problem

Write down the conditions of the micro-problem *(do not use technology specific jargon):*

- A technological system for *(specify the purpose of the system)* that includes *(a list main elements of the system).*
- Technical contradiction 1: *(formulate).*
- Technical contradiction 2: *(formulate).*
- It is required to achieve *(specify desirable result)* without incurring *(specify the undesirable result)* with minimal changes or complications introduced into the system.

Note: Technical contradictions are defined using nouns for the elements in the system and action verbs describing the interaction between them.

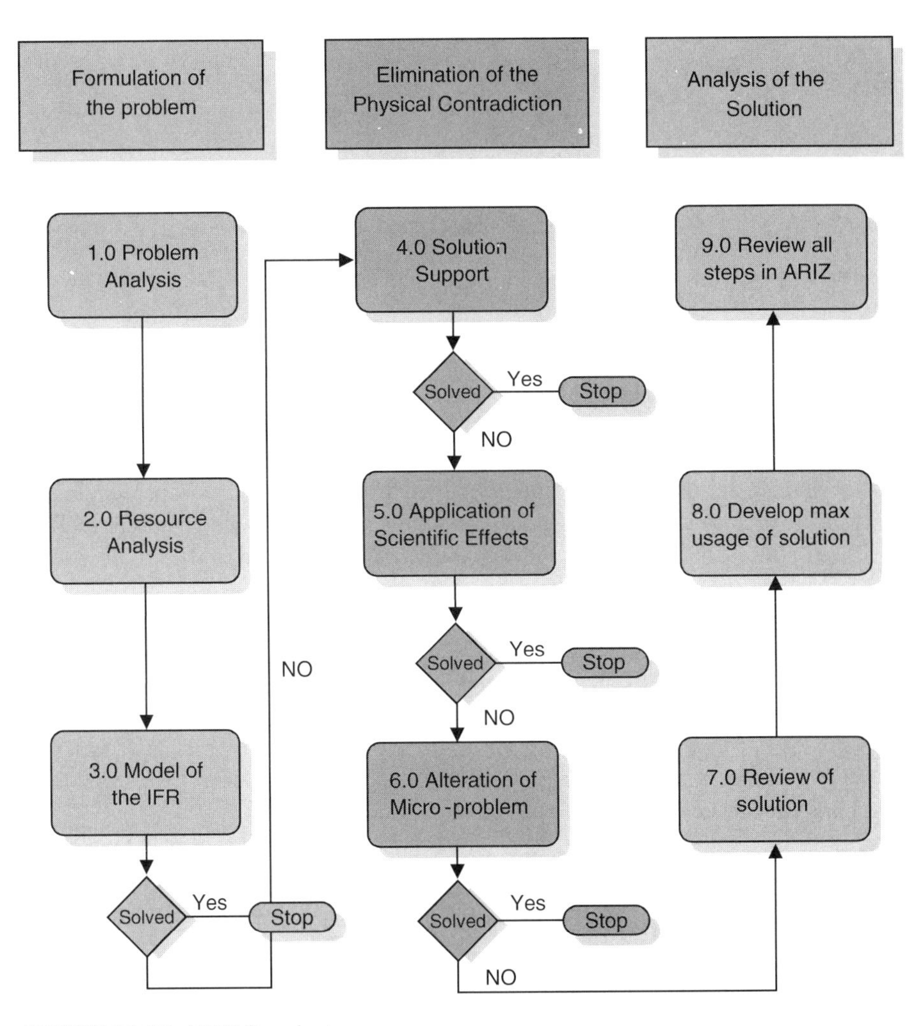

FIGURE 19.16 ARIZ flowchart.

19.5.2.2.2.2 Conflicting Elements

Identify the conflicting elements: an article and a tool

1. If an element can be in two states, point out both of them.
2. An article is an element that is to be processed or improved. A tool is an element that has an immediate interaction with the article.
3. If there is more than one pair of the identical conflicting elements, it is sufficient to analyze just one pair.

19.5.2.2.2.3 Conflict Intensification

Formulate the intensified technical contradiction (ITC) by showing an extreme state of the elements.

19.5.2.2.2.4 Conflict Diagrams

Compile diagrams of the intensified technical contradictions:

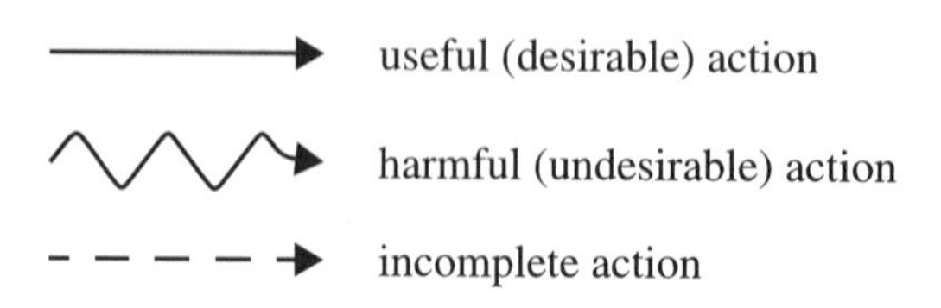

19.5.2.2.2.5 Selection of the Conflict

Select from one of the two conflict diagrams for further analysis:

1. Select a diagram that better emphasizes the main (primary) function.
2. If intensification of the conflicts resulted in the impossibility of performing the main function, select a diagram that is associated with an absent tool.
3. If intensification of the conflicts resulted in elimination of the article, use a 95% principle.
4. Select a diagram that better emphasizes the main function, but reformulate an associated technical contradiction by showing not extreme, but very close to extreme, states of the elements.

19.5.2.2.2.6 Model of Solution

Develop a model of the solution by specifying actions of an X-resource capable of resolving the selected ITC:

- Finding an X-resource that would preserve (*specify the useful action*) while eliminating (*specify harmful action*) with minimal changes or complications introduced into the system is required.

19.5.2.2.2.7 Model of Solution Diagram

Construct a diagram of the model of the solution.

19.5.2.2.2.8 Substance-Field Analysis

Compile a substance-field diagram that models the solution.

- Compile a substance-field model representing a selected ITC
- Compile a desirable substance-field model illustrating resolution of the conflict
- Select the appropriate standard solution and compile the complete substance field transformation

19.5.2.2.3 Resource Analysis

19.5.2.2.3.1 Conflict Domain

Define the space domain within which the conflict develops.

19.5.2.2.3.2 Operation Time

Define the period of time within which the conflict should be overcome.

- Operation Time is associated with the time resources available:
  - Pre-conflict time T1
  - Conflict time T2
  - Post-conflict time T3
- It is always preferable to overcome a conflict during T1 and/or T2.

19.5.2.2.3.3 Substance and Energy Resources

List the substance and energy resources of the system and its environment.

- The substance and energy resources are physical substances and fields that can be obtained or produced easily within the system or its environment. These resources can be of three types:
  - In-system resources
    a. Resources of the tool
    b. Resources of the article
  - Environmental resources
    a. Resources of the environment that are specific to the system
    b. General resources that are natural to any environment, such as magnetic or gravitation fields of the earth
  - Overall-system resources
    a. Side-products: waste products of any system or any cheap or free foreign objects

19.5.2.2.4 Model of Ideal Solution

19.5.2.2.4.1 Selection of the X-resource

Select one of the resources for further modification.

1. Select in-system resources in the conflict domain first
2. Modification of the tool is more preferable than modification of the article

19.5.2.2.4.2 First Ideal Final Result (IFR)

The IFR can be formulated as follows:

The X-resource, without any complications or any harm to the system, terminates *(specify the undesirable action)* during the operation time within the conflict domain, while providing the *(specify the useful action).*

19.5.2.2.4.3 Physical Contradiction

Formulate a physical contradiction:

- To terminate *(specify the undesirable action)*, the X-resource within the conflict domain and during the operation time must be *(specify the physical state P)*

- To provide (specify the desirable action), the X-resource within the conflict domain and during the operation time must be (specify the opposite physical state –P)

19.5.2.2.4.4 Elimination of Physical Contradiction Macro
Use methods for elimination of physical contradictions:

- Separation of opposite physical properties in time
- Separation of opposite physical properties in space
- Separation of opposite physical properties between system and its components
- Separation of opposite properties upon conditions
- Combination of the above methods

Note: When applying the separation principles (use one or a combination of the following techniques):

- Separation in time
  - Think of ways to make the X-resource to have property P before or after the conflict and property –P during the conflict
  - Use high-speed processes
  - Explore various phenomena possible for the X-resource developed during phase transitions
  - Change the parameters or characteristics of the X-resource using a field
  - Explore using phenomena associated with decomposition of the X-resource into its basic elementary structure and then its recovery, e.g., ionization, recombination, dissociation, association
- Separation in space
  - Divide the X-resource into two parts having properties P and –P with one part in the conflict domain and the other outside of the conflict domain
  - Combine the X-resource with a void, porosity, foam, bubbles, etc.
  - Combine X-resource with other resources
  - Combine X-resource with a derivative of another resource (e.g. hydrogen and oxygen are derivatives of water)
- Separation between the system and its components
  - Divide the X-resource into several components in a way that one component has property P while the other has property –P
  - Decompose the X-resource into elementary particles, granules, flexible rods, shells, etc.
  - Explore using the phenomena associated with the decomposition of the X-resource into its base elements

19.6 CAVEAT

ARIZ is a highly developed complex tool and should not be used on typical straightforward engineering problems. Also, becoming proficient with ARIZ takes time and practice. As a general rule of thumb, it is recommended that an individual solve ten problems with ARIZ before claiming a layman's level of competency with the tool.

19.7 CONCLUSION

TRIZ is a powerful comprehensive problem-solving tool. It is the product of a massive analytical study of the output of the world's best inventors and the world's most creative inventions. The fundamental underlying principle of TRIZ is Ideality. The ideality principle holds that over time systems evolve to higher levels of functionality through the elimination of internal contradictions and the efficient utilization of available resources.

In time, the study of inventions by Altshuller and others yielded a number of knowledge-based and analytical tools. Knowledge-based tools include the contradiction matrix, the 40 inventive principles and the laws of systems evolution. Analytical tools include substance field analysis and the algorithm for inventive problem solving (ARIZ).

REFERENCES

Altshuller, G.S., *Creativity as an Exact Science* (Translated by Anthony Williams), Gordon & Breach, New York, NY, 1988.

Altshuller, G.S., *The Innovation Algorithm* (Translated by Lev Shulyak and Steve Rodman), Technical Innovation Center, Inc., Worcester, MA, 1999.

Clarke, D.W., Sr., *TRIZ: Through the Eyes of an American Specialist*, Ideation International, Inc., Southfield, MI, 1997.

Covey, S.R., *The 7 Habits of Highly Effective People*, Simon & Schuster, New York, NY, 1990.

Kaplan, S., *An Introduction to TRIZ*, Ideation International, Inc., Southfield, MI, 1996.

Kuhn, T.S., *The Structure of Scientific Revolutions*, 3rd ed., University of Chicago Press, Chicago, IL, 1996.

Terninko, J., Zusman, A., and Zlotin, B., *Systematic Innovation*, St. Lucie Press, New York, NY, 1998.

Ungvari, S., *TRIZ Two Day Workshop Manual*, Strategic Product Innovations, Inc., Brighton, MI, 1998.

Ungvari, S., *TRIZ Refresher Course*, Strategic Product Innovations, Inc., Brighton, MI, 1999.

Ungvari, S., *TRIZ Problem Solving Guidebook*, Strategic Product Innovations, Inc., Brighton, MI, 1999.

Index

D

E

F

G

H

I

J

K

L

M

N

O

P

T